TEST METHODS AND DESIGN ALLOWABLES FOR FIBROUS COMPOSITES

A symposium
sponsored by ASTM
Committee D-30 on
High Modulus Fibers and
Their Composites
AMERICAN SOCIETY FOR
TESTING AND MATERIALS
Dearborn, Mich., 2,3 Oct. 1979

ASTM SPECIAL TECHNICAL PUBLICATION 734
C. C. Chamis, NASA Lewis Research Center,
editor

ASTM Publication Code Number (PCN)
04-734000-33

AMERICAN SOCIETY FOR TESTING AND MATERIALS
1916 Race Street, Philadelphia, Pa. 19103

NOTE

The Society is not responsible, as a body,
for the statements and opinions
advanced in this publication.

Printed in Baltimore, Md.
May 1981

Foreword

The symposium on Test Methods and Design Allowables for Fibrous Composites was held in Dearborn, Michigan, 2,3 October 1979. The symposium was sponsored by the American Society for Testing and Materials' through its Committee D-30 on High Modulus Fibers and Their Composites. C. C. Chamis, NASA Lewis Research Center, presided as chairman of the symposium and editor of this publication.

Related
ASTM Publications

Commercial Opportunities for Advanced Composites, STP 704 (1980), $13.50, 04-704000-33

Nondestructive Evaluation and Flaw Criticality for Composite Materials, STP 696 (1979), $34.50, 04-696000-33

Composite Materials: Testing and Design (Fifth Conference), STP 674 (1970), $52.50, 04-674000-33

Advanced Composite Materials—Environmental Effects, STP 658 (1978), $26.00, 04-658000-33

Fatigue of Filamentary Composite Materials, STP 636 (1977), $26.50, 04-636000-33

Composite Materials: Testing and Design (Fourth Conference), STP 617 (1977), $51.75, 04-617000-33

Environmental Effects on Advanced Composite Materials, STP 602 (1976), $10.00, 04-602000-33

A Note of Appreciation
to Reviewers

This publication is made possible by the authors and, also, the unheralded efforts of the reviewers. This body of technical experts whose dedication, sacrifice of time and effort, and collective wisdom in reviewing the papers must be acknowledged. The quality level of ASTM publications is a direct function of their respected opinions. On behalf of ASTM we acknowledge with appreciation their contribution.

ASTM Committee on Publications

Editorial Staff

Contents

Introduction

The composites structures community recognizes that fiber composites offer a multitude of desirable properties to meet diverse and competing design requirements in a cost-effective manner. This multitude of desirable properties, however, requires special test methods to quantify material properties for design. It also requires well-defined procedures in selecting and setting design allowables for composites. Over the past 15 years researchers have continuously sought test methods to measure composite material properties with an acceptable degree of repeatability. At the same time, designers have sought well-defined procedures to select and establish design allowables that are compatible with current practice margins. Special test methods to measure composite properties and procedures for setting design allowables go through a long period of peer evaluation and critique prior to their adaptation on a general consensus basis.

In order to shorten the peer evaluation and critique period, the specialists symposium, covered in this publication, was organized, 2 and 3 October 1979, in Dearborn, Michigan, with the objective to provide a forum for the discussion of: (1) special test methods for setting design allowables and reporting results; (2) selecting and establishing design allowables; (3) promoting an understanding of the procedures for establishing design allowables; and (4) developing new test methods streamlined for setting design allowables which will eventually lead to improved methodology for reliable composite structures design. Papers presented in response to the symposium objective are included in this volume. These papers are grouped into four major areas: New and Special Test Methods, Special Test Methods and Analysis, Design Allowables, and Design Allowables for Special Applications.

The papers grouped in New and Special Test Methods cover test methods for transverse tensile strength, shear modulus, longitudinal compression, and impact resistance. Those in Special Test Methods and Analysis include off-axes and angle-plied tension and compression, fracture and fatigue, and biaxial stress states. They also cover an evaluation of test methods for longitudinal compression, short-beam-shear, and sandwich beam. The papers under Design Allowables describe procedures for setting/ selecting design allowables, in general, for composite aircraft structures, for graphite/polyimide bolted joints, for cost-effective mechanical property characterization, for statistical considerations, for intraply hybrids, and for compression/compression fatigue. Those under Design Allowables for Special Applications describe procedures for buckling of cylindrical components, proof load of pressure vessels, strength of bolted composite joints, environmental effects and creep.

The papers in each major area provide a valuable source as to where the emphasis was at the time of the symposium with respect to special testing and procedures for selecting design allowables. In addition, the papers provide a good perspective, with suitable reference, of the test methods and procedures for setting design allowables that have been proposed, or are used, including ranges of applications and limitations. Furthermore, the papers in each major area offer specific recommendations for future research which will lead to improved test methods and procedures for establishing design allowables for composite structures. Lastly, the papers collectively provide the researcher, analyst, and designer with a wealth of information and data on a broad spectrum of composites testing, design allowables, and design procedures. The papers are not, nor should they be expected to be, inclusive in any one area. However, they do constitute a first step toward an integrated source of test methods and procedures for setting design allowables for composites and composite structures.

C. C. Chamis,

Aerospace and composite structures engineer,
NASA-Lewis Research Center, Cleveland,
Ohio 44135; editor.

New and Special Test Methods

R. L. Foye[1]

A Single-Ply Transverse Tension Strength Test for Unidirectional Composites

REFERENCE: Foye, R. L., **"A Single-Ply Transverse Tension Strength Test for Unidirectional Composites,"** *Test Methods and Design Allowables for Fibrous Composites, ASTM STP 734,* C. C. Chamis, Ed., American Society for Testing and Materials, 1981, pp. 5–20.

ABSTRACT: Moisture content is a factor in the mechanical response of many composites. To evaluate this effect, it is necessary to achieve a controlled and uniform moisture distribution within each test specimen. Thin specimens reach moisture equilibrium quicker than thick ones, leading to shorter and more economical test programs. This paper presents the results of an initial attempt to develop a single-ply transverse tension test for unidirectional composites. A number of these tests were performed on unidirectional graphite composites. The transverse strengths were close to the thick specimen values. The stress distributions and the tensile buckling characteristics of the specimens were analyzed using NASTRAN.

KEY WORDS: composite materials, testing, evaluation, quality control, moisture effects, transverse tension, thin specimens, graphite/epoxy, graphite/polyimide

Absorbed moisture is generally regarded as having a significant effect on the mechanical properties of many graphite/epoxy composites. This water absorption process takes place so slowly that thicker laminates may take months or even years of exposure time to approach their equilibrium moisture states. Thin specimens, by virtue of their ability to reach moisture equilibrium much faster than thick ones, are more practical in any test program. Thus, it is desirable to develop quality-control and material property tests which can use the thinnest possible specimens. The single-ply laminate constitutes the thinnest practical graphite/epoxy specimen. Single-ply longitudinal tension tests are not difficult to perform. However, there does

[1]Aerospace engineer, HQ, U.S. Army R&T Laboratories (AVRADCOM), NASA Ames Research Center, Moffett Field, Calif. 94035.

not appear to be a generally acceptable single-ply transverse tension or shear test. This paper presents the results of an investigation of a simple and practical single-ply transverse tension strength test for unidirectional graphite/epoxy composites. Its extension to other unidirectional materials is straightforward. The test is based on a long-standing method of quality-control testing of woven textile fabrics wherein a swatch of cloth is stretched flat and then clamped between the grips of a tension testing machine that is equipped with hydraulic or pneumatic grips (Fig. 1). Care is taken to properly align the warp and fill threads in the region between the grips. The excess cloth that extends beyond the grips is permitted to drape freely. A tension strength test is then performed. Fabric strengths thus obtained are not significantly influenced by the presence of the excess draped material because woven cloth generally has a very low shear modulus, which minimizes the transfer of tensile load beyond the region bounded by the sides of the grips. Also, the wrinkled or postbuckled state of the excess material diminishes its tensile load-carrying capability; that is, the tensile strains primarily serve to smooth out wrinkles. These mechanisms may work to a similar advantage in a single-ply composite test. The shear stiffness of unidirectional material is low, tending to confine the tensile load to the immediate area between the grips. Also, the single-ply laminate is generally in a postbuckled state from either initial curvature or its own deadweight loading, or from both. Thus, there is reason to expect that a similar composite test may give transverse strength results that are reasonable when compared with results from thicker machined specimens.

Such a test is attractive for another reason. It eliminates the need for machining the sides of the specimen. This machining is expensive and it can introduce critical flaws in the test section.

This paper reports the results of a number of single-ply transverse tests performed on unidirectional graphite composites. The tests were done using 7.6 by 25.4-cm (3 by 10 in.) molded strips of material (Fig. 2). These strips were clamped directly in the testing machine without the use of tabs. The resulting mean strengths and variabilities were acceptable for engineering purposes. The two main difficulties encountered were the need for a uniform prepreg material without gaps, waviness, or large variations in local resin content and the need for close thickness control in molding the test specimens.

NASTRAN calculations were performed in order to establish the nature of the stress distributions throughout the entire specimen and the stress concentrations in the vicinity of the grips. These calculations were repeated for rigid grips and grips which had thin flexible rubber pads attached. Account was also taken of the effect of grip clamping pressure, which tends to reduce the effective shear stiffness of the rubber pads. The same analysis also predicted the onset of tensile buckling outside the test region at loads significantly

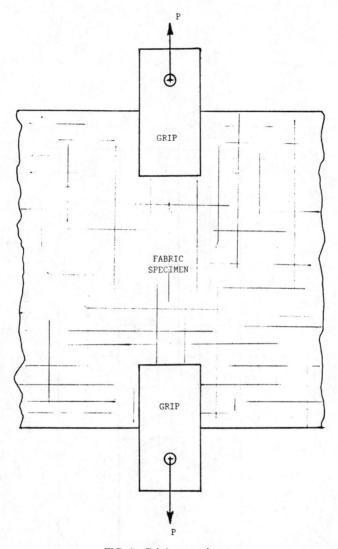

FIG. 1—*Fabric strength test*.

lower than the observed specimen strengths. The presence of rubber pads on the grips significantly influenced the critical buckling loads.

The specimens were fabricated and tested at the National Aeronautics and Space Administration (NASA) Langley Research Center, Hampton, Virginia. The analysis was done at NASA Ames Research Center, Moffett Field, California.

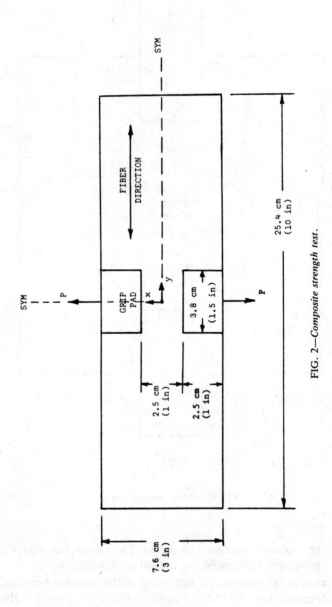

FIG. 2—Composite strength test.

Test Specimen

An individual test specimen consisted of a flat single-ply rectangular strip of unidirectional material measuring 7.6 cm (3 in.) in the transverse direction and 25.4 cm (10 in.) in the longitudinal direction. Prior to curing, the specimen was cut with scissors from a 7.6-cm-wide (3 in.) roll of prepreg tape. The longitudinal dimension was chosen somewhat arbitrarily. The uncured strip was laid up on a flat ground steel surface which had been treated with mold release. Similarly treated shim stock of the appropriate single-ply thickness was placed alongside the specimen. Another flat plate was then placed on top of the specimen. The mold was then placed in a heated platten press and the prescribed thermal cure cycle applied. Sufficient pressure was applied to close the mold and maintain thickness control with a minimum of distortion to the faceplates. Generally, a number of specimens were laid up and cured at the same time. When the cure cycle was complete, the mold was allowed to cool slowly, the face plates carefully separated, and the specimens carefully peeled off the plates. Each specimen was visually checked for excessive fiber washing or waviness. Holding a specimen in front of a bright light would clearly show resin-rich gaps, cracks, and similar nonuniformities. As a final quality check, each specimen thickness was measured with a micrometer at a dozen or more points on the surface. A thickness variation of more than 0.013 mm (0.0005 in.) was considered excessive.

Specimens were made from three different graphite prepregs. They were Narmco 5209/T300, Narmco 5208/T300, and PMR-15/T300 materials. The 5209/T300 tape was uniform and resulted invariably in good-quality specimens with a thickness of about 0.18 mm (0.007 in.) after curing and a fiber content of about 70 percent. However, all of these specimens contained about 3.2 mm (0.125 in.) of maximum out-of-plane distortion or initial deformation as measured from a flat surface to the point of maximum laminate departure from that surface. Most of this distortion appeared to be in the form of anticlastic curvature plus simple curvature over the 25.4-cm (10 in.) longitudinal length.

The 5208/T300 material produced very poor-quality specimens on the first and only attempt to use this material. The outer fibers in the tape washed nonuniformly in the transverse direction, resulting in about a 10.2-cm-wide (4 in.) specimen after curing. It was felt that this problem could be easily resolved by placing an appropriately sized strip of bleeder material on both sides of the uncured tape to absorb the excess resin and prevent migration of the fibers during cure. However, this remedy was not applied and the 5208/T300 specimens were discarded.

A minimum number of specimens were fabricated using a polyimide matrix PMR-15/T300 material. These specimens were acceptable in appearance and uniform in thickness, but did not look as good as the 5209/T300 material with regard to uniformity of fiber distribution. Their average thickness was about 0.25 mm (0.010 in.).

Test Method

All the tests were run on an Instron tester using a constant head motion of 1.27 mm/min (0.05 in./min). The pneumatic grips had rubber pads 3.8 cm (1.5 in.) wide and 2.5 cm (1 in.) in height which made contact with the test specimen (Fig. 2). The pads had a thickness of about 0.8 mm (0.03 in.) with a rubber hardness in the 55 to 60 range on the Shore Durometer A hardness scale. This translates into a shear modulus of about 0.86 MPa (125 psi) [1].[2] The grips operated with 0.55 MPa (80 psi) of air pressure acting on a 6.3-cm-diameter (2.5 in.) cylinder linked to the movable grip blocks. This led to a clamping pressure on the order of 0.69 MPa (100 psi). When the grips were positioned at the centers of the longer sides of a test specimen (Fig. 2) about 2.5 cm (1 in.) separated the grips. Thus, the effective gage length was about 2.5 cm (1 in.) plus a small amount of specimen pullout. The load was measured by a 890-N (200 lb) load cell and plotted against head motion by means of a continuous x-y recorder. A typical load/displacement curve is shown in Fig. 3. The initial curvilinear portion of the plot at very low load levels is due to straightening of the specimen and initial misalignment of the grips. The second curved portion of the plot occurring at about 20 percent of ultimate load is due to some slipping or pullout of the specimen in the grips. The curve was relatively linear beyond this point to ultimate strength. The small residual load-carrying capability was due to occasional incomplete fracture near the edges of the specimen.

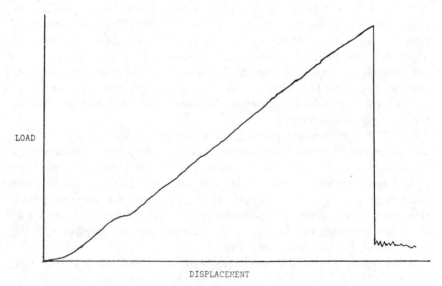

LOAD

DISPLACEMENT

FIG. 3—*Typical load/displacement plot.*

---

[2]The italic numbers in brackets refer to the list of references appended to this paper.

Experimental Data

The fracture features were similar for both of the materials tested. The specimens all cracked perpendicular to the direction of the applied load and parallel to the fiber direction. The fracture usually occurred near one of the grips. The broken specimens appeared to be undamaged away from the immediate vicinity of the longitudinal crack. The values of the apparent breaking stresses are given in Table 1. This calculation was based on the assumption that the area of the test specimen located between the sides of the grips carries all the load (as shown in Fig. 4). The average apparent strength was 69.7 MPa (10.1 ksi) for the graphite/epoxy specimens with a coefficient of variation of 14.1 percent. The apparent strength for the graphite/polyimide specimens was 59.5 MPa (8.6 ksi). Its coefficient of variation, based on just four specimens, was 4.6 percent. The apparent strengths are significantly higher than the corresponding values from thick specimen tests; namely, 60.0 MPa (8.7 ksi) for 5209/T300 [2] and 33.5 MPa (4.9 ksi) for PMR-15/T300 [3]. These differences will be discussed later. Coefficients of variation in the transverse strength of thicker graphite/epoxy laminates have been reported to be on the order of 5 to 7 percent [4].

There was no attempt to hygrothermally condition any of the test specimens. They were all permitted to remain unprotected in a standard test laboratory environment [21°C (70°F) and 50 percent relative humidity] for several months before testing. All testing was carried out in a short time span at the storage conditions.

TABLE 1—*Specimen apparent strength.*

Specimen No.	Specimen Material	Thickness, mm (in.)	Breaking Load, N (lb)	Apparent Strength MPa (ksi)
1	graphite/epoxy	0.170 (0.0067)	455 (102)	70.1 (10.15)
2	graphite/epoxy	0.178 (0.0070)	521 (117)	76.8 (11.13)
3	graphite/epoxy	0.172 (0.0068)	431 (97)	65.5 (9.50)
4	graphite/epoxy	0.178 (0.0070)	431 (97)	63.6 (9.23)
5	graphite/epoxy	0.172 (0.0068)	360 (81)	54.8 (7.94)
6	graphite/epoxy	0.170 (0.0067)	556 (125)	85.9 (12.44)
7	graphite/epoxy	0.172 (0.0068)	405 (91)	61.5 (8.93)
8	graphite/epoxy	0.180 (0.0071)	486 (109)	70.5 (10.21)
9	graphite/epoxy	0.187 (0.0074)	455 (102)	63.3 (9.18)
10	graphite/epoxy	0.183 (0.0072)	636 (143)	91.5 (13.25)
11	graphite/epoxy	0.175 (0.0069)	472 (106)	70.6 (10.24)
12	graphite/epoxy	0.170 (0.0067)	396 (89)	61.1 (8.86)
13	graphite/polyimide	0.257 (0.0101)	606 (136)	62.0 (8.98)
14	graphite/polyimide	0.254 (0.0100)	591 (133)	61.2 (8.87)
15	graphite/polyimide	0.243 (0.0097)	521 (117)	55.5 (8.05)
16	graphite/polyimide	0.238 (0.0095)	543 (122)	59.0 (8.56)

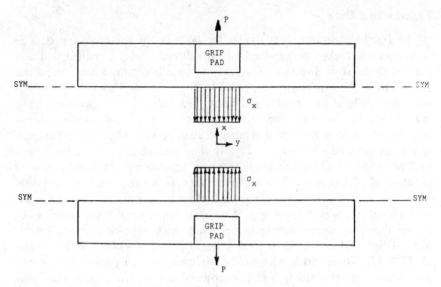

FIG. 4—*Assumed stress distribution.*

Stress Analysis

For the purposes of finite-element stress analysis, the geometry of the test specimen is defined in Fig. 2. The thickness is assumed to have a constant value of 0.18 mm (0.007 in.). Since the geometry, loading, and material properties are doubly symmetric, only one quadrant of Fig. 2 needs to be analyzed. Figure 5 shows the finite-element grid comprised primarily of rectangular bending elements, with triangular ones used only to facilitate grid mesh refinement in the vicinity of the grip pads. The material is assumed to be linearly elastic, homogeneous, and orthotropic with the following properties

$$
\begin{aligned}
E_L &= 139.0 \, \text{GPa} \, (20.1 \times 10^6 \, \text{psi}) \\
E_T &= 8.3 \, \text{GPa} \, (1.2 \times 10^6 \, \text{psi}) \\
G_{LT} &= 2.1 \, \text{GPa} \, (0.3 \times 10^6 \, \text{psi}) \\
\gamma_{TL} &= 0.3 \\
\gamma_{LT} &= 0.018
\end{aligned}
$$

Infinite shear stiffness is assumed in planes normal to the X-Y plane of Fig. 5.

The rubber grip pads were modeled by a pair of discrete shear springs attached to each node point which was covered by the pads. These springs resist relative motion of the covered node point in the X- and Y-directions of Fig. 5. These same nodes were constrained from any motion normal to the X-Y plane. The stiffnesses of the springs were assumed to be proportional to

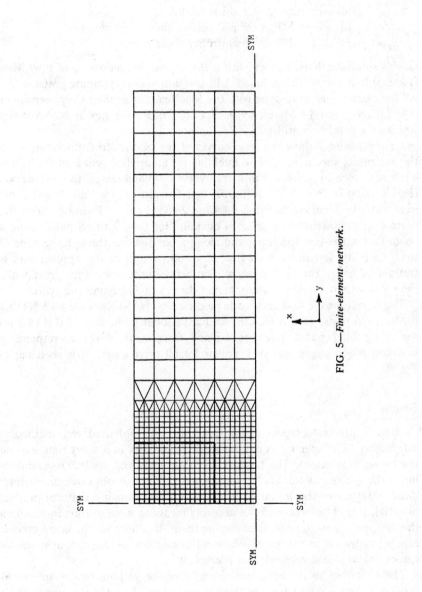

FIG. 5—*Finite-element network.*

both the rectangular area of rubber pad surrounding each covered node and the shear modulus of the rubber. The spring stiffnesses were inversely proportional to the rubber thickness. Three pad concepts were considered:

1. rigid pads—infinite rubber shear modulus,
2. hard pads—0.86-MPa (125 psi) rubber shear modulus, and
3. soft pads—0.17-MPa (25 psi) rubber shear modulus.

Case 3 simulates hard rubber with a shear modulus reduction of 0.69 MPa (100 psi) to account for the loss of stiffness due to grip clamping pressure [5]. All the calculations were done with the MacNeal-Schwendler Corp. version of NASTRAN dated 14 March 1979. A CDC 7600 computer at NASA Ames Research Center was utilized.

Figures 6 and 7 show the level curves of the planar stress distributions in the composite specimen prior to buckling for an applied grip load P which is equivalent to a hypothetical unit transverse normal stress in the test section (Fig. 4); that is, $P = 0.047$ N (0.0105 lb). The stresses for Case 3 (soft pads) were virtually identical to those for Case 2 (hard pads). Figure 8 shows the normal stress distributions prior to buckling for Case 2 (hard pads) along a centerline of the test specimen and along a parallel line through the edge of the grip pads. It can be seen that 55 to 60 percent of the applied load is transferred across the central test section between the sides of the grips, while the remainder is carried in the material that extends beyond the grips.

The lowest tensile buckling loads predicted by NASTRAN are 80.5 N (18.1 lb) for rigid pads, 37.8 N (8.5 lb) for hard rubber pads, and 37.0 N (8.3 lb) for soft rubber pads. The out-of-plane component of the corresponding buckling mode shape was very similar for all three cases. It is sketched in Fig. 9.

Results

From a practical experimental viewpoint, the proposed test method is satisfactory. The specimens can be made in batches in a short time and no machining is required. The test can be run quickly and easily. Preconditioning of the specimens can also be done rapidly. The only obvious experimental disadvantages are the initial requirement for a reasonably uniform prepreg material, free of defects, and the necessity for using some caution in handling the thin specimens to prevent damaging them. It is felt that the layup process can be improved to the point where all acceptable unidirectional materials can be made into acceptable test specimens.

The variability in the apparent strengths of the graphite/epoxy specimens is greater than is customary for thicker specimens. Some of this variability is accountable to inherent stress nonuniformity and volume differences between the proposed test and the traditional, thicker specimen, transverse strength test. However, much of it is due to increased variability of thickness

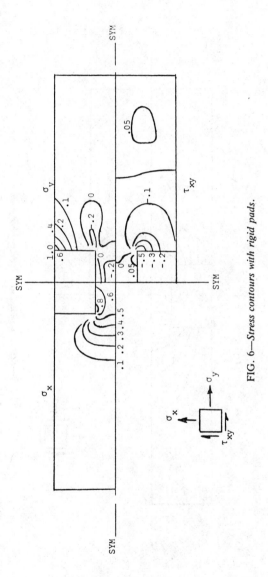

FIG. 6—*Stress contours with rigid pads.*

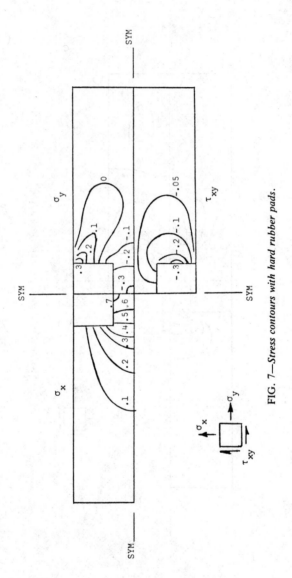

FIG. 7—Stress contours with hard rubber pads.

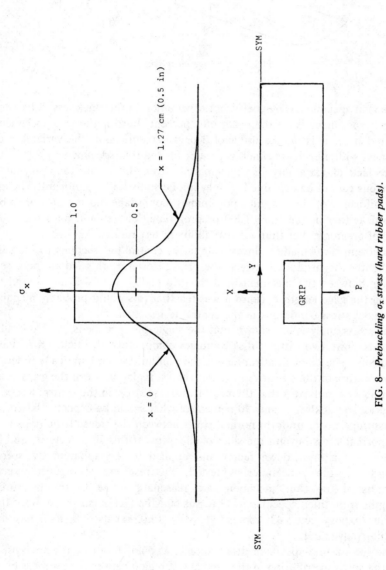

FIG. 8—*Prebuckling σ_x stress (hard rubber pads).*

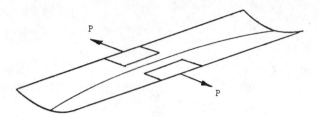

FIG. 9—*Buckling mode shape.*

in the thin specimen. The coefficient of variation in the thickness of batches of thick specimens is on the order of 1 percent, but for the thin specimens reported here, it is on the order of 3 or 4 percent. Also, the variability of thickness within each specimen is greater for the thin specimen.

The lack of variability in the apparent strengths of the four polyimide specimens cannot be explained. It may not be statistically significant. The initial uniformity of the graphite/polyimide prepreg tape did not appear to be as good as that of the 5209/T300 prepreg tape. This should have been the cause of greater rather than less variability in test results.

An attempt was made to investigate larger central test sections by increasing slightly the distance between the grips. However, this led to increased slipping between the specimens and the grip pads. There was no attempt to shorten the gage length because it was felt that this would probably magnify the biaxial stress conditions in the central test section.

The stress analysis has shown that the single-ply specimens, with the rubber pads, have two distinct disadvantages compared with thicker machined specimens. The chief disadvantage is the nonuniform and statically indeterminate nature of the stresses in the central test region between the grips. The linear analysis indicates that the peak transverse stress in the central test section prior to buckling is only 70 percent of what would be expected based on the assumptions of uniform normal stress between the sides of the grips and zero normal stress beyond the sides of the grips. If this 70 percent is used in the form of a knockdown factor and applied to the apparent transverse strengths of the two materials tested, the resulting average transverse strengths of the thin specimens are reasonably close to the measured strengths from thick specimens [49 versus 60 MPa (7.1 versus 8.7 ksi) for the graphite/expoxy and 42 versus 34 MPa (6.0 versus 4.9 ksi) for the graphite/polyimide].

The second thin-specimen disadvantage, as pointed out by the analysis, is that the stress distribution in the central test region between the grips is biaxial. However, this may not be such a serious drawback as it first appears because the magnitude of the more significant shear stress component is small in magnitude within the central test section. The larger longitudinal stress component probably does not have a significant effect on the transverse ultimate strength level. This is intuitively credible because, for most unidirectional materials, the reinforcing fibers carry the major share of the longitudinal stresses while the matrix and fiber/matrix interface are the critical elements in transverse tension.

The stress distributions indicate that the thin specimen test would be of little value without the flexible rubber pads. Rigid grips would introduce unacceptable stress concentrations at their corners. The rubber pads virtually eliminate these stress concentrations. The stresses are not sensitive to the rubber pad thickness or stiffness for the range of material properties considered here.

Conclusions

The thin-specimen test has advantages and disadvantages over the conventional thicker-specimen test. It is cheaper, quicker, and easier to perform, and it is not subject to machining damage. However, a significant portion of the transverse load is transferred beyond the central test region between the grips, and the stresses within that test region are neither uniform, uniaxial, nor statically determinate. A knockdown factor of 70 percent applied to the apparent strengths appears to provide a reasonable correction that brings the single-ply transverse strengths into agreement with thick-specimen results. However, these uncertainties place this new test in the category of a potential quality-control and material acceptance test rather than in the category of a basic property test, as is the case for the short-beam shear and three-point bending tests. Nevertheless, there is a continuing need for this type of test method development, and the advantages of the thin specimen method may make it attractive for many purposes.

There is considerable room for improvement and refinement of the single-ply test. Other specimen and grip pad geometries should be investigated. A positive means for initial alignment of the specimen and the grips should be developed. Improvements in specimen fabrication should improve the resulting test data and extend the range of material application.

This paper should be considered as a preliminary study or point of departure for subsequent improvements in a test method that has potential applications in quality control, acceptance testing, and the assessment of moisture effects.

Acknowledgments

The author wishes to express his appreciation to Mr. Mladen Chargin of NASA Ames Research Center for his assistance in the NASTRAN analysis and to Messrs. John Gleason and Todd Hodges of the U.S. Army AVRADCOM Structures Laboratory at NASA Langley Research Center for their assistance in the testing.

References

[1] Burton, W. E., *Engineering With Rubber*, McGraw-Hill, New York, 1949.
[2] Private correspondence with Narmco Materials, Costa Mesa, Calif.
[3] Shuart, M. J., "An Evaluation of the Sandwich Beam as a Compressive Test Method for Composites," MS Thesis in Engineering Mechanics, Virginia Polytechnic Institute and State University, Blacksburg, Va., Aug. 1978.
[4] Sandhu, R. S., "Analytical-Experimental Correlation of the Behavior of 0°, ±45°, 90° Family of AS/3501-5 Graphite Epoxy Composite Laminates Under Uniaxial Tensile Loading," AFFDL-TR-79-3064, Air Force Flight Dynamics Laboratory, Dayton, Ohio, May 1979.
[5] Foye, R. L., "Compression Strength of Unidirectional Composites," AIAA 3rd Aerospace Sciences Meeting, American Institute of Aeronautics and Astronautics, Paper No. 66-143, 24-26 Jan. 1966.

L. B. Greszczuk[1]

Application of Four-Point Ring-Twist Test for Determining Shear Modulus of Filamentary Composites

REFERENCE: Greszczuk, L. B., **"Application of Four-Point Ring-Twist Test for Determining Shear Modulus of Filamentary Composites,"** *Test Methods and Design Allowables for Fibrous Composites, ASTM STP 734,* C. C. Chamis, Ed., American Society for Testing and Materials, 1981, pp. 21-33.

ABSTRACT: A new test technique for measuring the shear moduli of isotropic and composite materials is described. The test technique employs circular rings subjected to out-of-plane four-point loading: four forces of equal magnitude, two upward at 0 and 180 deg and two downward at 90 and 270 deg. The pertinent equation relating ring deformation to applied loading, ring-section properties, and elastic properties is given. It is shown that by selecting appropriate specimen geometry, the deflection of the ring subjected to four-point loading is governed primarily by shear deformations. The reliability and accuracy of the test method are demonstrated by conducting tests on metallic rings as well as on rings made of various composite materials. Room as well as elevated-temperature shear-moduli data are measured using the ring tests. The results from the four-point ring-twist tests are compared with data presented in the literature. The test is simple and fast, with no requirement for elaborate instrumentation or setup. Moreover, the test is applicable for measuring the shear moduli at room temperature, cryogenic temperatures, and elevated temperatures.

KEY WORDS: composite materials, testing, shear modulus, ring tests, elevated temperature

Although a number of test methods for experimentally determining the shear moduli of composite materials exist [1,2],[2] not all lend themselves to determining the shear moduli at elevated or cryogenic temperatures. Moreover, the conventional test methods employ either flat plates or torsion bars or tubes and are difficult to apply when the specimen configuration for measuring the shear modulus is constrained to a ring, to a section of a cylinder, or to a section of the cone. For this reason, two new test methods employing circular rings have been developed. They are the Douglas Split-

---

[1] McDonnell Douglas Astronautics Co., Huntington Beach, Calif.
[2] The italic numbers in brackets refer to the list of references appended to this paper.

Ring Test described in Ref *3* and the four-point ring-twist test, which is the subject of this paper.

Basic Relationships

If a circular ring is subjected to out-of-plane four-point loading as shown in Fig. 1, the resultant vertical deflection at the point of the application of the load can be derived using the energy method [4], and is[3]

$$\delta = \frac{PR^3(\pi - 2)}{8EI} + \frac{PR^3(\pi - 3)}{4GJ} \tag{1}$$

where the first and second terms on the right-hand side of the equation denote, respectively, the deflections due to bending and torsion. In addition to the terms defined in Fig. 1, E is the Young's modulus in the hoop direction, G is the longitudinal shear modulus (G_{LT} in the case of filament-wound composite ring), I is the moment of inertia, and J is the section constant

$$J = \beta a b^3 \tag{2}$$

where β is a numerical factor that is a function of the a/b ratio and is tabulated in Table 1.

If Eq 1 is expressed as

$$\delta = \delta_B + \delta_T \tag{3}$$

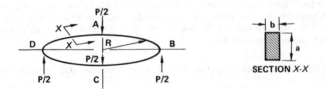

FIG. 1—*Ring subject to out-of-plane loading.*

[3]A more rigorous expression for the deflection, including the shear deformation term, can also be obtained using the energy method, and is

$$\delta = \frac{PR^3(\pi - 2)}{8EI} + \frac{PR^3(\pi - 3)}{4GJ}\left[1 + \frac{\pi J}{2abR^2(\pi - 3)}\right]$$

For typical test specimen configurations described later on in this paper it can be shown that $\pi J/[2abR^2(\pi - 3)] \ll 1$; that is, the deflections due to shear are negligible compared with torsional deflections. Moreover, as can readily be established from Ref 9, for the case of filament-wound rings or bars subjected to torsion, $G_{13} \equiv G_{23} \equiv G_{LT}$, where subscripts 1, 2 and 3 denote the radial, axial (perpendicular to plane of ring), and circumferential directions, respectively, in the ring and thus $\beta \neq f(G_{23}, G_{13})$, but is identical to values given in Table 1.

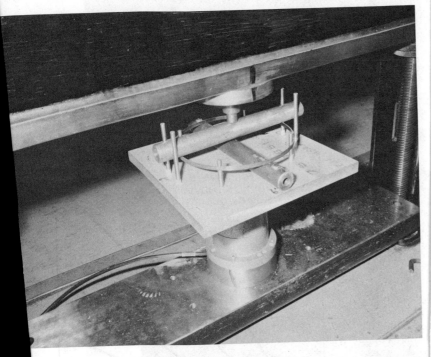

FIG. 3—Overall view of test setup.

pecimens was monitored with thermocouples attached to the speci-
After the specimen reached the desired temperature, it was soaked
at temperature for 10 min prior to test. Figure 4 shows the deflected
on of a ring subjected to the four-point twist test.

ical load-deflection curves for aluminum and graphite/epoxy rings
ted to four-point loading are shown in Figs. 5-8. Figure 6 shows the
eflection curve for a Thornel 300/epoxy ring at 21°C (70°F), and
7 and 8 show similar results for Pitch 75/epoxy rings tested at 21 an
(70 and 250°F). The resin system used for making the composit
onsisted of APCO resin 2447 and APCO hardener 2343. Before th
int flexure tests, the various rings were tested in diametrical com
n to obtain their Young's moduli for use in Eq 6.

he aluminum ring, the shear modulus obtained from the four-poi
st test was 28.3 GPa (4.1 $\times$ 10^6 psi), as compared with the han
lue for the shear modulus of $G = 26.2$ GPa (3.8 $\times$ 10^6 psi). In t
composites, one of the variables investigated was specimen geo

TABLE 1—Values of β as a function of a/b [5].

a/b	1.0	1.2	1.5	1.75	2.0	2.5	3.0	4.0	5.0	6.0	8.0	10.0	∞
β	0.141	0.166	0.196	0.214	0.229	0.249	0.263	0.281	0.291	0.299	0.307	0.313	0.333

where δ_B is the deflection due to bending

$$\delta_B = \frac{PR^3 (\pi - 2)}{8EI} \qquad (4)$$

and δ_T is the deflection due to torsion

$$\delta_T = \frac{PR^3 (\pi - 3)}{4GJ} \qquad (5)$$

then the relative magnitudes of the two components of the ring deflection can be compared by plotting the δ_B/δ_T ratio. Such a plot is shown in Fig. 2, where δ_B/δ_T is plotted as a function of G/E, the ratio of shear modulus to Young's modulus for various values of a/b. From the results presented in Fig. 2, it is apparent that for low values of G/E the deflection of the ring subjected to four-point loading is primarily due to twisting. This test, therefore, is especially suitable for measuring the shear moduli of filamentary composites.

Solving Eq 1 for G, the following expression is obtained for determining the shear modulus from the four-point ring-twist test

$$G = \frac{\pi - 3}{J \left[\dfrac{4\delta}{PR^3} - \dfrac{\pi - 2}{EI} \right]} \qquad (6)$$

If $E \gg G$ or the specimen configuration (a/b ratio) is chosen such that $\delta_B/\delta_T \ll 1$, the second term in the denominator of Eq 6 can be neglected, thus giving

$$G = \frac{(\pi - 3) PR^3}{4\delta J} \qquad (7)$$

Equation 6 can be used as is, if the Young's modulus, E, of the ring is first evaluated by subjecting the ring to two diametrically opposite concentrated loads and calculating the modulus from the following well-known equation [6]

$$E = \frac{0.149 \, FR^3}{\delta_F I} \qquad (8)$$

where F is the applied load and δ_F the deflection at the load application point.

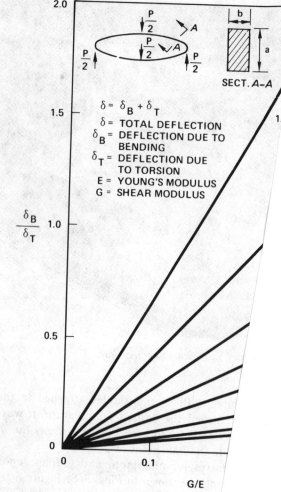

FIG. 2—*Effect of material properties and ring geome* *ponents of ring deflection.*

Test Procedure

The test setup used for conducting the four in Fig. 3. The loading and support bars were and held in position by alignment pins. T Instron test machine at a loading rate of (load versus machine-head travel was record and unloading. In the elevated-temperatu fixture were placed inside the temperatur

the s
men.
at th
positi
Typ
subje
load-
Figs.
121°C
rings
four-p
pressio

Results

For t
ring-twi
book va
case of

FIG. 4—*Deflections in graphite/epoxy ring subjected to four-point loading.*

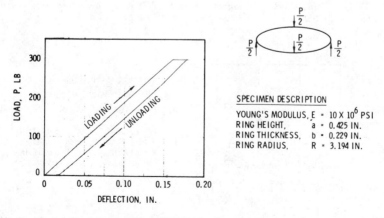

SPECIMEN DESCRIPTION

YOUNG'S MODULUS, E = 10×10^6 PSI
RING HEIGHT, a = 0.425 IN.
RING THICKNESS, b = 0.229 IN.
RING RADIUS, R = 3.194 IN.

FIG. 5—*Load-deflection curve for a 6061-T6 aluminum ring subjected to four-point loading (1 in. = 25.4 mm; 1 psi = 6.895 Pa; 1 lb = 4.45 N).*

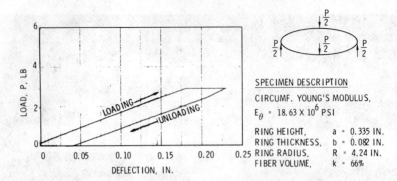

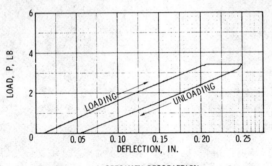

FIG. 6—*Typical load-deflection curve for unidirectional Thornel 300/epoxy ring subjected to four-point twist test (1 in. = 25.4 mm; 1 psi = 6.845 Pa; 1 lb = 4.45 N).*

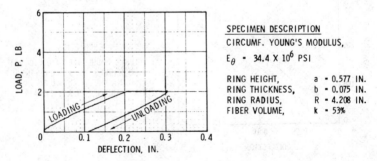

FIG. 7—*Load-deflection curve for unidirectional Pitch 75/epoxy ring subjected to four-point twist test at 21°C (70°F) (1 in. = 25.4 mm; 1 psi = 6.895 Pa; 1 lb = 4.45 N).*

FIG. 8—*Load-deflection curve for unidirectional Pitch 75/epoxy ring subjected to four-point twist test at 121°C (250°F) (1 in. = 25.4 mm; 1 psi = 6.895 Pa; 1 lb = 4.45N).*

TABLE 2—*Influence of specimen geometry on shear modulus of unidirectional T300/epoxy composites.*

Specimen	Ring Thickness, b, in.[a]	Ring Height, a, in.	a/b	Ring Radius, R, in.	Young's Modulus, $E \times 10^{-6}$ psi[b]	Shear Modulus,[c] $G \times 10^{-6}$ psi	$G^{*[d]} \times 10^{-6}$ psi	Percent Difference[e]
1	0.0826	0.168	2.04	4.241	18.90	0.822	0.716	12.9
2	0.0825	0.253	3.07	4.241	19.08	0.845	0.788	6.8
3	0.0825	0.335	4.06	4.241	18.63	0.830	0.793	4.5
4	0.0820	0.500	6.10	4.239	18.82	0.873	0.855	2.1
Avg	...	...	...	...	...	0.842	...	...

[a] 1 in. = 25.4 mm.
[b] 1 psi = 6.895 Pa.
[c] Based on Eq 6.
[d] Based on Eq 7.
[e] Percent difference = $[(G - G^{*})/G \times 100]$.

TABLE 3—*Room-temperature and elevated-temperature shear Moduli for unidirectional graphite/epoxy rings.*

Specimen	Fiber Material	Test Temperature, °F[a]	Ring Thickness, b, in.[b]	Ring Height, a, in.	a/b	Ring Radius, R, in.	Young's Modulus, $E \times 10^{-6}$ psi[c]	Shear Modulus, $G \times 10^{-6}$ psi
5	Thornel 300	70	0.0816	0.311	3.81	4.238	19.04	0.879
6A	Thornel 300	70	0.0812	0.500	6.16	4.241	19.04	0.826
6B	Thornel 300	250	0.0812	0.500	6.16	4.241	18.43	0.531
6C	Thornel 300	350	0.0812	0.500	6.16	4.241	...	0.059
7	Pitch 75	70	0.0762	0.246	3.23	4.204	38.32	0.510
8A	Pitch 75	70	0.075	0.500	7.69	4.208	40.72	0.600
8B	Pitch 75	250	0.075	0.500	7.69	4.208	34.44	0.343

[a] $°C = \dfrac{5}{9}(°F - 32)$.

[b] 1 in. = 25.4 mm.

[c] 1 psi = 6.895 Pa.

TABLE 4—Comparison of shear moduli measured using four-point ring-twist test with data given in literature.[a]

Material (Fiber/Resin)	Fiber Volume Fraction, k (%)	Shear Modulus $G \times 10^{-6}$ psi[b]	G/k $\times 10^{-6}$ psi	Test Method	Reference
T300/934	63	0.66	1.05	not specified	[7]
T300/934	62	0.66	1.06	not specified	[7]
T300/934	60	0.72	1.20	not specified	[7]
T300/313	61	0.78	1.28	±45-deg coupon	[7]
T300/5208	63	0.93	1.47	rail shear	[7]
T300/5208	63	1.04	1.65	±45-deg coupon	[7]
T300/5208	60.7	0.79	1.30	±45-deg coupon	[8]
T300/APCO 2447-2343	66	0.84	1.27	four-point ring twist	present paper
GY70/5208	57.2	0.61	1.17	±45-deg coupon	[8]
Pitch 75/APCO 2447-2343	53.0	0.56	1.06	four-point ring twist	present paper

[a]All data are for unidirectional composites.
[b]1 psi = 6.895 Pa.

etry. Table 2 shows the shear moduli measured using rings with various a/b ratios; it contains values of shear moduli obtained from Eqs 6 and 7. If $a/b \geq 5$, the error resulting from use of Eq 7 rather than Eq 6 would be of the same magnitude as the scatter in test data. The fiber content of composites for which data were presented in Table 2 was $k = 66$ percent by volume. Room and elevated-temperature shear-moduli data for unidirectional Thornel 300/epoxy and Pitch 75/epoxy composites are presented in Table 3. Specimens 6 and 8 were tested at both room and elevated temperatures. Except for Specimen 6c, all values of the shear moduli were calculated from Eq 6. The shear modulus of Specimen 6c was calculated from Eq 7, because no measurement of E at 177°C (350° F) was made. Following the 177°C (350°F) four-point ring-twist test, the specimen exhibited permanent deformation. The permanent deformation and the low value of shear modulus at 177°C (350°F) could be indicative of the improper postcure of the specimen. The amount of permanent deformation following the 121°C (250°F) tests was negligible.

A comparison of shear moduli data obtained from the four-point ring-twist tests with data given in literature for similar types of composites is presented in Table 4. Because the data presented in the literature were for composites with fiber contents different than those of the specimens described in this paper, Table 4 also gives the normalized results (G/k). No shear moduli data for composites made with Pitch 75 fibers could be found in the literature. Consequently, the data for Pitch 75 composites are compared with the literature data for GY70/epoxy composites, because the GY70 graphite fibers have mechanical properties very similar to those of Pitch 75 graphite fibers.

Conclusions

The four-point ring-twist test described in this paper appears to be a simple, accurate, and economical way to determine the shear moduli of isotropic and composite materials. The test can be used at room temperature as well as at elevated and cryogenic temperatures.

References

[1] Greszczuk, L. B. in *Composite Materials: Testing and Design ASTM STP 460*, American Society for Testing and Materials, 1969, pp. 140–149.
[2] Adams, D. F. and Thomas, R. L. in *Proceedings*, Twelfth National Conference of the Society of Aerospace Materials and Process Engineers, Anaheim, Calif. 10–12 Oct. 1967, p. AC-5.
[3] Greszczuk, L. B. in *Proceedings*, 23rd Annual Technical and Management Conference, Society of Plastics Industry, Washington, D.C., Feb. 1968.
[4] Greszczuk, L. B. in *Proceedings*, 24th National Symposium of the Society of Aerospace Materials and Process Engineers, San Francisco, Calif., May 1979.
[5] Timoshenko, S. and Goodier, J. N., *Theory of Elasticity*, McGraw-Hill, New York, 1951.
[6] Roark, R. J., *Formula for Stress and Strain*, McGraw-Hill, New York, 1954.

TABLE 1—Values of β as a function of a/b [5].

a/b	1.0	1.2	1.5	1.75	2.0	2.5	3.0	4.0	5.0	6.0	8.0	10.0	∞
β	0.141	0.166	0.196	0.214	0.229	0.249	0.263	0.281	0.291	0.299	0.307	0.313	0.333

where δ_B is the deflection due to bending

$$\delta_B = \frac{PR^3 (\pi - 2)}{8EI} \tag{4}$$

and δ_T is the deflection due to torsion

$$\delta_T = \frac{PR^3 (\pi - 3)}{4GJ} \tag{5}$$

then the relative magnitudes of the two components of the ring deflection can be compared by plotting the δ_B/δ_T ratio. Such a plot is shown in Fig. 2, where δ_B/δ_T is plotted as a function of G/E, the ratio of shear modulus to Young's modulus for various values of a/b. From the results presented in Fig. 2, it is apparent that for low values of G/E the deflection of the ring subjected to four-point loading is primarily due to twisting. This test, therefore, is especially suitable for measuring the shear moduli of filamentary composites.

Solving Eq 1 for G, the following expression is obtained for determining the shear modulus from the four-point ring-twist test

$$G = \frac{\pi - 3}{J \left[\dfrac{4\delta}{PR^3} - \dfrac{\pi - 2}{EI} \right]} \tag{6}$$

If $E \gg G$ or the specimen configuration (a/b ratio) is chosen such that $\delta_B/\delta_T \ll 1$, the second term in the denominator of Eq 6 can be neglected, thus giving

$$G = \frac{(\pi - 3) PR^3}{4\delta J} \tag{7}$$

Equation 6 can be used as is, if the Young's modulus, E, of the ring is first evaluated by subjecting the ring to two diametrically opposite concentrated loads and calculating the modulus from the following well-known equation [6]

$$E = \frac{0.149 \, FR^3}{\delta_F I} \tag{8}$$

where F is the applied load and δ_F the deflection at the load application point.

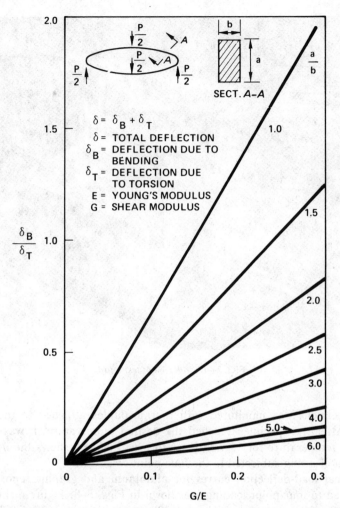

FIG. 2—*Effect of material properties and ring geometry on bending and torsional components of ring deflection.*

Test Procedure

The test setup used for conducting the four-point ring-twist tests is shown in Fig. 3. The loading and support bars were placed at 90 deg to each other and held in position by alignment pins. The tests were conducted in an Instron test machine at a loading rate of 0.0118 m/s (0.5 in./min). The load versus machine-head travel was recorded continuously during loading and unloading. In the elevated-temperature tests, the specimen and test fixture were placed inside the temperature chamber. The temperature of

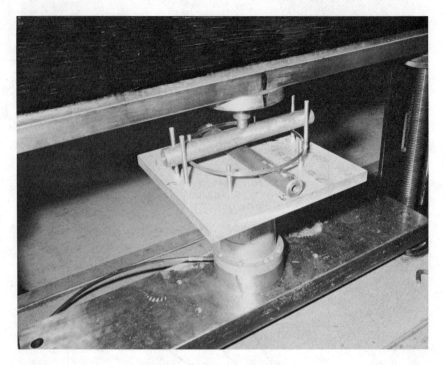

FIG. 3—*Overall view of test setup.*

the specimens was monitored with thermocouples attached to the specimen. After the specimen reached the desired temperature, it was soaked at that temperature for 10 min prior to test. Figure 4 shows the deflected position of a ring subjected to the four-point twist test.

Typical load-deflection curves for aluminum and graphite/epoxy rings subjected to four-point loading are shown in Figs. 5–8. Figure 6 shows the load-deflection curve for a Thornel 300/epoxy ring at 21°C (70°F), and Figs. 7 and 8 show similar results for Pitch 75/epoxy rings tested at 21 and 121°C (70 and 250°F). The resin system used for making the composite rings consisted of APCO resin 2447 and APCO hardener 2343. Before the four-point flexure tests, the various rings were tested in diametrical compression to obtain their Young's moduli for use in Eq 6.

Results

For the aluminum ring, the shear modulus obtained from the four-point ring-twist test was 28.3 GPa (4.1 $\times$ 10^6 psi), as compared with the handbook value for the shear modulus of $G = 26.2$ GPa (3.8 $\times$ 10^6 psi). In the case of composites, one of the variables investigated was specimen geom-

FIG. 4—*Deflections in graphite/epoxy ring subjected to four-point loading.*

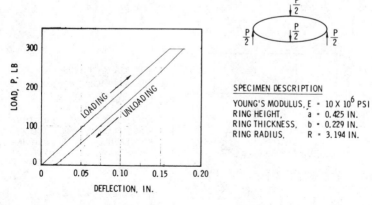

SPECIMEN DESCRIPTION

YOUNG'S MODULUS, E = 10×10^6 PSI
RING HEIGHT, a = 0.425 IN.
RING THICKNESS, b = 0.229 IN.
RING RADIUS, R = 3.194 IN.

FIG. 5—*Load-deflection curve for a 6061-T6 aluminum ring subjected to four-point loading (1 in. = 25.4 mm; 1 psi = 6.895 Pa; 1 lb = 4.45 N).*

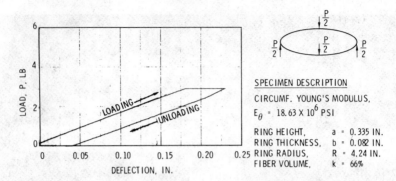

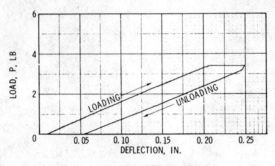

FIG. 6—*Typical load-deflection curve for unidirectional Thornel 300/epoxy ring subjected to four-point twist test (1 in. = 25.4 mm; 1 psi = 6.845 Pa; 1 lb = 4.45 N).*

FIG. 7—*Load-deflection curve for unidirectional Pitch 75/epoxy ring subjected to four-point twist test at 21°C (70°F) (1 in. = 25.4 mm; 1 psi = 6.895 Pa; 1 lb = 4.45 N).*

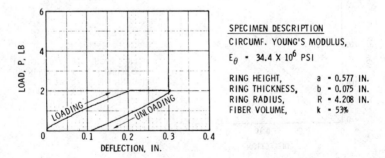

FIG. 8—*Load-deflection curve for unidirectional Pitch 75/epoxy ring subjected to four-point twist test at 121°C (250°F) (1 in. = 25.4 mm; 1 psi = 6.895 Pa; 1 lb = 4.45N).*

TABLE 2—*Influence of specimen geometry on shear modulus of unidirectional T300/epoxy composites.*

Specimen	Ring Thickness, b, in.[a]	Ring Height, a, in.	a/b	Ring Radius, R, in.	Young's Modulus, $E \times 10^{-6}$ psi[b]	Shear Modulus,[c] $G \times 10^{-6}$ psi	G^{*}[d] $\times 10^{-6}$ psi	Percent Difference[e]
1	0.0826	0.168	2.04	4.241	18.90	0.822	0.716	12.9
2	0.0825	0.253	3.07	4.241	19.08	0.845	0.788	6.8
3	0.0825	0.335	4.06	4.241	18.63	0.830	0.793	4.5
4	0.0820	0.500	6.10	4.239	18.82	0.873	0.855	2.1
Avg	...	...	...	...	...	0.842	...	...

[a] 1 in. = 25.4 mm.
[b] 1 psi = 6.895 Pa.
[c] Based on Eq 6.
[d] Based on Eq 7.
[e] Percent difference = $[(G - G^{*})/G \times 100]$.

TABLE 3—*Room-temperature and elevated-temperature shear Moduli for unidirectional graphite/epoxy rings.*

Specimen	Fiber Material	Test Temperature, °F[a]	Ring Thickness, b, in.[b]	Ring Height, a, in.	a/b	Ring Radius, R, in.	Young's Modulus, E × 10⁻⁶ psi[c]	Shear Modulus, G × 10⁻⁶ psi
5	Thornel 300	70	0.0816	0.311	3.81	4.238	19.04	0.879
6A	Thornel 300	70	0.0812	0.500	6.16	4.241	19.04	0.826
6B	Thornel 300	250	0.0812	0.500	6.16	4.241	18.43	0.531
6C	Thornel 300	350	0.0812	0.500	6.16	4.241	...	0.059
7	Pitch 75	70	0.0762	0.246	3.23	4.204	38.32	0.510
8A	Pitch 75	70	0.075	0.500	7.69	4.208	40.72	0.600
8B	Pitch 75	250	0.075	0.500	7.69	4.208	34.44	0.343

[a] °C = $\frac{5}{9}$(°F − 32).

[b] 1 in. = 25.4 mm.

[c] 1 psi = 6.895 Pa.

TABLE 4—Comparison of shear moduli measured using four-point ring-twist test with data given in literature.[a]

Material (Fiber/Resin)	Fiber Volume Fraction, k (%)	Shear Modulus $G \times 10^{-6}$ psi[b]	$G/k \times 10^{-6}$ psi	Test Method	Reference
T300/934	63	0.66	1.05	not specified	[7]
T300/934	62	0.66	1.06	not specified	[7]
T300/934	60	0.72	1.20	not specified	[7]
T300/313	61	0.78	1.28	±45-deg coupon	[7]
T300/5208	63	0.93	1.47	rail shear	[7]
T300/5208	63	1.04	1.65	±45-deg coupon	[7]
T300/5208	60.7	0.79	1.30	±45-deg coupon	[8]
T300/APCO 2447-2343	66	0.84	1.27	four-point ring twist	present paper
GY70/5208	57.2	0.61	1.17	±45-deg coupon	[8]
Pitch 75/APCO 2447-2343	53.0	0.56	1.06	four-point ring twist	present paper

[a] All data are for unidirectional composites.
[b] 1 psi = 6.895 Pa.

etry. Table 2 shows the shear moduli measured using rings with various a/b ratios; it contains values of shear moduli obtained from Eqs 6 and 7. If $a/b \geq 5$, the error resulting from use of Eq 7 rather than Eq 6 would be of the same magnitude as the scatter in test data. The fiber content of composites for which data were presented in Table 2 was $k = 66$ percent by volume. Room and elevated-temperature shear-moduli data for unidirectional Thornel 300/epoxy and Pitch 75/epoxy composites are presented in Table 3. Specimens 6 and 8 were tested at both room and elevated temperatures. Except for Specimen 6c, all values of the shear moduli were calculated from Eq 6. The shear modulus of Specimen 6c was calculated from Eq 7, because no measurement of E at 177°C (350° F) was made. Following the 177°C (350°F) four-point ring-twist test, the specimen exhibited permanent deformation. The permanent deformation and the low value of shear modulus at 177°C (350°F) could be indicative of the improper postcure of the specimen. The amount of permanent deformation following the 121°C (250°F) tests was negligible.

A comparison of shear moduli data obtained from the four-point ring-twist tests with data given in literature for similar types of composites is presented in Table 4. Because the data presented in the literature were for composites with fiber contents different than those of the specimens described in this paper, Table 4 also gives the normalized results (G/k). No shear moduli data for composites made with Pitch 75 fibers could be found in the literature. Consequently, the data for Pitch 75 composites are compared with the literature data for GY70/epoxy composites, because the GY70 graphite fibers have mechanical properties very similar to those of Pitch 75 graphite fibers.

Conclusions

The four-point ring-twist test described in this paper appears to be a simple, accurate, and economical way to determine the shear moduli of isotropic and composite materials. The test can be used at room temperature as well as at elevated and cryogenic temperatures.

References

[1] Greszczuk, L. B. in *Composite Materials: Testing and Design ASTM STP 460*, American Society for Testing and Materials, 1969, pp. 140–149.
[2] Adams, D. F. and Thomas, R. L. in *Proceedings*, Twelfth National Conference of the Society of Aerospace Materials and Process Engineers, Anaheim, Calif. 10–12 Oct. 1967, p. AC-5.
[3] Greszczuk, L. B. in *Proceedings*, 23rd Annual Technical and Management Conference, Society of Plastics Industry, Washington, D.C., Feb. 1968.
[4] Greszczuk, L. B. in *Proceedings*, 24th National Symposium of the Society of Aerospace Materials and Process Engineers, San Francisco, Calif., May 1979.
[5] Timoshenko, S. and Goodier, J. N., *Theory of Elasticity*, McGraw-Hill, New York, 1951.
[6] Roark, R. J., *Formula for Stress and Strain*, McGraw-Hill, New York, 1954.

[7] *Advanced Composite Design Guide,* Vol. 1, Book 1, 3rd ed. (first draft), Prepared by North American Rockwell Corp. under USAF Contract No. F33615-71-C-1362.

[8] Greszczuk, L. B. and Ashizawa M., "Advanced Composite Foil Test Component (Tapered Box Beam)," Final Report prepared under U.S. Naval Ship Systems Command Contract N00024-74-C-5441, May 1977.

[9] Lekhnitskii, S. G. *Theory of Elasticity of an Anisotropic Elastic Body* (translation from a 1950 Russian edition), Holden-Day, San Francisco, 1963, pp. 197–204.

R. K. Clark[1] *and W. B. Lisagor*[1]

Compression Testing of Graphite/Epoxy Composite Materials

REFERENCE: Clark, R. K. and Lisagor, W. B., **"Compression Testing of Graphite/Epoxy Composite Materials,"** *Test Methods and Design Allowables for Fibrous Composites, ASTM STP 734,* C. C. Chamis, Ed., American Society for Testing and Materials, 1981, pp. 34–53.

ABSTRACT: The focus of this investigation was to provide results that will support the selection of a reliable method of compression testing coupon specimens of filament-reinforced polymer-matrix composite materials. Three schemes were examined for testing graphite/epoxy (Narmco T300/5208) composite material specimens to failure in compression, including an adaptation of the Illinois Institute of Technology Research Institute (IITRI) "wedge grip" compression fixture, a face-supported compression fixture, and an end-loaded-coupon fixture. The effects of specimen size, specimen support arrangement, and method of load transfer on compressive behavior of graphite/epoxy were investigated.

Compression tests with the modified IITRI and face-supported fixture were conducted on specimens of 12.5, 25, and 50-mm widths, of 8, 16, and 24-ply thicknesses, and of [0], [±45], and [0/±45/90] fiber orientations. The end-loaded coupon fixture was used to test 16-ply [0/±45/90] specimens.

Compressive stress-strain, strength, and modulus data obtained with the three fixtures are presented with evaluations showing the effects of all test parameters, including fiber orientation. The IITRI fixture has the potential to provide good stress/strain data to failure for unidirectional and quasi-isotropic laminates. The face-supported fixture was found to be the most desirable for testing $[\pm 45]_s$ laminates.

KEY WORDS: composite materials, mechanical properties, compression testing, graphite/epoxy, compressive strength, compressive modulus, failure modes

The efficient use of filament-reinforced composite materials in aerospace applications requires that their thermal, physical, and mechanical properties be established accurately. Because of the inhomogeneity and brittle nature of these composites, the properties measured are more sensitive to testing equipment and procedures than are those for isotropic, homogeneous materials possessing some ductility. Reliable compressive properties for composite materials are the most difficult of all mechanical properties to

[1] Research scientists, NASA-Langley Research Center, Hampton, Va. 23665.

acquire because of the sensitivity of compression tests to a range of factors, including test method, quality of material, and uneven loading of specimens [1].[2] The importance of good compression test methods is related to the fact that compression loads are often a dominant factor when composites degrade under cyclic loading and environmental exposure [2] and the fact that compressive strength is the property most severely affected when composites experience environmental degradation.

Compression test methods currently in use generally are of three types: the sandwich beam compression test method, unsupported compression coupon test methods, and supported compression coupon test methods. Reference 1 gives an evaluation of the sandwich beam compression test method, which has been highly regarded as a dependable means of testing composite materials in compression, although questions have been raised regarding the effects of the honeycomb on performance of the laminate. Drawbacks include the relatively high cost of the sandwich beam specimen and its general unsuitability for environmental testing. A host of fixtures have been employed to test specimens that are unsupported in the gage length. Reference 3 presents a good description of the Illinois Institute of Technology Research Institute (IITRI) wedge-grip compression text fixture which is perhaps the most widely used fixture of this class. The supported compression coupon test methods include those procedures in which the specimen is fully supported in the gage length to prevent buckling during loading. References 4 and 5 contain good descriptions of fixtures of this class.

Three American Society for Testing and Materials (ASTM)-approved standard test procedures for compression tests of composite materials include Test for Compressive Properties of Rigid Plastics (D 695-69), Test for Compressive Properties of Oriented Fiber Composites (D 3410-75), and Flexure Test of Flat Sandwich Constructions (C 393-62).

This paper presents results from an evaluation of three schemes for compression testing coupons of graphite/epoxy composite material. These results are presented for the purpose of identifying the sensitivity of individual test techniques to laminate, specimen, and test parameters and comparing results from the three test schemes. The test fixtures utilized included an adaptation of the IITRI compression fixture, a face-supported compression fixture, and an end-loaded-coupon fixture.

Procedure

Compression tests of Narmco T300/5208 graphite-fiber reinforced epoxy-resin matrix (graphite/epoxy) composite material specimens were conducted using three test fixtures. Table 1 defines the number and type of speci-

---

[2] The italic numbers in brackets refer to the list of references appended to this paper.

TABLE 1—*Number and type of specimens tested.*

Fiber Orientation	[0]			[0/±45/90]			[±45]		
No. of Plies	8	16	24	8	16	24	8	16	24
Type/No. of Specimens									
IITRI:									
12.5 mm	...	6	...	5	5	5	...	5	...
25 mm	...	4	...	5	5	5	...	6	...
50 mm	...	5	...	6	5	6	...	5	...
Face-supported:									
12.5 mm	...	...	...	...	6	...	...	...	...
25 mm	...	6	...	5	5	6	...	5	...
50 mm	...	4	...	...	4	...	...	5	...
End loaded:									
25 mm	...	...	...	...	10	...	...	...	...

mens tested. Preliminary tests were conducted with each fixture, using 2024-T4 aluminum alloy sheet specimens to verify the experimental procedures. All tests were conducted at a nominal strain rate of $17 \times 10^{-5}\,\mathrm{s}^{-1}$.

IITRI Compression Test Fixture

The IITRI compression test fixture, shown in Fig. 1, was modified to permit testing of 12.5, 25, and 50-mm-wide specimens. Figure 1 also shows

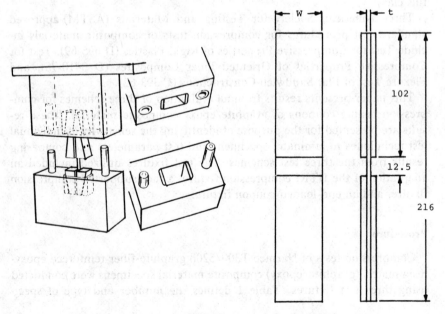

FIG. 1—*Schematic of IITRI fixture and sketch of specimen.*

a sketch of the IITRI specimen. Tabs for the specimen were fabricated from a glass-reinforced epoxy-matrix material (fiber glass) and were bonded to the specimen using a 392 K cure adhesive. The wedge grips are bolted to each tabbed end of the specimen. This prestressed the tabs transverse to the plane of the specimen and prevented slippage of the tabs under low axial loads. The outer surfaces of the wedge grips react with mating surfaces in the upper and lower bolsters to transmit compressive loads to the specimen. The lower bolster has two parallel alignment shafts that fit into two roller bushings in the upper bolster to ensure lateral alignment of the upper and lower units. Axial alignment of the upper and lower units is verified by gaging the parallelism of the matching surfaces. Axial alignment was adjusted as necessary by shimming between the contact surfaces of the test machine and the bolsters.

Considerable attention to detail was directed toward achieving precision in fabricating specimens for the IITRI fixture. One of the most critical details was to ensure that the opposing tab surfaces which are gripped during loading are flat within ± 25 μm.

Face-Supported Compression Test Fixture

Figure 2 shows an exploded view of the face-supported compression fixture with a 50-mm-wide specimen and a sketch of the specimen. The specimen was mounted in the fixture with about 0.1 mm clearance between the specimen and the inner platens. Strain gages were positioned on the specimen such that when the specimen is mounted in the fixture the gages are located within the gap in the inner platens. Tests with this fixture were conducted in a universal hydraulic testing machine with hydraulic grips. Compression loads are transmitted to the specimen through the specimen tabs. Great care was taken in installing the fixture and specimen in the testing machine to ensure alignment of the specimen and testing machine axes.

End-Loaded-Coupon Compression Text Fixture

The end-loaded-coupon compression fixture (Fig. 3) consists of two end blocks with provisions to anchor the ends of coupon specimens and a guide cylinder which ensures alignment of the end blocks. The mating surfaces of the guide cylinder and the end blocks are lubricated to minimize frictional loading of the cylinder. Load is transmitted to the specimen by the end blocks. The end blocks have retainers for anchoring the specimens to the end blocks which also provide support of the specimen transverse to the load axis and prevent failure of the specimen by "brooming." Specimen width up to 25 mm may be accommodated.

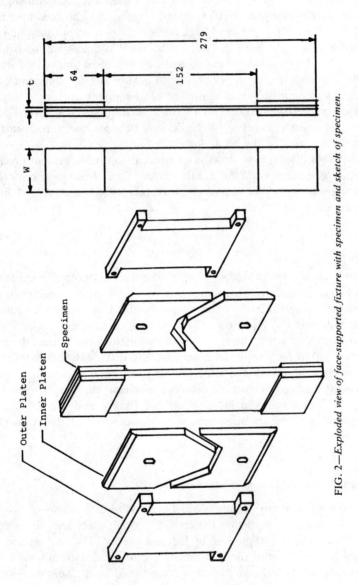

FIG. 2.—*Exploded view of face-supported fixture with specimen and sketch of specimen.*

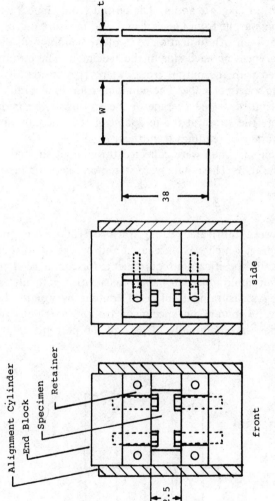

FIG. 3—*Sketch of end-loaded-coupon fixture and specimen.*

Instrumentation and Data Collection

Load-strain data were obtained for each specimen throughout the test by monitoring the output of a load-cell mounted in the load train of the testing machine and by monitoring the output from resistance strain gages positioned on the specimen as shown in Fig. 4. Note the numbering sequence used to identify the gages—Gages 1 and 3 were on opposite sides of the specimen from Gages 2 and 4. The output from Gage 2 was compared with the average output from Gages 1 and 3 to measure out-of-plane bending in each specimen. An indication of in-plane bending was obtained from the strain differences on each side of the specimen. The output from Gage 4 was compared with the output from other gages to identify large strain gradients at the specimen edge. The nominal strain in a specimen at a given load was determined as the average of the strains at the midpoint of the two sides, where the strain at the midpoint of the side containing Gages 1 and 3 is the average of readings from those gages.

Data for each specimen were collected throughout the test using an on-line digital computer. These data were stored on magnetic tape.

Materials

Specimens were fabricated using Narmco T300/5208 graphite/epoxy composite prepreg, which was purchased and processed to conform to prepreg and laminate specification requirements for commercial aircraft applications [6]. The quality of each panel was verified with ultrasonic C-scan and by making measurements of resin content by weight, fiber volume, density, and void content on specimens from two regions of the panel. These properties were nominally 27 percent, 66 percent, 1600 kg/m^3, and 0.66 percent, respectively.

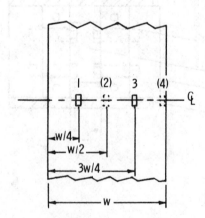

FIG. 4—*Location of strain gages on specimen.*

Specimens were stored in a laboratory environment, 21 to 27°C and about 70 percent relative humidity, for 6 to 12 months prior to being tested.

Results and Discussion

Results are presented from compression tests of coupons of T300/5208 graphite/epoxy composite material with the IITRI compression test fixture, a face-supported compression test fixture, and an end-loaded-coupon compression test fixture. Data are analyzed to identify sensitivities of techniques to laminate, specimen, and test parameters.

Modulus data from fixture checkout tests on 2024-T4 aluminum alloy sheet specimens tested in compression with the three fixtures were within 6 percent of the published data [7] and the coefficient of variation of the data was 0.9, 0.4, and 4 percent, respectively, for the IITRI, face-supported, and end-load-coupon compression test fixtures. These results suggest that test procedures for each fixture were satisfactory.

Table 2 gives average compression test results for each series of T300/5208 graphite/epoxy composite material specimens. The ultimate compressive strength was based on the maximum load applied to the specimen. The ultimate strain was the highest strain indicated by any gage at maximum load. The secant modulus is the secant of the stress-strain curve at an average compressive strain of 0.004, which is in the range of strain encountered in applications of graphite/epoxy composites. The strain variations due to out-of-plane bending and in-plane bending were also determined at an average compressive strain of 0.004.

Uniformity of Load Transfer

The effect of nonuniform load transfer during compression testing is to induce out-of-plane bending or in-plane bending with accompanying strain variations. In extreme cases the bending can result in failure by buckling. Table 2 gives the average strain variations due to out-of-plane bending and in-plane bending for each series of specimens tested. Average out-of-plane and in-plane strain variations were as high as 12 and 28 percent, 21 and 4 percent, and 6 and 10 percent for groups of specimens tested with the IITRI, face-supported, and end-loaded-coupon fixtures, respectively.

The sensitivity of IITRI-specimen stress-strain curves to flatness and parallelism of opposing tab surfaces of specimens was confirmed early in the program. Figure 5 shows stress-strain curves for a 25-mm-wide 24-ply quasi-isotropic IITRI specimen. The data in Figure 5a, which were obtained by preloading the specimen in the as-fabricated condition, show strong evidence of out-of-plane and in-plane bending. In view of these data the tabs of the specimen were ground to be "flat and parallel" and the data in Fig. 5b were obtained. Examination of the specimen disclosed sig-

TABLE 2—Average compression test results for T300/5208 graphite/epoxy composite material.

| Test Series | No. of Tests | Specification/Test Type | | | | Ultimate Strength/Standard Deviation, MPa | Ultimate Strain/Standard Deviation, % | Elastic Modulus/Standard Deviation, GPa | % Strain Variation | |
		Fixture	Width, mm	Thickness, No. of Plies	Laminate				Out-of-Plane	In-Plane
1	6	IITRI	12.5	16	UNI[d]	1413/68	1.3/0.12	132/5.1	8	4
2[a]	4	IITRI	25.	16	UNI	1593/130	1.7/0.31	131/2.5	5	7
3	5	IITRI	50	16	UNI	1517/87	1.5/0.17	137/6.6	9	28
4	5	IITRI	12.5	8	QI[e]	483/78	1.2/0.26	52/1.5	4	4
5[a]	5	IITRI	25	8	QI	490/26	1.3/0.18	49/1.4	12	10
6	6	IITRI	50	8	QI	490/38	1.3/0.15	52/2.3	10	21
7[a]	5	IITRI	12.5	16	QI	531/79	1.3/0.21	48/2.8	11	5
8[a]	5	IITRI	25	16	QI	558/62	1.5/0.11	48/1.3	7	20
9[a]	5	IITRI	50	16	QI	572/25	1.4/0.12	50/1.6	5	10
10[a]	5	IITRI	12.5	24	QI	586/46	1.4/0.15	50/2.5	4	5
11[a]	5	IITRI	25	24	QI	579/48	1.5/0.05	47/3.8	5	5
12	6	IITRI	50	24	QI	579/31	1.4/0.09	54/0.30	8	19
13[a]	5	IITRI	12.5	16	CP[f]	192/6	3.3/0.23	16.5/0.76	2	3
14[a]	6	IITRI	25	16	CP	217/6	3.8/0.34	17.2/0.90	11	14
15	5	IITRI	50	16	CP	331/28	4.9/0.09	19.3/0.90	5	8
16	6	FS[b]	25	16	UNI	1317/54	1.2/0.10	130/3.9	7	3
17	4	FS	50	16	UNI	1179/33	1.1/0.22	132/3.5	21	2
18	5	FS	25	8	QI	421/74	1.0/0.21	50/0.90	2	4
19	6	FS	12.5	16	QI	510/31	1.3/0.11	47/1.2	2	1
20	5	FS	25	16	QI	517/28	1.3/0.17	50/2.8	4	2
21	4	FS	50	16	QI	510/25	1.3/0.05	46/3.7	4	3
22	6	FS	25	24	QI	503/56	1.3/0.15	48/1.0	4	2
23	5	FS	25	16	CP	196/8	2.2/0.23	17.9/0.76	4	1
24	5	FS	50	16	CP	205/5	2.3/0.25	17.9/1.1	4	3
25	10	EL[c]	25	16	QI	531/40	1.4/0.16	50/0.90	6	10

[a] Tabs ground to be parallel.
[b] Face-supported fixture.
[c] End-loaded-coupon fixture.
[d] [0] laminate.
[e] [0/±45/90] laminate.
[f] [±45/∓45] laminate.

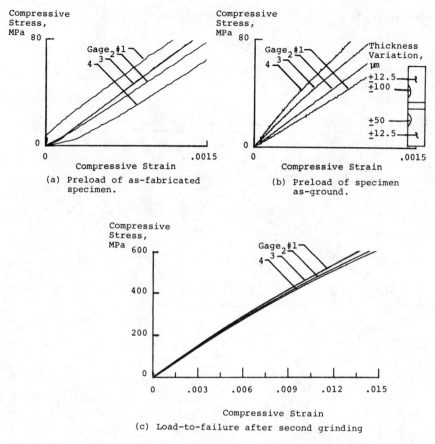

FIG. 5—*Compressive stress-strain curves for 25-mm-wide 24-ply-thick quasi-isotropic specimen tested with the IITRI fixture.*

nificant variations in flatness of the tabs, indicated by the inset in Fig. 5b, which resulted from grinding error. These variations were sufficient to produce the significant strain differences across the specimen shown here. The specimen was ground a second time to be flat within ± 25 μm. Figure 5c shows the stress-strain curves for the specimen loaded to failure. Note the uniformity of strain across the specimen.

In light of results like those in Fig. 5, the tabs of a number of IITRI specimens were machined to be flat and parallel. In Table 2 the groups of specimens that were machined are indicated with a subscript "a" by their test series number.

Figure 6 shows representative "best" and "worst" stress-strain curves for the three fixtures. Figure 6a shows representative stress-strain curves for the specimens tested with the IITRI fixture. Figures 6b and 6c show similar

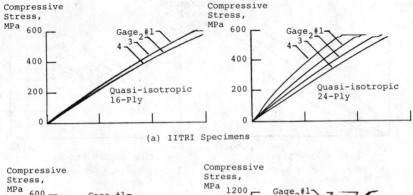

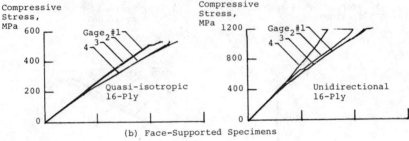

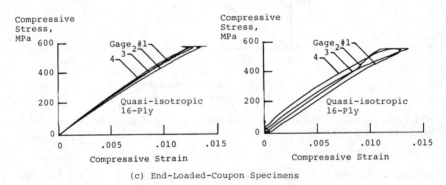

FIG. 6—*Representative best and worst compressive stress-strain curves for graphite/epoxy composite material specimens.*

curves for specimens tested with the face-supported fixture and the end-loaded-coupon fixture, respectively. The characteristics of the curves shown here are typical in that stress-strain curves for IITRI specimens were generally continuous from start of testing to failure, while the curves for face-supported specimens frequently exhibited large changes in slope particularly as the stress in the specimens neared ultimate. This characteristic with the face-supported fixture is no doubt related to the fact that the longer gage length specimens, even when supported by the fixture, experience some out-of-plane buckling. This buckling, which might not precipitate catastrophic failure, produces asymmetry in the specimen and may result in total failure at an average stress less than the compressive ultimate stress. The stress-strain curves for end-loaded-coupon specimens were similar to those for the IITRI specimens.

Figure 7 shows the effect of total strain variation on compressive strength of graphite/epoxy composite specimens tested with the IITRI fixture, where total strain is the sum of the in-plane strain variation and the out-of-plane strain variation. The data points in this figure represent single specimens comprising the test series in Table 2. The data here show a consistent trend toward lower strength with higher strain variations. Note the linear relationship between strength and strain variation for 12.5- and 50-mm-wide unidirectional specimens, with higher strength occurring at lower strain variation. Also note the strong effect of specimen width. Similar data for 25-mm-wide unidirectional specimens did not show this trend. The quasi-isotropic data show the trend toward lower strength with greater strain variation; however, the effects of specimen width and thickness were not evident. Compressive strength versus total strain variation data for specimens tested with the end-loaded-coupon fixture and the face-supported fixture do not show the consistent trends noted for the IITRI fixture data, which is probably due to the lesser precision of these fixtures in producing compressive strength failures in composite material specimens.

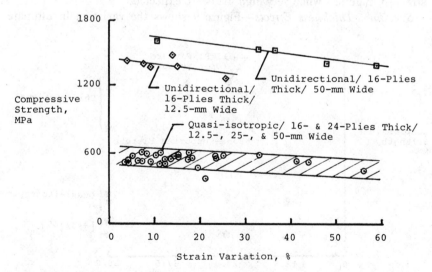

FIG. 7—*Effect of strain variation on compressive strength of graphite/epoxy specimens tested with the IITRI fixture.*

Ultimate Compressive Strength

Specimen Width Effects—Figure 8 shows ultimate compressive strength as a function of specimen width for 16-ply graphite/epoxy composite material in three fiber orientations. Data are shown for the IITRI fixture and the face-supported fixture. The most noteworthy point here is the difference in results from the two fixtures for the $[\pm 45/\mp 45]_{2s}$ laminate. The ultimate compressive strength results from the face-supported fixture are nearly constant at about 200 MPa compared with the results from the IITRI fixture, which vary linearly from 190 to 330 MPa over a range of specimen widths from 12.5 to 50 mm. The wide range in strength of $[\pm 45/\mp 45]_{2s}$ specimens tested with the IITRI fixture probably result from the biaxial state of stress present in low-aspect-ratio specimens of high Poisson's ratio under load.

The ultimate compressive strength of quasi-isotropic specimens tested with the IITRI fixture and the face-supported fixture was independent of specimen width, with the IITRI fixture producing consistently higher strength results than the face-supported fixture (Fig. 8).

Note the substantially lower strength of the 50-mm-wide unidirectional specimens tested with the face-supported fixture compared with other data for unidirectional specimens (Fig. 8). The face-supported fixture was examined after testing the 50-mm-wide specimens and was found to be bent. This explains the lower strength obtained for that series of specimens and points to the importance of fabricating the test fixture from high-yield-strength materials when very high loads are expected.

Specimen Thickness Effects—Figure 9 shows the variation in ultimate

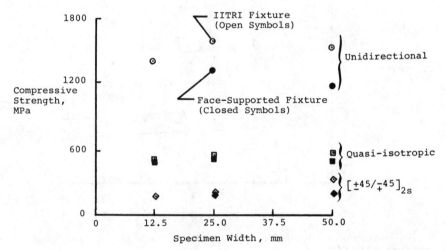

FIG. 8—*Average compressive strength of 16-ply graphite/epoxy composite material as a function of specimen width.*

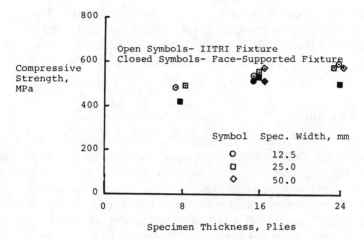

FIG. 9—*Average compressive strength of quasi-isotropic graphite/epoxy composite material as a function of specimen thickness.*

compressive strength with specimen thickness for quasi-isotropic specimens tested with the face-supported fixture and for quasi-isotropic specimens tested with the IITRI fixture. The significant point here is the lower average strengths of 8-ply specimens tested with each fixture. Examination of stress-strain data for individual specimens represented by the data in Fig. 9 showed some evidence of buckling in 8-ply specimens tested in each fixture.

Comparison of Strength by Fixtures—Figure 10 shows a comparison of ultimate compressive strengths for 16-ply quasi-isotropic 25-mm-wide specimens tested with the end-loaded-coupon fixture, the IITRI fixture, and the face-supported fixture. The shaded region of each bar represents the range of values. The IITRI fixture produced the highest average strength at 552

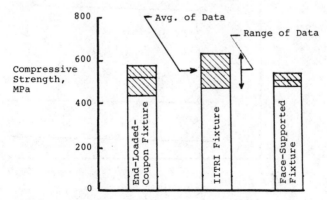

FIG. 10—*Compressive strength of 25-mm-wide 16-ply quasi-isotropic graphite/epoxy specimens tested with the end-loaded, IITRI, and face-supported fixtures.*

MPa followed by the end-loaded-coupon fixture at 531 MPa. The data obtained with the face-supported fixture exhibits less scatter than the data with the other fixtures. The higher strength of specimens tested with the IITRI fixture is the result of greater precision in alignment and loading of specimens with that fixture compared with the other two fixtures.

Compressive Stiffness

Figures 11a and 11b show measured secant modulus as a function of specimen width and thickness for specimens tested with the IITRI fixture and the face-supported fixture. The noteworthy fact shown here is that modulus is independent of specimen width for every case except for the [±45/∓45] specimens tested with the IITRI fixture where modulus is linear with width and shows a 17 percent change over the width range from 12.5 to 50 mm. As noted earlier in discussion of the strength data for [±45/∓45] specimens tested with the IITRI fixture, the change in modulus with specimen width is probably due to the biaxial state of stress present in low-aspect-ratio specimens of high Poisson's ratio under load.

Figure 12 shows a comparison of secant modulus for 16-ply quasi-isotropic 25-mm-wide specimens tested with the end-loaded-coupon fixture, the IITRI fixture, and the face-supported fixture. The shaded region of each bar represents the range of values. The average modulus was approximately the same for all fixtures. Variability in modulus for the specimens tested was satisfactorily low; even the face-supported fixture data which have the most scatter have a coefficient of variation of only 5.6 percent.

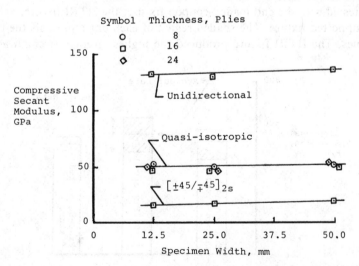

FIG. 11a—*Secant modulus at 0.004 strain of graphite/epoxy composite material as a function of specimen width and thickness for specimens tested with the IITRI fixture.*

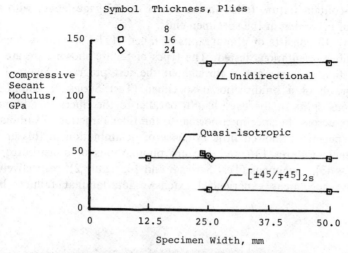

FIG. 11*b*—*Secant modulus at 0.004 strain of graphite/epoxy composite material as a function of specimen width and thickness for specimens tested with the face-supported fixture.*

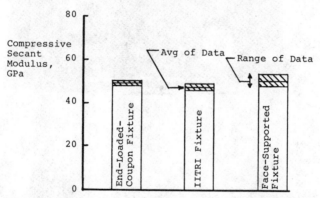

FIG. 12—*Secant modulus at 0.004 strain of 25-mm wide 16-ply-thick quasi-isotropic graphite/epoxy specimens tested with the end-loaded, IITRI, and face-supported fixtures.*

Failure Modes

Failures in the quasi-isotropic and $[\pm45/\mp45]_s$ specimens tested with the IITRI fixture were always centered in the gage length of the specimens, whereas fractures of the unidirectional specimens were generally located nearer the tab ends. Failures in specimens tested with the face-supported fixture were generally 25 mm or more away from the tab ends; however, the unidirectional specimen tended to fail nearer the tab ends than did the other specimen types. The quasi-isotropic specimens tested with the end-loaded-coupon fixture failed in the center of the gage length with no evidence of brooming at the specimen ends.

Figure 13 consists of photographs of failed IITRI specimens having the three fiber orientations tested. The types of failure shown here are typical of the failures experienced throughout the test program. Figure 13a is a photograph of a unidirectional specimen (Test Series 2, Table 2) with numerous splits in the gage length parallel to the fibers in addition to a fracture across the specimen normal to the fiber direction. Examination of failed specimens showed little evidence of delamination in this or similar specimens. Figures 13b and 13c are photographs of quasi-isotropic and $[\pm 45/\mp 45]_s$ specimens (Test Series 8 and 14, Table 2), respectively. Both of these specimens experienced extensive interlaminar failures between

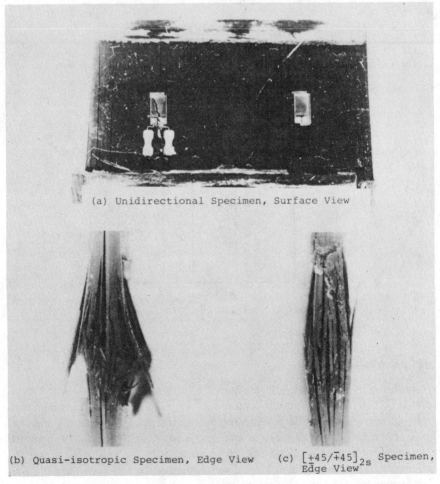

(a) Unidirectional Specimen, Surface View

(b) Quasi-isotropic Specimen, Edge View   (c) $[+45/\overline{+}45]_{2s}$ Specimen, Edge View

FIG. 13—Photographs of failed 16-ply 25-mm-wide specimens tested in IITRI fixture.

neighboring dissimilar plies. In general, the failed $[\pm 45/\mp 45]_s$ specimens showed very little evidence of fiber fractures, whereas the failed quasi-isotropic specimens showed extensive evidence of fiber fractures.

To gain more insight into the mode of failure of specimens tested in this program, consider that Ref 8 states that since the microbuckling strength of graphite/epoxy composite material is approximately equal to the shear modulus of the composite and that specimens tested in compression fail at a fraction of that level, the mode of failure must be something other than microbuckling. Reference 8 states further that the critical parameter which is responsible for the low compressive strength of graphite/epoxy composites is the transverse tensile strength and that, if composite materials have sufficiently high transverse tensile strength, the compressive strength of unidirectional composites approaches their tensile strength. Note that the average ultimate compressive strength of all unidirectional specimens tested with the IITRI fixture was 1500 MPa compared with an ultimate tensile strength of 1450 MPa [9]. Furthermore, following the assumption of Ref 8 that the tensile and compressive strength and stiffness of graphite fibers are the same, the average of all strain-to-failure data for unidirectional specimens tested with the IITRI fixture is 1.43 percent compared with the 1.32 percent strain-to-failure for T300 graphite fibers obtained from published data [10]. These points indicate that unidirectional specimens tested with the IITRI fixture failed by compressive strength failure rather than by microbuckling or general buckling, that compressive strength failure of unidirectional graphite/epoxy composite material is governed by fiber behavior, and that the fiber failure mode is similar to the tensile failure of the fiber. Additionally, these results suggest that the IITRI fixture produces near maximum compressive strength data for unidirectional graphite/epoxy composite material.

Comparison of compressive strength and stiffness data for unidirectional face-supported specimens with tensile data for graphite/epoxy strength and fiber stiffness was less favorable than for the IITRI specimens. These specimens failed at lower stress levels than the IITRI specimens as a result of the greater instability of the longer-gage-length specimens.

The average strain-to-failure of 16-ply and greater thickness quasi-isotropic specimens tested with the IITRI fixture, the face-supported fixture, and the end-loaded-coupon fixture compares favorably with the estimated fiber maximum strain-to-failure. This suggests that the compressive behavior of quasi-isotropic composites tested in these fixtures is governed primarily by the unidirectional fibers.

Failure of the $[\pm 45/\mp 45]_{2s}$ specimens occurred in the form of delamination of the plies. In light of this fact, the maximum fiber strain for the $[\pm 45/\mp 45]_{2s}$ specimens should be less than the strains encountered in the unidirectional and quasi-isotropic specimens. The strain-to-failure of specimens tested with the face-supported fixture was quite constant at about 2.2

percent, which, assuming uniform compression over the length and width of the specimen, is equivalent to 0.22 percent fiber strain or about one-sixth the maximum fiber strain encountered in the unidirectional and quasi-isotropic specimens.

The strain-to-failure of the $[\pm 45/\mp 45]_s$ specimens tested with the IITRI fixture varied approximately linearly with specimen width, ranging from 3.3 percent at a width of 12.5 mm to 4.9 percent at a width of 50 mm. This wide range of failure strain and accompanying failure stresses is probably the result of the biaxial state of stress that exists in loading low-aspect-ratio specimens of high Poisson's ratio.

Concluding Remarks

Compression tests of T300/5208 graphite/epoxy composite material were performed using a modified IITRI fixture, a face-supported fixture, and an end-loaded-coupon fixture to determine the effects of loading, specimen width, specimen thickness, and fiber orientation on the compressive behavior of the material. Each specimen was instrumented with four longitudinal strain gages to determine the extent of strain variation in the specimen during the test.

Based on results reported herein, no single test fixture appears universally adequate for compression testing. However, each of the three fixtures has the potential to provide reliable compressive properties data in certain instances. For example, the IITRI fixture provided the most consistent data for unidirectional composite specimens, while the face-supported fixture provided the most consistent results for $[\pm 45/\mp 45]_s$ specimens.

The IITRI fixture was found to be sensitive to flatness and parallelism of opposing tab surfaces of specimens. Specimen variances of this type produced significant strain variations in the specimens. Specimens experiencing large strain variations during testing had lower compressive strengths than did more uniformly strained specimens. Tests of 16- and 24-ply unidirectional and quasi-isotropic specimens using the IITRI fixture produced high strength values, with specimen failures governed by 0-deg fibers in both cases. Tests of 8-ply specimens showed evidence of failure by buckling and correspondingly lower strengths.

The strength data from $[\pm 45/\mp 45]_s$ specimens tested with the IITRI fixture showed a strong dependence on specimen width, which is probably the result of the biaxial state of stress that exists in low-aspect-ratio specimens of high Poisson's ratio under axial load. These specimens failed by delamination of the plies.

Unidirectional and quasi-isotropic specimens tested with the face-supported fixture experienced a small amount of strain variation at low loads; at loads approaching failure, however, the long-gage-length specimens experienced varying amounts of general instability which resulted in failure at lower stresses than were achieved in specimens tested with the IITRI fixture. This strength differential is greatest for unidirectional specimens. These data were independent of specimen width, but dependent on specimen thickness to the extent that 8-ply specimens failed at much lower stress levels than did thicker specimens.

Tests of 16-ply $[\pm 45/\mp 45]_s$ specimens with the face-supported fixture produced results independent of width. Failure of these specimens was by delamination of plies.

Modulus data from the three fixtures were not significantly different except for tests of $[\pm 45/\mp 45]$ specimens with the IITRI fixture which showed a strong variation with specimen width. The coefficient of variation of the modulus data was less than 9 percent for every series of test.

Data obtained with the end-loaded-coupon fixture are not substantially different from the data obtained with the IITRI and face-supported fixtures. In view of this and the simplicity of the specimen and the fixture, further study of the end-loaded-coupon fixture is justified.

References

[1] Shuart, Mark J. and Herakovich, Carl T., "An Evaluation of the Sandwich Beam in Four-point Bending as a Compressive Test Method for Composites," NASA TM 78783, National Aeronautics and Space Administration, Washington, D.C., Sept. 1978.

[2] Haskins, J. F., Kerr, J. R., and Stein, B. A. in *Proceedings,* SCAR Conference, NASA CP-001, NASA-Langley Research Center, Hampton, Va., 9-12 Nov. 1976.

[3] Hofer, K. E., Jr., and Rao, P. N., *Journal of Testing and Evaluation,* Vol. 5, No. 4, July 1977, pp. 278-283.

[4] Ryder, J. T. and Black, E. D. in *Composite Materials: Testing and Design (Fourth Conference), ASTM STP 617,* American Society for Testing and Materials, 1977, pp. 170-189.

[5] Lauraitis, K. N., "Effect of Environment on the Compressive Strengths of Laminated Epoxy Matrix Composites," Interim Technical Quarterly Report, LR 28508-1, Lockheed California Company, Burbank, Calif., Feb. 1975.

[6] Douglas Material Specification, DMS1936C, McDonnell Douglas Corp., Long Beach, Calif., Sept. 23, 1975.

[7] *Metals Handbook, Vol. 1 Properties and Selection of Metals,* American Society for Metals, 1961.

[8] Greszczuk, L. B., "Failure Mechanics of Composites Subjected to Compressive Loading," AFML-TR-72-107, Air Force Materials Laboratory, Dayton, Ohio, Aug. 1972.

[9] *Advanced Composites Design Guide,* Vol. 4, 3rd ed., Third Revision, Jan. 1977.

[10] Lauraitis, K. N., "Effect of Environment on the Compressive Strengths of Laminated Epoxy Matrix Composites," Interim Technical Quarterly Report, LR 28508-3, Lockheed California Company, Burbank, Calif., June 1978.

A. V. Sharma[1]

Low-Velocity Impact Tests on Fibrous Composite Sandwich Structures

REFERENCE: Sharma, A. V., **Low-Velocity Impact Tests on Fibrous Composite Sandwich Structures,"** *Test Methods and Design Allowables for Fibrous Composites, ASTM STP 734*, C. C. Chamis, Ed., American Society for Testing and Materials, 1981, pp. 54–70.

ABSTRACT: An experimental investigation was conducted to assess the damage tolerance of composite sandwich structures subjected to low-velocity projectile impact. Sandwich-type specimens fabricated with graphite/epoxy materials having laminate configurations such as $(\pm 45,0_4)_s$, $(\pm 45,90,0)_s$, and $(90, \pm 45,0)_s$ were tested. Impact tests were performed at low energy levels to assess the strength degradation of composite laminates when compared with their ultimate strengths. Low-energy projectile impact was considered to simulate typically the damage caused by runway debris and the accidental dropping of hand tools during servicing and other similar objects on secondary aircraft structures fabricated with composites. The preload and impact energy combination necessary to cause catastrophic failure was determined. The residual strength of specimens that survived the impact damage was also measured.

Based on the experimental results, a faired curve indicating the lower bound of the failure threshold for each of the laminate configurations tested in compression and tension as a function of the projectile impact energy is shown. Further, the strength degradation due to impact is found to be dependent on the laminate configuration and the fiber/matrix combination.

KEY WORDS: composite materials, damage, impact strength, projectile energy, failure threshold, tension tests, compression tests, honeycomb panels, residual strength

The use of graphite/epoxy composites in the design of the secondary structural components of aircraft is increasing. Some of these components have a honeycomb-type core sandwiched between laminated composite facings. In the normal operational mode of the aircraft, these composite facings may be exposed to foreign object damage (FOD) resulting from the dropping of hand tools during servicing, runway debris, hail, etc. Consequently, it is of interest

---

[1]Professor of Mechanical Engineering, North Carolina A. & T. State University, Greensboro, N. C. 27411.

FIG. 2—*A typical specimen with loading frame (tension mode).*

apparatus. The load and the resulting strains in the specimen were measured using standard strain-gage techniques. Two strain gages, oriented to measure the longitudinal strains, were bonded to the specimen equidistantly [2.5 cm (1 in.) from the geometric center of the test section]. The ultimate static load and the corresponding strains were determined using undamaged specimens. With the results of earlier tests in the background [5,6], the specimens were loaded and impacted by releasing the projectile to assess the damage tolerance of the specimens. Depending on the magnitude of the initial load (preload) and the projectile kinetic energy, the specimen either survived or failed catastrophically upon impact. The preload applied to those specimens that survived the projectile impact is shown as a "no failure" point on the graphs. Further, the loading on these (survived) specimens subsequent to the impact was continued to assess their residual strength.

Following Rhodes [2], the term "failure threshold" used in subsequent sections is defined as the lowest static load which precipitated catastrophic failure in the face sheet of a sandwhich beam specimen at a given impact energy. The stress ratio, σ/σ_{ult}, used in this paper is defined as the ratio of the stress in the specimen prior to impact, or the residual strength of the specimen, to the ultimate static strength of the virgin specimen.

Experimental Results and Analyses

A limited number of specimens (often about 10) were tested in each of the loading (tension/compression) modes to observe the trends in the variation of the failure thresholds for various laminates. The sum of the number of stars and open circles in the graphs would generally indicate the total number of specimens tested in each of the loading modes.

to study the impact damage caused by low-velocity projectiles to the composite facings. Various theoretical approaches to study the behavior of the notched composites based on linear elastic fracture mechanics and other methods have been reviewed by Yeow et al [1].[2] Experimental verification of the theoretical models was also common in many of the studies reviewed [1]. The experimental studies typically deal with implanted flaws such as holes and slots of known dimensions in the composite materials. The study of the effect of variables such as the lamina configuration (ply orientation and stacking sequence), fiber/matrix combination, specimen width to projectile diameter ratio in the FOD-type studies, and specimen width to the implanted flaw size ratio on the behavior of the composite materials is typical of experimental work. Some of the investigations by Rhodes [2,3], Adsit and Waszczak [4], and Sharma [5,6] were involved in the development of graphite/epoxy composite failure thresholds leading to some design implications.

The objective of the present investigation is to evaluate the effect of low-velocity projectile impact on the load-carrying ability of the composite sandwich structural components. The preload and the impact energy combinations necessary to initiate the catastrophic failure of the sandwich specimens were determined. The residual strengths of the specimens that survived the projectile impact were found. As a result of the projectile impact, the degradation of the ultimate strength of the composites was evaluated as a function of the kinetic energy of the impacting projectile.

Specimens and Experimental Arrangement

Three different specimen configurations were evaluated in this investigation (Table 1). The material and the stacking sequence in the fabrication of the specimens were Thornel 300/Rigidite 5208 $(\pm 45,0_4)_s$, Thornel 300/Rigidite 5208 $(\pm 45,90,0)_s$, and Thornel 300/Fiberite 934 $(90,\pm 45,0)_s$. Each lamina in the panel had a nominal cured thickness of 140×10^{-6} m (0.0055 in.).[3] The honeycomb sandwich specimens were fabricated using the composite laminate as face sheet (test surface) and a steel plate as the back surface. The nominal dimensions of the specimen were 56 cm (22 in.) long by 7.5 cm (3 in.) wide with a honeycomb core thickness of 2.5 cm (1 in.). The area of the core in the center test section, where a uniform stress field was imposed through a four-point loading apparatus, was 7.5 by 7.5 cm (3 by 3 in.). All the specimens had a 0.32-cm (1/8 in.) cell, 130-kg/m^3 (8.1 lbm/ft^3) aluminum honeycomb core in the test section. A dense aluminum core was used in the end sections of the beam, where high shear loads exist. Large face panels were laid at predetermined ply orientations and cured. These cured panels were cut to size and bonded adhesively to the core material.

---

[2]The italic numbers in brackets refer to the list of references appended to this paper.
[3]Original measurements were in English units.

TABLE 1—Specimen description and some experimental results.

Series	Material	Laminate Configuration	Ultimate Tensile			Ultimate Compressive		
			Strength. σ_u GPa (ksi)	Strain	Strength Retention,[a] % of σ_u	Strength. σ_u GPa (ksi)	Strain	Strength Retention,[a] % of σ_u
A	Thornel 300/Rigidite 5208	$(\pm45, 0_4)_s$	1.09(158)	0.010	58	1.11 (161)	0.013	38
B	Thornel 300/Rigidite 5208	$(\pm45, 90, 0)_s$	0.57(83)	0.010	55	0.60 (87)	0.012	51
C	Thornel 300/Fiberite 934	$(90, \pm45, 0)_s$	0.68(99)	0.011	33	0.54 (78)	0.011	38

[a] This is the ratio of the residual strength, σ (asymptotic value of the failure threshold or the midpoint value of the failure threshold), to the laminate ultimate strength, σ_u, expressed as a percent.

schematic diagram of the projectile firing mechanism is shown in Fig. 1. ... is bled from a supply line into a cylindrical reservoir until the ...4-mm-thick (1 mil) Mylar diaphram ruptures. The air escapes through ... hole (the size of which was predetermined based on the desired projec-... velocity) located in the center of the orifice plate and propels the projec-... toward the specimen through a gun barrel. The projectile is an aluminum ... 1.27 cm (0.5 in.) in diameter. In the present investigation, the aspect ratio (specimen width to projectile diameter) was about 6. The strength degradation that results with smaller aspect ratios should be considered when evaluating the residual strength of the impact-damaged laminates. A velocity measuring device is located at about 7.5 cm (3 in.) in front of the test specimen. As the projectile travels through this device, light beams emitted by photodiodes are interrupted and an electronic counter is triggered. The average velocity of the projectile is calculated from the distance between the photodiodes and the time required by the projectile to traverse this distance. This arrangement for measuring the velocity is similar to the technique used in Ref 2. The specimens were subjected to the projectile impact velocities from about 17 m/s (56 ft/s) to about 64 m/s (210 ft/s).

Static bending loads were applied to the specimens through a specially built four-point loading apparatus. The tensile or the compressive loads were applied through a whiffletree arrangement connected to a screw jack in the rear. The whiffletree arrangement consists of a linkage mechanism that pulls in or pushes out the central section of the sandwich specimen with respect to the end posts (supports). A photograph of a typical specimen with the loading apparatus is shown in Fig. 2. A load cell was built into the loading

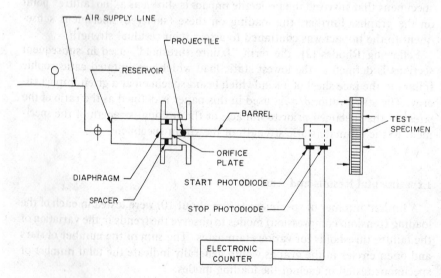

FIG. 1—Schematic diagram of firing mechanism.

Series A: Graphite/Epoxy, (±45,0₄)ₛ

The composite face sheet in this series of specimens has eight unidirectional plies and four angle plies.

Tension-Loaded Laminates—The sandwich specimens were subjected to tensile loads using the four-point beam-loading apparatus. The variation of the nondimensional stress ratio (σ/σ_{ult}) as a function of the kinetic energy of the projectile just before impact is shown in Fig. 3. The solid line in this figure and in similar figures shown elsewhere is a faired curve drawn on the basis of the observed experimental results. The faired curve is defined as the locus of the experimental data and is drawn as a smooth curve just above the data points that represent the stress in the laminate at impact ("no failure" points in the graphs). This curve is designated as the failure threshold curve. The significance of the failure threshold is that if the sum of the preload (strain) energy and the kinetic energy of the impacting projectile exceeds a certain value (the numerical value of which was not established but which could be found through further experimentation), the material would fail catastrophically. Otherwise, the material survives the impact, exhibiting residual strength. Since one of the objectives of this experimental program was to assess the effect of low-velocity projectile impact on the laminate strength, the projectile impact energy in all the tests was confined to a value less than 6 J (53 in.-lb). The graph in Fig. 3 shows that the tension-loaded laminates retain 58 percent of their ultimate strength upon impact in the range of 2 to 6 J

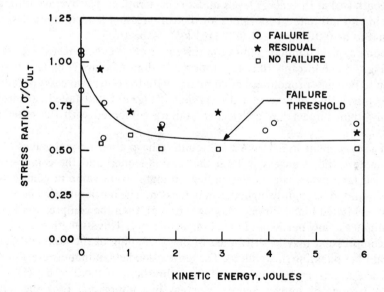

FIG. 3—*Graphite/epoxy composite, (±45,0₄)ₛ, loaded in tension; stress ratio versus kinetic energy of projectile.*

(18 to 53 in.-lb) of impact energy. The asymptotic trend of the failure threshold curve with an increase in the impact energy may be seen in Fig. 3. The average ultimate-strain and ultimate-stress values in tension for this series of laminates were found to be 0.010 and 1.09 GPa (158 ksi), respectively.

Typical laminate tensile failures are shown in Fig. 4. The intraply and interply delaminations may be seen in this figure. The intraply delamination due to the maximum static load (Fig. 4a) is more pronounced as compared with the failure modes in the other two cases (Figs. 4b and 4c). The failure of the zero-degree plies may also be seen clearly in Fig. 4. In general, it was visually observed that the crack lines have penetrated through the laminate thickness. Some of the specimens exhibited separation of the face sheet from the honeycomb core in the central testing portion of the sandwich beam.

Compression-Loaded Laminates—Some of the specimens in this series were subjected to the compressive loads. The variation of the stress ratio (σ/σ_{ult}) with the changing projectile kinetic energy of impact is shown in Fig. 5. The failure threshold curve as shown is varying gradually with an increase in the kinetic energy of the impacting projectile. In other words, the residual strength of the impact-damaged specimens was found to decrease with an increase in the projectile energy. In order to study the variation of the damage strength among various laminates, the failure threshold strength of the compression-loaded laminates is estimated at 38 percent (a value taken at about the midrange of the kinetic energy axis) of the compressive ultimate static strength (σ_u) at the energy levels under consideration. The average ultimate-strain and ultimate-stress values for the compression-loaded laminates were found to be 0.013 and 1.11 GPa (161 ksi), respectively.

Some typical laminate failures in this series of specimens tested are shown in Fig. 6. The intraply fiber separation at ultimate failure (Fig. 6a) may be seen to be more pronounced as compared with the other two cases of failure. It is interesting to note that the specimen (Fig. 6c) that failed catastrophically upon impact has a "cleaner" failure compared with the other two cases.

The specimens in Series A were found to have almost equal values of the ultimate static strengths in both the tension-loaded and the compression-loaded laminates. The corresponding ultimate strain value in compression was found to be slightly higher than in tension. The tension-loaded laminates have exhibited higher impact damage strength than the compression-loaded laminates at any particular level of impact energy. This strength degradation in compression may be attributed to the weakening of the laminate due to local fiber buckling (instability). The tension-loaded laminates were observed to have a nonvarying (constant) failure threshold in the 2 to 6-J (18 to 53 in.-lb) range of impact energy, whereas in compression tests the failure threshold strength was found to decrease uniformly with an increase in the kinetic energy of the impacting projectile.

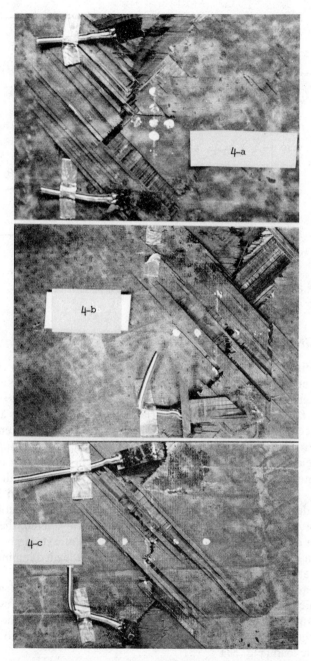

FIG. 4—*Fracture modes of graphite/epoxy laminates, tension, $(\pm 45, 0_4)_s$. (a) Specimen that failed at maximum static load. (b) Specimen that sustained local impact damage and subsequently failed upon further loading. (c) Specimen that failed catastrophically upon impact.*

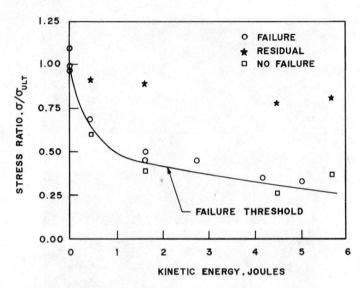

FIG. 5—*Graphite/epoxy composite, (±45,0₄)ₛ, loaded in compression; stress ratio versus kinetic energy of projectile.*

Series B: Graphite/Epoxy, (±45,90,0)ₛ

The specimens in this series have two unidirectional plies, two crossplies, and four angle plies. The tensile and the compressive loads were applied to the specimens using the procedure decribed earlier.

Tension-Loaded Laminates—The variation of the stress ratio as a function of the projectile impact energy is shown in Fig. 7. The projectile energy was varied from 1 to 4 J (9 to 36 in.-lb). The shape of the initial part of the failure threshold curve, shown as a broken line here and in subsequent graphs, is approximate and is drawn based on the trends observed in similar earlier tests [2,3]. The strength retention of the specimens subjected to impact is seen to be decreasing monotonically with an increase in the projectile impact energy (Fig. 7). In the range of impact energy levels considered, the strength retention is estimated at 55 percent of the ultimate static strength. The ultimate-strain and ultimate-stress values of the tension-loaded laminates were found to be 0.010 and 0.57 GPa (83 ksi), respectively.

Two photographs showing specimen failure at the ultimate load and at the catastrophic impact load are shown in Figs. 8a and 8b, respectivey. The interply delamination in Figure 8b (impact) appears to spread more widely from the center test section than in the case of the failure due to the maximum static load (Fig. 8a).

Compression-Loaded Laminates—The eight-ply laminates of this series were tested in compression also. The strengths of the undamaged and the impact damaged specimens were evaluated. A failure threshold curve showing

to study the impact damage caused by low-velocity projectiles to the composite facings. Various theoretical approaches to study the behavior of the notched composites based on linear elastic fracture mechanics and other methods have been reviewed by Yeow et al [1].[2] Experimental verification of the theoretical models was also common in many of the studies reviewed [1]. The experimental studies typically deal with implanted flaws such as holes and slots of known dimensions in the composite materials. The study of the effect of variables such as the lamina configuration (ply orientation and stacking sequence), fiber/matrix combination, specimen width to projectile diameter ratio in the FOD-type studies, and specimen width to the implanted flaw size ratio on the behavior of the composite materials is typical of experimental work. Some of the investigations by Rhodes [2,3], Adsit and Waszczak [4], and Sharma [5,6] were involved in the development of graphite/epoxy composite failure thresholds leading to some design implications.

The objective of the present investigation is to evaluate the effect of low-velocity projectile impact on the load-carrying ability of the composite sandwich structural components. The preload and the impact energy combinations necessary to initiate the catastrophic failure of the sandwich specimens were determined. The residual strengths of the specimens that survived the projectile impact were found. As a result of the projectile impact, the degradation of the ultimate strength of the composites was evaluated as a function of the kinetic energy of the impacting projectile.

Specimens and Experimental Arrangement

Three different specimen configurations were evaluated in this investigation (Table 1). The material and the stacking sequence in the fabrication of the specimens were Thornel 300/Rigidite 5208 $(\pm 45,0_4)_s$, Thornel 300/Rigidite 5208 $(\pm 45,90,0)_s$, and Thornel 300/Fiberite 934 $(90,\pm 45,0)_s$. Each lamina in the panel had a nominal cured thickness of 140×10^{-6} m (0.0055 in.).[3] The honeycomb sandwich specimens were fabricated using the composite laminate as face sheet (test surface) and a steel plate as the back surface. The nominal dimensions of the specimen were 56 cm (22 in.) long by 7.5 cm (3 in.) wide with a honeycomb core thickness of 2.5 cm (1 in.). The area of the core in the center test section, where a uniform stress field was imposed through a four-point loading apparatus, was 7.5 by 7.5 cm (3 by 3 in.). All the specimens had a 0.32-cm (1/8 in.) cell, 130-kg/m^3 (8.1 lbm/ft^3) aluminum honeycomb core in the test section. A dense aluminum core was used in the end sections of the beam, where high shear loads exist. Large face panels were laid at predetermined ply orientations and cured. These cured panels were cut to size and bonded adhesively to the core material.

---

[2]The italic numbers in brackets refer to the list of references appended to this paper.
[3]Original measurements were in English units.

TABLE 1—*Specimen description and some experimental results.*

Series	Material	Laminate Configuration	Ultimate Tensile			Ultimate Compressive		
			Strength. σ_u GPa (ksi)	Strain	Strength Retention,[a] % of σ_u	Strength. σ_u GPa (ksi)	Strain	Strength Retention,[a] % of σ_u
A	Thornel 300/Rigidite 5208	(± 45, 0,)$_s$	1.09(158)	0.010	58	1.11 (161)	0.013	38
B	Thornel 300/Rigidite 5208	(± 45, 90, 0)$_s$	0.57(83)	0.010	55	0.60 (87)	0.012	51
C	Thornel 300/Fiberite 934	(90, ± 45, 0)$_s$	0.68(99)	0.011	33	0.54 (78)	0.011	38

[a]This is the ratio of the residual strength, σ (asymptotic value of the failure threshold or the midpoint value of the failure threshold), to the laminate ultimate strength, σ_u, expressed as a percent.

A schematic diagram of the projectile firing mechanism is shown in Fig. 1. Air is bled from a supply line into a cylindrical reservoir until the 0.0254-mm-thick (1 mil) Mylar diaphram ruptures. The air escapes through a tiny hole (the size of which was predetermined based on the desired projectile velocity) located in the center of the orifice plate and propels the projectile toward the specimen through a gun barrel. The projectile is an aluminum sphere 1.27 cm (0.5 in.) in diameter. In the present investigation, the aspect ratio (specimen width to projectile diameter) was about 6. The strength degradation that results with smaller aspect ratios should be considered when evaluating the residual strength of the impact-damaged laminates. A velocity measuring device is located at about 7.5 cm (3 in.) in front of the test specimen. As the projectile travels through this device, light beams emitted by photodiodes are interrupted and an electronic counter is triggered. The average velocity of the projectile is calculated from the distance between the photodiodes and the time required by the projectile to traverse this distance. This arrangement for measuring the velocity is similar to the technique used in Ref 2. The specimens were subjected to the projectile impact velocities from about 17 m/s (56 ft/s) to about 64 m/s (210 ft/s).

Static bending loads were applied to the specimens through a specially built four-point loading apparatus. The tensile or the compressive loads were applied through a whiffletree arrangement connected to a screw jack in the rear. The whiffletree arrangement consists of a linkage mechanism that pulls in or pushes out the central section of the sandwich specimen with respect to the end posts (supports). A photograph of a typical specimen with the loading apparatus is shown in Fig. 2. A load cell was built into the loading

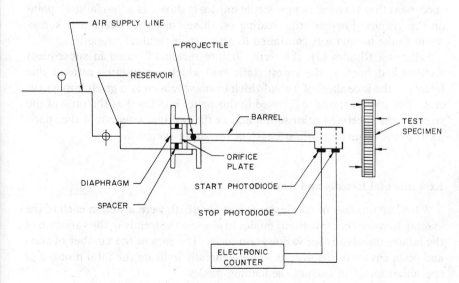

FIG. 1—*Schematic diagram of firing mechanism.*

FIG. 2—*A typical specimen with loading frame (tension mode).*

apparatus. The load and the resulting strains in the specimen were measured using standard strain-gage techniques. Two strain gages, oriented to measure the longitudinal strains, were bonded to the specimen equidistantly [2.5 cm (1 in.) from the geometric center of the test section]. The ultimate static load and the corresponding strains were determined using undamaged specimens. With the results of earlier tests in the background [5,6], the specimens were loaded and impacted by releasing the projectile to assess the damage tolerance of the specimens. Depending on the magnitude of the initial load (preload) and the projectile kinetic energy, the specimen either survived or failed catastrophically upon impact. The preload applied to those specimens that survived the projectile impact is shown as a "no failure" point on the graphs. Further, the loading on these (survived) specimens subsequent to the impact was continued to assess their residual strength.

Following Rhodes [2], the term "failure threshold" used in subsequent sections is defined as the lowest static load which precipitated catastrophic failure in the face sheet of a sandwhich beam specimen at a given impact energy. The stress ratio, σ/σ_{ult}, used in this paper is defined as the ratio of the stress in the specimen prior to impact, or the residual strength of the specimen, to the ultimate static strength of the virgin specimen.

Experimental Results and Analyses

A limited number of specimens (often about 10) were tested in each of the loading (tension/compression) modes to observe the trends in the variation of the failure thresholds for various laminates. The sum of the number of stars and open circles in the graphs would generally indicate the total number of specimens tested in each of the loading modes.

Series A: Graphite/Epoxy, (±45,0₄)ₛ

The composite face sheet in this series of specimens has eight unidirectional plies and four angle plies.

Tension-Loaded Laminates—The sandwich specimens were subjected to tensile loads using the four-point beam-loading apparatus. The variation of the nondimensional stress ratio (σ/σ_{ult}) as a function of the kinetic energy of the projectile just before impact is shown in Fig. 3. The solid line in this figure and in similar figures shown elsewhere is a faired curve drawn on the basis of the observed experimental results. The faired curve is defined as the locus of the experimental data and is drawn as a smooth curve just above the data points that represent the stress in the laminate at impact ("no failure" points in the graphs). This curve is designated as the failure threshold curve. The significance of the failure threshold is that if the sum of the preload (strain) energy and the kinetic energy of the impacting projectile exceeds a certain value (the numerical value of which was not established but which could be found through further experimentation), the material would fail catastrophically. Otherwise, the material survives the impact, exhibiting residual strength. Since one of the objectives of this experimental program was to assess the effect of low-velocity projectile impact on the laminate strength, the projectile impact energy in all the tests was confined to a value less than 6 J (53 in.-lb). The graph in Fig. 3 shows that the tension-loaded laminates retain 58 percent of their ultimate strength upon impact in the range of 2 to 6 J

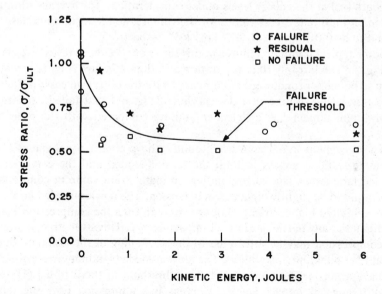

FIG. 3—*Graphite/epoxy composite, (±45,0₄)ₛ, loaded in tension; stress ratio versus kinetic energy of projectile.*

(18 to 53 in.-lb) of impact energy. The asymptotic trend of the failure threshold curve with an increase in the impact energy may be seen in Fig. 3. The average ultimate-strain and ultimate-stress values in tension for this series of laminates were found to be 0.010 and 1.09 GPa (158 ksi), respectively.

Typical laminate tensile failures are shown in Fig. 4. The intraply and interply delaminations may be seen in this figure. The intraply delamination due to the maximum static load (Fig. 4a) is more pronounced as compared with the failure modes in the other two cases (Figs. 4b and 4c). The failure of the zero-degree plies may also be seen clearly in Fig. 4. In general, it was visually observed that the crack lines have penetrated through the laminate thickness. Some of the specimens exhibited separation of the face sheet from the honeycomb core in the central testing portion of the sandwich beam.

Compression-Loaded Laminates—Some of the specimens in this series were subjected to the compressive loads. The variation of the stress ratio ($\sigma/\sigma_{\text{ult}}$) with the changing projectile kinetic energy of impact is shown in Fig. 5. The failure threshold curve as shown is varying gradually with an increase in the kinetic energy of the impacting projectile. In other words, the residual strength of the impact-damaged specimens was found to decrease with an increase in the projectile energy. In order to study the variation of the damage strength among various laminates, the failure threshold strength of the compression-loaded laminates is estimated at 38 percent (a value taken at about the midrange of the kinetic energy axis) of the compressive ultimate static strength (σ_u) at the energy levels under consideration. The average ultimate-strain and ultimate-stress values for the compression-loaded laminates were found to be 0.013 and 1.11 GPa (161 ksi), respectively.

Some typical laminate failures in this series of specimens tested are shown in Fig. 6. The intraply fiber separation at ultimate failure (Fig. 6a) may be seen to be more pronounced as compared with the other two cases of failure. It is interesting to note that the specimen (Fig. 6c) that failed catastrophically upon impact has a "cleaner" failure compared with the other two cases.

The specimens in Series A were found to have almost equal values of the ultimate static strengths in both the tension-loaded and the compression-loaded laminates. The corresponding ultimate strain value in compression was found to be slightly higher than in tension. The tension-loaded laminates have exhibited higher impact damage strength than the compression-loaded laminates at any particular level of impact energy. This strength degradation in compression may be attributed to the weakening of the laminate due to local fiber buckling (instability). The tension-loaded laminates were observed to have a nonvarying (constant) failure threshold in the 2 to 6-J (18 to 53 in.-lb) range of impact energy, whereas in compression tests the failure threshold strength was found to decrease uniformly with an increase in the kinetic energy of the impacting projectile.

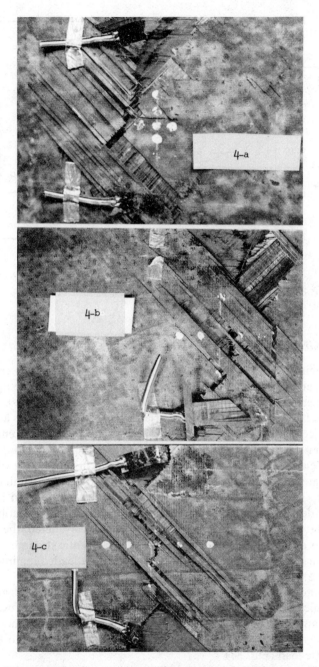

FIG. 4—*Fracture modes of graphite/epoxy laminates, tension, $(\pm 45, 0_4)_s$. (a) Specimen that failed at maximum static load. (b) Specimen that sustained local impact damage and subsequently failed upon further loading. (c) Specimen that failed catastrophically upon impact.*

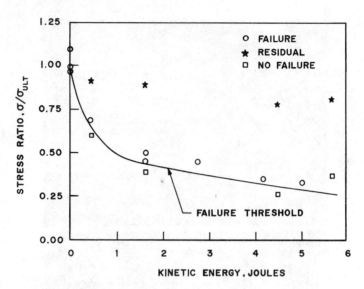

FIG. 5—*Graphite/epoxy composite,* $(\pm 45, 0_4)_s$, *loaded in compression; stress ratio versus kinetic energy of projectile.*

Series B: Graphite/Epoxy, $(\pm 45, 90, 0)_s$

The specimens in this series have two unidirectional plies, two crossplies, and four angle plies. The tensile and the compressive loads were applied to the specimens using the procedure decribed earlier.

Tension-Loaded Laminates—The variation of the stress ratio as a function of the projectile impact energy is shown in Fig. 7. The projectile energy was varied from 1 to 4 J (9 to 36 in.-lb). The shape of the initial part of the failure threshold curve, shown as a broken line here and in subsequent graphs, is approximate and is drawn based on the trends observed in similar earlier tests [2,3]. The strength retention of the specimens subjected to impact is seen to be decreasing monotonically with an increase in the projectile impact energy (Fig. 7). In the range of impact energy levels considered, the strength retention is estimated at 55 percent of the ultimate static strength. The ultimate-strain and ultimate-stress values of the tension-loaded laminates were found to be 0.010 and 0.57 GPa (83 ksi), respectively.

Two photographs showing specimen failure at the ultimate load and at the catastrophic impact load are shown in Figs. 8a and 8b, respectivey. The interply delamination in Figure 8b (impact) appears to spread more widely from the center test section than in the case of the failure due to the maximum static load (Fig. 8a).

Compression-Loaded Laminates—The eight-ply laminates of this series were tested in compression also. The strengths of the undamaged and the impact damaged specimens were evaluated. A failure threshold curve showing

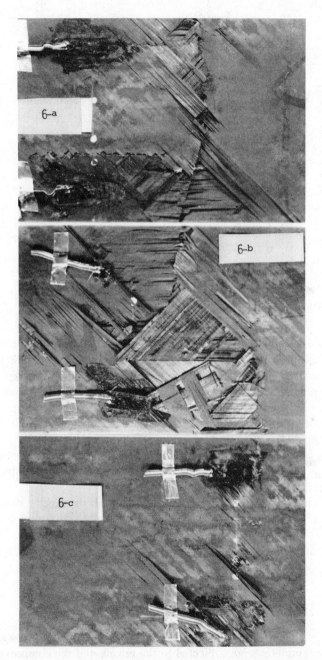

FIG. 6—*Fracture modes of graphite/epoxy laminates, compression, $(\pm 45, 0_4)_s$. (a) Specimen that failed at maximum static load. (b) Specimen that sustained local impact damage and subsequently failed upon further loading. (c) Specimen that failed catastrophically upon impact.*

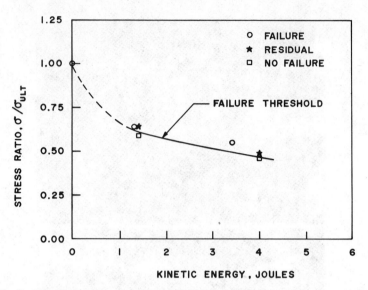

FIG. 7—*Graphite/epoxy composite,* $(\pm 45,90,0)_s$, *loaded in tension; stress ratio versus kinetic energy for projectile.*

the stress ratio as a function of the impact energy of the projectile is shown in Fig. 9. Based on the limited number of specimens tested in this series, the strength of the impact damaged laminates appears to be 51 percent of the ultimate strength. The failure threshold curve is fairly constant over the range of the projectile impact energy, 1 to 5 J (9 to 45 in.-lb). The compressive ultimate-strain and ultimate-stress values of the laminates tested in this series were found to be 0.012 and 0.60 GPa (87 ksi), respectively.

Two typical views of the ultimate failure mode and the catastrophic failure mode (due to a combination of the preload and the projectile impact energy) are shown in Figs. 10a and 10b, respectively.

The ultimate strength of the laminates in both the tension and the compression tests of this series was found to be very close. However, the ultimate strain was noted to be higher in compression than in tension. The failure lines in both the tension and the compression tests (Figs. 8a and 10a, 8b and 10b) appear to be almost identical.

Series C: Graphite/Epoxy, $(90, \pm 45, 0)_s$

The epoxy matrix used in this eight-ply series of laminates was Fiberite 934. The laminates were subjected to the tensile and the compressive loads.

Tension-Loaded Laminates—The stress ratio (σ/σ_{ult}) as a function of the impact energy of the projectile in tension-loaded laminates is shown in Fig. 11. The failure threshold is seen to be fairly constant in the projectile kinetic

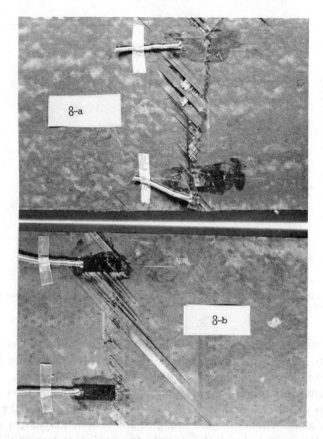

FIG. 8—*Fracture modes of graphite/epoxy laminates, tension, (±45,90,0)ₛ. (a) Specimen that failed at maximum static load. (b) Specimen that failed catastrophically upon impact.*

energy range of 1 to 4 J (9 to 36 in.-lb). The stress level at which catastrophic failure would result upon impact is found to be 33 percent of the laminate ultimate strength. The laminate ultimate-strain and ultimate-stress values were found to be 0.011 and 0.68 GPa (99 ksi), respectively.

Some typical views of the failure modes under three different loading conditions are shown in Fig. 12. It is interesting to note the relatively "clean" catastrophic failure as compared with the failures at the other two loading conditions.

Compression-Loaded Laminates—The compression-loaded laminates have exhibited relatively more scatter in the limited amount of the experimental data. However, some trends in the failure threshold may be seen by observing Fig. 13. The stress ratio is decreasing with an increase in the impact energy. The strength retention under impact is estimated at about 38 percent of the ultimate static strength. The average ultimate-strain and

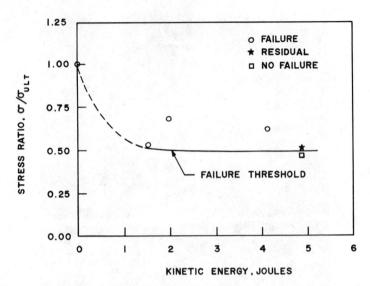

FIG. 9—*Graphite/epoxy composite,* $(\pm 45, 90, 0)_s$, *loaded in compression; stress ratio versus kinetic energy of projectile.*

ultimate-stress values were found to be 0.011 and 0.54 GPa (78 ksi), respectively. The failed surfaces in compression were observed to have the same pattern of failures as in the case of the tension-loaded laminates.

An examination of all the experimental results in Table 1 shows that the specimens in Series A and Series B having ±45-deg plies near the test surface have almost equal ultimate strengths and considerably larger strains in compression compared with the corresponding values in tension. The Series C with a crossply on the test surface has not revealed any similar trends. This may be interpreted to mean that locating a 45-deg ply on the laminate surface would help in achieving higher strains at failure in compression than in tension. On the other hand, in designs where the retention of higher residual strength (improved impact resistance) is desired, the tension-loaded laminates of Series A and B have shown better impact resistance than the corresponding compression-loaded laminates.

The failure surfaces of the specimens in Series A and B reveal pronounced intraply fiber separation under static (tensile or compressive) loads than under catastrophic loads (impact failure). This phenomenon may be observed by comparing Figs. 4a and 4c, Figs. 6a and 6c, and to a less-pronounced extent Figures 8a and 8b and Figs. 10a and 10b. The interply delaminations may be seen in all the specimens to a varying degree.

An examination of the specimens that failed under the maximum static loads (Figs. 4a, 6a, 8a, and 10a) in Series A and B shows a distinct crack line in the lateral (that is, in the width) direction whereas the specimens that

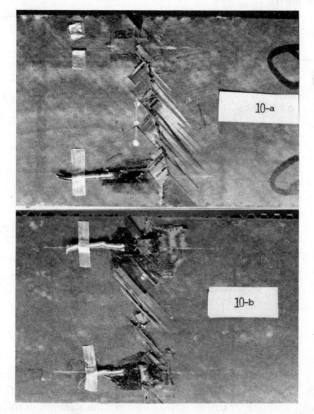

FIG. 10—*Fracture modes of graphite/epoxy laminates, compression, $(\pm 45,90,0)_s$. (a) Specimen that failed at maximum static load. (b) Specimen that failed catastrophically upon impact.*

failed catastrophically under impact (Figs. 4c, 6c, 8b, and 10b) do not in general show similar lines.

In general, it may be seen that the laminates (Series A) having more unidirectional plies (in the direction of load) with angle plies (45-deg orientation to the load direction) on the test surface show more extensive delamination and fiber separation at failure (Figures 4 and 6) than the laminates (Series B, Figs. 8 and 10) that have fewer unidirectional fibers. It is also interesting to observe that the laminates in Series C (having a crossply on the surface) have not shown extensive delamination (Fig. 12) and fiber separation.

Conclusions

An experimental investigation was conducted to assess the damage tolerance of the laminated graphite/epoxy composite materials subjected to low-

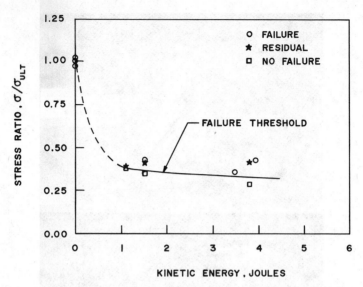

FIG. 11—*Graphite/epoxy composite,* $(90, \pm 45, 0)_s$, *loaded in tension; stress ratio versus kinetic energy of projectile.*

velocity projectile impact. The sandwich-type beam specimens were fabricated with graphite/epoxy face sheets, aluminum honeycomb core, and a steel (back) plate. Using a four-point beam-loading apparatus, the ultimate strength and ultimate strain, as well as the residual strength of the composite laminates that survived the projectile impact, were found. Based on the limited number of tests conducted in this investigation, the following observations can be made:

1. The laminates (Series A) having angle plies as face sheets exhibit almost equal ultimate strength values in both the tension and compression tests. The Series B laminates also show similar results. The corresponding strains in compression were found to be higher than in tension. On the other hand, the failure threshold was noted to be higher under tension than in compression. The specimens of Series C having a crossply on the test surface apparently show an opposite pattern of strength retention upon impact.

2. The graphs of the failure thresholds for the various laminates tested show either asymptotic, or monotonically decreasing residual strength values with an increase in the projectile impact energy.

3. The laminates having more angle plies near the impact surface and unidirectional plies elsewhere appear to show extensive interply and intraply fiber delaminations at failure relative to the laminates with a crossply on the impact surface.

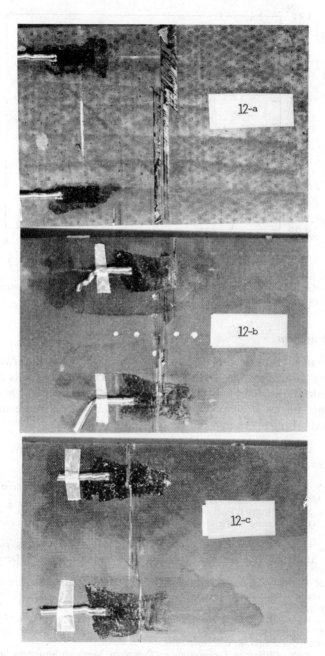

FIG. 12—*Fracture modes of graphite/epoxy laminates, tension.* $(90, \pm 45, 0)_s$. *(a) Specimen that failed at maximum static load. (b) Specimen that sustained local impact damage and subsequently failed upon further loading. (c) Specimen that failed catastrophically upon impact.*

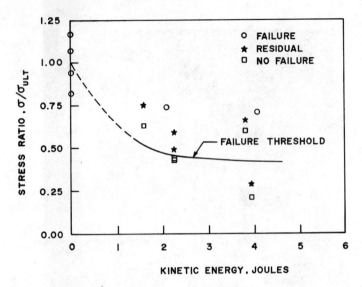

FIG. 13—*Graphite/epoxy composite, (90, ±45, 0)ₛ, loaded in compression; stress ratio versus kinetic energy of projectile.*

Acknowledgment

The financial support provided for this work by the National Aeronautics and Space Administration (NASA) through a Grant NSG-1296 is gratefully acknowledged. Further, sincere appreciation is extended to Marvin D. Rhodes of Langley Research Center, NASA, for his assistance in the experimental arrangement and for many valuable suggestions.

References

[1] Yeow, Y. T., Morris, D. H., and Brinson, H. F., *Journal of Testing and Evaluation*, Vol. 7, No. 2, March 1979.

[2] Rhodes, M. D., "Impact Tests on Fibrous Composite Sandwich Structures," NASA TM 78719, National Aeronautics and Space Administration, Langley Research Center, Hampton, Va., Oct. 1978.

[3] Rhodes, M. D., "Impact Fracture of Composite Sandwich Structures" in *Proceedings*, ASME/AIAA/SAE 16th Structural Dynamics and Materials Conference, Denver, Colo., May 1975.

[4] Adsit, N. R. and Waszczak, J. P., "Investigation of Damage Tolerance of Graphite/Epoxy Structures and Related Design Implications," Final Report, Convair Division, General Dynamics, NADC-76387-30, San Diego, Calif., Dec. 1976.

[5] Sharma, A. V., "Effect of Low Velocity Projectile Impact on the Strength of Composite Sandwich Structures" in *Proceedings*, Fifth Inter-American Conference on Materials Technology, Sao Paulo, Brazil, Nov. 1978.

[6] Sharma, A. V., "Effect of Temperature on Composite Sandwich Structures Subjected to Low Velocity Projectile Impact," Paper No. 78-WA/Aero-2, Winter Annual Meeting, The American Society of Mechanical Engineers, San Francisco, Calif., Dec. 1978.

Special Test Methods and Analysis

J. F. Mandell,[1] *A. Y Darwish,*[1] *and F. J. McGarry*[1]

Fracture Testing of Injection-Molded Glass and Carbon Fiber-Reinforced Thermoplastics

REFERENCE: Mandell, J. F., Darwish, A. Y., and McGarry, F. J., **"Fracture Testing of Injection-Molded Glass and Carbon Fiber-Reinforced Thermoplastics,"** *Test Methods and Design Allowables for Fibrous Composites, ASTM STP 734,* C. C. Chamis, Ed., American Society for Testing and Materials, 1981, pp. 73-90.

ABSTRACT: A study is reported of the application of linear elastic fracture toughness testing techniques to short glass and carbon fiber-reinforced injection-molded thermoplastics. The materials were typical of high-temperature engineering thermoplastics and included polycarbonate, polysulfone, nylon 6/6, polyphenylene sulfide, and poly-(amide-imide). Tests were run on specimens of various sizes and shapes with machined notches and fatigue cracks. Results are reported for modes of crack growth, applicability of linear elastic fracture mechanics, and test peculiarities. Attempts to correlate the matrix toughness and fiber length with composite toughness are discussed.

KEY WORDS: composite materials, thermoplastics, injection molding, fracture (materials)

Injection-molded fiber-reinforced thermoplastics are typified by short dispersed fibers whose orientation is determined by the flow characteristics of the melt during molding. The fibers increase the strength, elastic modulus, and temperature resistance, relative to the unreinforced polymer, but the mechanical properties may vary with direction and location depending on the mold geometry and the molding conditions [1,2].[2] Usually, the impact resistance increases with brittle matrices and decreases with ductile ones [2]. Carbon fibers generally give higher modulus and lower impact strength than E-glass fibers [3]. Although short fiber composites have lower mechanical properties than continuous fiber systems, they possess advantages of lower cost, more variety in matrix thermal, mechanical, and chemical properties, and they can be molded into very complex shapes at high production rates.

The purpose of this study was to investigate the crack resistance of a va-

---

[1]Research associate, graduate student, and professor, respectively, Department of Materials Science and Engineering, Massachusetts Institute of Technology, Cambridge, Mass. 02139.
[2]The italic numbers in brackets refer to the list of references appended to this paper.

riety of reinforced, high-temperature-engineering thermoplastics. Several areas have been studied: (1) the general modes and mechanisms of crack initiation and propagation, (2) the applicability of linear elastic fracture mechanics (LEFM), and (3) the comparative fracture toughness of various matrices and reinforcements. The materials used and their sources are listed in Table 1. In most cases, the unreinforced polymer is produced by one manufacturer, then combined with fibers, usually by an extrusion method, by a second. Finally, the material is molded by an injection molder; in the present case, all the specimens were molded by Keyon Materials of Lord Corp. The matrix materials listed in Table 1 include two standard engineering thermoplastics, semicrystalline nylon 6/6 (N66) and amorphous polycarbonate (PC) along with polysulfone (PSUL), which is a higher-temperature amorphous polymer; these three are nominally ductile in uniaxial tension tests but are also notch-sensitive under some conditions. The semicrystalline polyphenylene sulfide (PPS) and the amorphous poly(amide-imide) (PAI) are higher-temperature polymers which are not ductile at room temperature, although the PAI may have a strain to failure as high as 10 percent [4]. The PPS is an extremely brittle matrix which is difficult to mold without voids in its unreinforced form.

Experimental Methods

All fracture tests were conducted on a servohydraulic testing machine under displacement control in an air-conditioned laboratory environment.

TABLE 1—*Materials and manufacturers.*[a]

Material Designation	Matrix	Fiber	Manufacturer or Compounder
N66	nylon 6/6	. . .	E.I. Du Pont de Nemours & Co. Inc.
N66/G	nylon 6/6	40% E-glass	LNP Corp.
N66/C	nylon 6/6	40% carbon	LNP Corp.
PC	polycarbonate	. . .	General Electric Co.
PC/G	polycarbonate	40% E-glass	LNP Corp.
PC/C	polycarbonate	40% carbon	LNP Corp.
PSUL	polysulfone	. . .	Union Carbide Corp.
PSUL/G	polysulfone	40% E-glass	LNP Corp.
PSUL/C	polysulfone	40% carbon	LNP Corp.
PPS	polyphenylene sulfide	. . .	Phillips Petroleum Co.
PPS/G	polyphenylene sulfide	40% E-glass	Phillips Petroleum Co.
PPS/C	polyphenylene sulfide	40% carbon	Fiberite Corp.
PAI/G	poly(amide-imide)	30% E-glass	Amoco Chemicals Corp.

[a] All materials were injection molded by Keyon Materials of The Lord Corp.
[b] Percent by weight; carbon fiber is PAN-based; Young's modulus = 207 GPa(30 × 10^6 psi).
[c] Unreinforced polymer obtained through LNP Corp., same batch as for reinforced material.

Ultimate strength and fatigue crack tests were run under load control. The testing rate for fracture tests was 0.5 cm/s unless otherwise noted. Most of the fracture specimens were of the single-edge-notched (SEN) type, loaded in uniaxial tension. Values for the fracture toughness were calculated from the maximum force using an isotropic K-calibration [5]. The study was exploratory in nature, with only three test replications in most cases, so no statistical treatment of the data was attempted.

Figures 1 and 2 show the geometry of specimens and moldings. The tensile bar is a standard ASTM D638 Type 1 steel. The single end gate results in flow of material down the length of the bar during molding. The rectangular plaque specimen in Fig. 2 is gated in the center of one surface, so the material flows in a circular pattern about the gate until the mold sides are reached, whereupon the flow continues down the mold lengthwise to the ends. The D638 bar without a notch was used to obtain ultimate tensile strength (UTS) values. Notches in fracture specimens were machined with a diamond-edged wheel of 0.46-mm thickness for specimens cut from plaques, or 0.30 mm for D638 specimens. In each case the radius at the notch root corners was approximately 0.06 mm. In some, the notches were further sharpened by razor blade cuts or fatigue cracks.

Results and Discussion

Modes of Crack Growth

As an illustration of the effects of processing on crack extension, Fig. 3 shows two failed specimens arranged as they were cut from a plaque. The specimens are of a pin-loaded compact tension geometry which was initially investigated. The cracks propagate in the form of a circle about the gate, which is also the geometry of the material boundary as it flows into the mold. This lack of control over the crack direction limited the use of the compact tension specimens to directions for which the crack propagation was colinear with the original notch axis. Even then, it was found that the cracks on the surface tended to resist growth in the direction favored in the interior, so that extensive crack extension on the interior would occur without full penetration to the surfaces, particularly in fatigue loading. As a result of this, the specimen geometry was changed to a SEN tension bar for which there is a greater tendency for colinear, through-thickness growth.

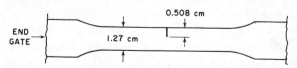

FIG. 1—*Tension bar specimens after ASTM D638 Type I, 0.138 cm thick, showing notch tip location.*

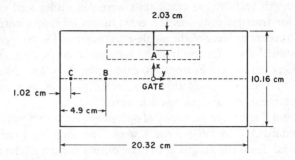

FIG. 2—*Plaque specimens, 0.64 cm thick; A,B, and C indicate crack-tip locations of test specimens.*

FIG. 3—*Failed compact-tension specimens as cut from plaque, PPS/G.*

Figure 4*a* shows two SEN specimens cut from plaques as in Fig. 2, with the notch tip at Position A; each specimen comprises half of one plaque, cut lengthwise along the dashed line through the gate. All specimens taken from this position tended to fail in a colinear fashion (bottom specimen, Fig. 4*a* except for reinforced PPS and a few carbon-reinforced N66 (N66/C). These failed with the crack at an angle to the notch axis as indicated in the upper photograph of Fig. 4*a*. No consistent effect of the differences in crack direction in Fig. 4*a* could be found in the fracture loads of materials for which both cases were observed.

Figure 4*b* is an enlarged view of a noncolinear failure in the PPS/G system. A small colinear crack is evident on the surface at the notch tip despite the gross separation in another direction. This branching of the initial crack was common in PPS systems and complicated efforts to sharpen the crack in fatigue. In all except reinforced PPS, fatigue cracks grew in a

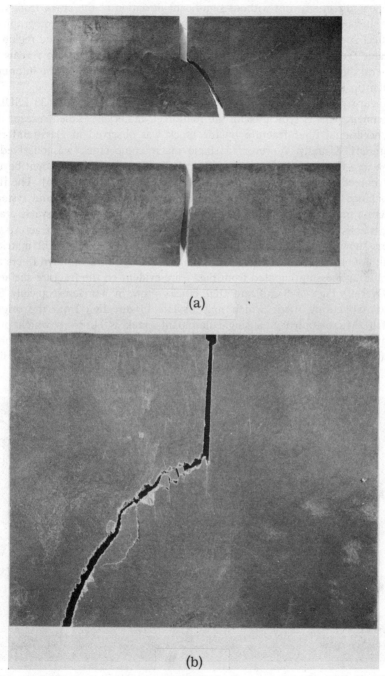

FIG. 4—Failed SEN specimens of (a) PPS/C showing general mode of failure, and (b) magnified view of a noncolinear failure in PPS/G.

colinear direction, and without significant deviation of the crack front from the surface to the interior. Figure 5 shows fracture surface photographs of N66/G specimens with and without fatigue cracks. The whitened region indicates the extent and geometry of the fatigue crack. Figure 6 reveals the general difference in fiber orientation near the surface and on the interior at notch tip Position A on the plaques.

Figure 7 is a micrograph of a fatigue crack growing in a D638 PSUL/C specimen. This is typical of all the reinforced materials studied, except that an occasional fiber fracture by the crack was observed in glass-reinforced materials. Usually, however, a single macroscopic crack was observed to grow in a fiber-avoidance mode with little damage or yielding evident beyond the immediate crack area. This has been described previously [6]. The fiber avoidance contrasts with continuous fiber or even chopped strand systems, where a much larger damage zone of ply delamination and transverse cracking is associated with a macroscopic crack tip [7]. The resulting fracture surfaces show significant local ductility for the N66, PC, and PSUL matrices, with a more brittle character for PPS and PAI (Fig. 8). Debonded fibers and local roughness anticipated from Fig. 7 are evident on the fracture surfaces.

Overall, Figs. 3–8 indicate that cracks grow in a macroscopically flat, brittle-appearing mode which may deviate in direction from the original notch. Microscopically, a single dominant crack is observed which propagates in a fiber-avoidance mode. Matrix ductility in these reinforced systems is limited to the local region adjacent to the crack surface, with little

FIG. 5—*Fracture surfaces of N66/G with initial fatigue crack (*top, *whitened region), and machined notch (*bottom).

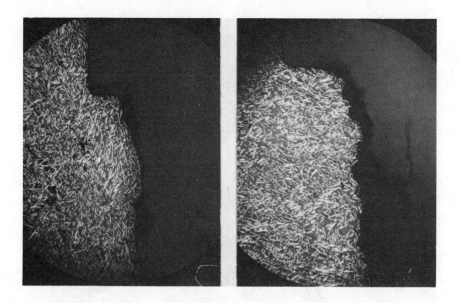

FIG. 6—*Fiber orientation at surface and interior of failed specimens with notch (*top*) at Position A in Fig. 2, PC/C.*

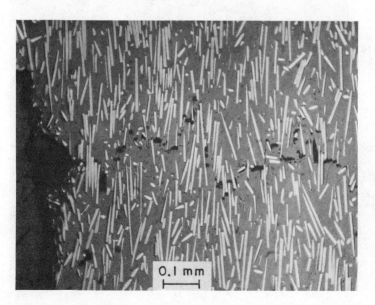

FIG. 7—*Fatigue crack growing left to right from notch in PSUL/C D638 specimen.*

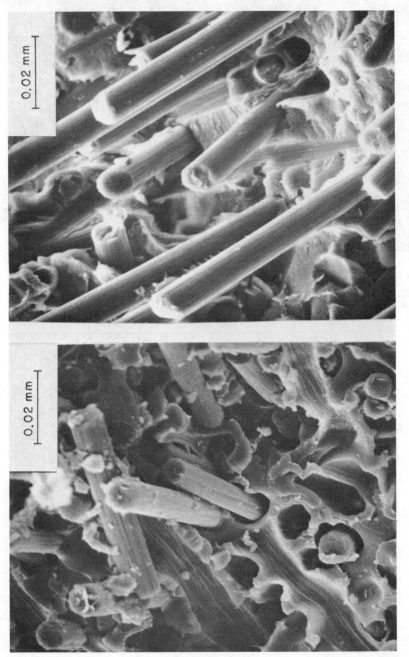

FIG. 8—Fracture surfaces of PSUL/C (left) and PPS/C (right).

apparent damage extending beyond the length of the longer fibers into the bulk material.

Applicability of Linear Elastic Fracture Mechanics

The brittle failure of the reinforced matrices suggests a small crack-tip process zone, typical of cases to which a brittle fracture criterion can be applied. However, the validity of a single parameter to define crack resistance, independent of geometry, must be demonstrated; here this was accomplished by testing specimens of different sizes. The effect of thickness has not been studied, but the fracture surfaces display no shear lips which accompany a thickness effect. Also, the variation of fiber orientation through the thickness, which depends upon mold geometry, made a good study of thickness effects difficult, if not impossible.

Exploratory fracture tests were run on notched D638 bars and on smaller strips cut from them. These are interesting because of their common use to determine fatigue and ultimate strength properties, coupled with attempts to rationalize these properties using LEFM models. In all SEN tests, the notch tip was located at the position in Fig. 1, with narrower specimens cut from around this point to avoid variations in material properties with position. The ratio c/w, crack length to specimen width, remained constant at 0.40. Test results for several materials and two specimen widths are listed in Table 2. In each case, the calculated fracture toughness is higher for the wider specimens. As will be discussed later, the N66/G material had the largest apparent process zone radius of all materials tested, approximately 0.5 mm, and the size effect was greatest for this material.

Other results from the D638 bars indicated that sharpening of the machined notch by a razor blade or by fatigue crack growth did not significantly lower the K_Q-value calculated. The data in Table 2 indicate the need for a two-parameter fracture criterion to account for the size effects in this range,

TABLE 2—*Effect of size on the fracture toughness of SEN specimens cut from D638 bars (c/w = 0.40 in all cases).*

Material	Specimen Width, cm	No. of Specimens	Avg K_Q, MNm$^{-3/2}$
N66/G	1.27	3	12.2
N66/G	0.64	3	9.6
N66/C	1.27	3	13.8
N66/C	0.64	2	12.5
PPS/G	1.27	3	8.5
PPS/G	0.64	4	6.3
PPS/C	1.27	2	7.4
PPS/C	0.64	4	6.3
PAI/G	1.27	4	7.8
PAI/G	0.64	4	7.1

as is commonly the case for other composites [8, 9]. Although it is difficult to assess the gross notch sensitivity due to variations in strength from point to point, these results, as well as similar double-edge-notch results, give a net section strength about half the unnotched bar UTS, so the D638 bars appear to be notch-sensitive.

The effects found with the D638 bars precluded the determination of a size-independent K_Q-value even for the largest notches which could be introduced. This led to the use of specimens taken from the plaque moldings shown in Fig. 2. For this part of the study, all specimens had a notch tip located at Position A, and were of the SEN geometry depicted by dashed lines, with the load applied in the y-direction. The specimen gage length was three times the width, and the crack length-to-width ratio was 0.40.

Figures 9 and 10 give K_Q versus specimen size for glass- and carbon-reinforced N66; each point represents a single test. These show no significant size effect. Further, both fatigue crack and razor crack initiation give values as high as the machined notch, indicating that the latter produces a valid test. Similar results for PPS in Figs. 11 and 12 show a trend to lower toughness for sizes below 3.5 cm width, but these materials have smaller crack-tip zones than the N66 materials, and hence they would be expected to show less size effect. The calculation of K_Q for the data in Figs. 9–12 as-

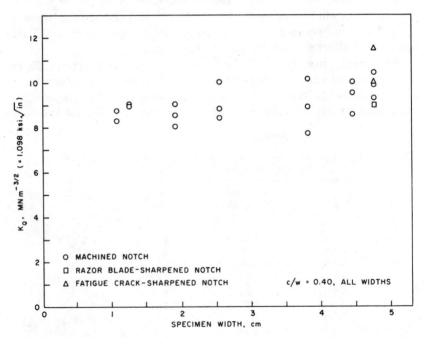

FIG. 9—*Fracture toughness calculated from maximum force versus specimen width for nylon with 40 percent glass fibers by weight.*

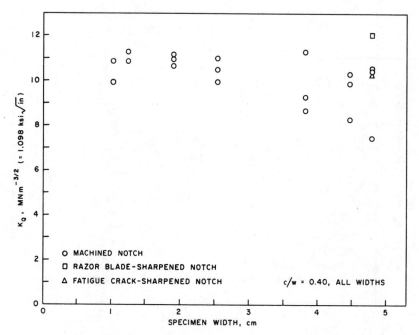

FIG. 10—*Fracture toughness calculated from maximum force versus specimen width for nylon with 40 percent carbon fibers by weight.*

sumed uniform isotropic elastic properties over the area of specimen. In fact, the fiber distribution follows a circular pattern from the gate outward, as discussed earlier, so the elastic property distribution would show a similar pattern. This could lead to errors in the calculation of K_I, particularly for the larger specimens, and these may contribute to the trend with size for the PPS materials, since they show a greater anisotropy. Some PPS specimens contained internal cracks parallel to the flow front prior to testing; this could also lead to inaccuracies in the K_Q calculation.

The force used to calculate K_Q was the maximum force during the test. Force versus crack-opening displacement (COD) curves given in Fig. 13 tend to be nearly linear to failure in most cases. This is typical of valid fracture toughness tests in metals [10], and it justifies the use of the maximum force in the K_Q calculation. A few tests with significant nonlinearity, usually in glass-reinforced systems, appeared to show some limited stable crack extension prior to fracture.

Variation of K_Q *with Position and Direction*

Table 3 gives the results of tests from several positions and directions on the plaque moldings; the positions A,B,C, and force directions x and y are

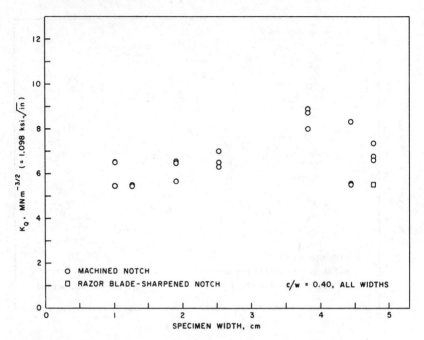

FIG. 11—*Fracture toughness calculated from maximum force versus specimen width for polyphenylene sulfide with 40 percent glass fibers by weight.*

indicated in Fig. 2. The combinations (A,y), (B,x), and (C,x) are for cracks growing normal to the material flow front. These give the highest K_Q and do not vary much from position to position. The combination (B,y), for cracks parallel to the flow, gives values much lower, less than half in the case of PPS/G. The 45-deg case shows intermediate toughness. The PPS/G system tested in the strong (B,x) combination failed away from the notch, where the circular flow pattern gave a weaker plane near the grips.

Comparison of Various Materials

SEN specimens of each material listed in Table 1 were tested to determine K_Q at Position A, force direction y in Fig. 2. All specimens were 4.75 cm wide with a c/w ratio of 0.40. Table 4 compares the values obtained in each case, and Table 5 gives the ultimate tensile strengths of D638 bars for comparison.

The higher toughness matrices, N66 and PC, show similar K_Q whether reinforced or unreinforced for this position and crack orientation. It should be recalled that the K_Q-value for N66/G with the crack parallel to the flow front is only approximately 60 percent of that listed in Table 4, so the K_Q for reinforced specimens can be much lower than that for the unreinforced

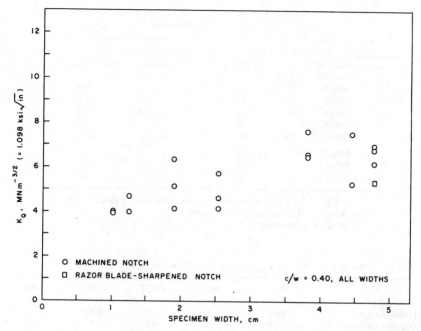

FIG. 12—*Fracture toughness calculated from maximum force versus specimen width for polyphenylene sulfide with 40 percent carbon fibers by weight.*

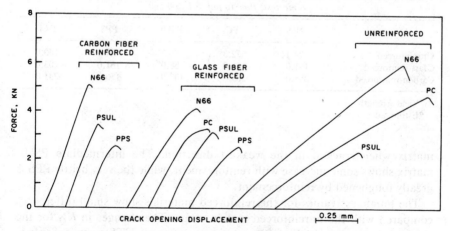

FIG. 13—*Force versus crack-opening displacement for various materials (specimen width = 5.08 cm; razor-sharpened notch; displacement rate = 0.05 cm/s).*

TABLE 3—*Average* K_Q *versus position and direction for plaque moldings.* [a]

Material	Position	Applied Stress Direction	K_Q(MNm$^{-3/2}$)
N66/G	A	y	9.1
N66/G	B	y	5.9
N66/G	B	x	10.3
N66/G	B	45 deg	8.2
N66/G	C	x	10.5
PPS/G	A	y	6.6
PPS/G	B	y	2.8
PPS/G	B	x	> 5.7[b]

[a]See Fig. 2 for positions and directions. Specimen width = 2.5 cm; crack length = 1.0 cm; replication factor = 3.
[b]Failure away from notch.

TABLE 4—*Average* K_Q *from plaques of various materials,* [a] *MNm* $^{-3/2}$.

	N66	PC	PSUL	PPS	PAI
Unreinforced	10.9	8.4	4.7	0.80[b]	...
Glass-reinforced	9.9	8.7	6.5	7.0	9.4
Carbon-reinforced	9.4	7.5	7.2	6.6	...

[a]Specimen width = 4.75 cm; crack length = 2.03 cm; Position A, Stress Direction y; replication factor = 3.
[b]For material with high void content.

TABLE 5—*Ultimate tensile strength of various materials, MPa (D638 specimen, Type 1, load ramp test; time to fail, 0.1 to 0.5 s).*

	N66	PC	PSUL	PPS	PAI
Unreinforced	74.1[a]	72.3[a]	76.5	34.5	140.0
Glass-reinforced	181.0	161.0	158.0[b]	181.0	203.0
Carbon-reinforced	256.0	203.0	197.0	156.0	231.0

[a]Yield strength.
[b]Estimated.

matrix when cracked in the weakest direction. The intermediate PSUL matrix shows some increase with reinforcement, while the very brittle PPS is greatly toughened by reinforcement.

The toughness values for the reinforced materials show small differences compared with the unreinforced matrix range. The changes in K_Q for the reinforced cases are similar to the relative variations in UTS given in Table 5. However, the UTS of all the matrices is substantially increased by fiber reinforcement, while the K_Q-value is increased only for the brittle matrices. The convergence of K_Q-values to similar levels for the reinforced systems tends to confirm reported experience for notched impact strength, for which a wider range of matrices has been tested (summary impact data may be found in Ref *11*).

Effect of Fiber Length on K_Q

To rationalize the fracture toughness results, the following observations appear relevant for these materials:

1. K_Q of reinforced systems does not necessarily correlate with K_Q of the matrix.

2. The local mode of crack propagation in the reinforced systems is primarily fiber avoidance, completely so in carbon fiber systems (Fig. 7).

3. Little or no damage or yielding is evident away from the immediate crack surface area, outside of a region on the order of the length of the longest fibers.

In the absence of larger-scale yielding or damage, a limiting condition would appear to be achieved when the local strength is exceeded at a distance from the crack tip equal to the length of the longer fibers; the crack is then free to propagate without breaking any fibers, instead breaking the material which surrounds them. If the UTS listed in Table 5 is taken as the local strength, then the radius of the process zone may be calculated, following Ref *12* for plane stress, as

$$r_p = \frac{1}{2\pi} \left(\frac{K_Q}{\text{UTS}} \right)^2 \tag{1}$$

If we equate r_p with the effective length of the longer fibers, l_f^*, then

$$l_f^* = \frac{1}{2\pi} \left(\frac{K_Q}{\text{UTS}} \right)^2 \tag{2}$$

and

$$K_Q = (\text{UTS})\,(2\pi l_f^*)^{1/2} \tag{3}$$

Thus, K_Q is predicted to be dependent on the matrix through the composite UTS, and on the fiber length. Figure 14 shows r_p versus fiber length for all the materials tested. With a distribution of fiber lengths, it was necessary to define some characteristic as a representative value and, since the local mode of crack growth appeared to be controlled by the longer fibers, the effective length was taken as that which was exceeded by 5 percent of the population, excluding short fragments with a length-to-diameter ratio less than 5. This

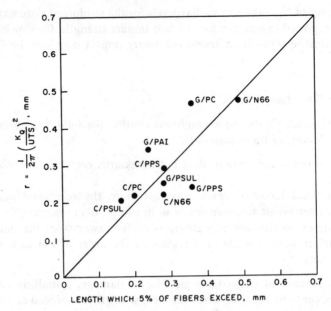

FIG. 14—*Calculated plastic zone radius versus fiber length exceeded by 5 percent of fibers; various carbon and glass fiber-reinforced thermoplastics with 30 to 40 percent fiber by weight.*

definition effectively excludes the very few longest fibers, which are rarely encountered, but other definitions might serve equally well. (The fiber lengths were determined by burnoff or solvent dissolution of the matrix, followed by direct measurement from a micrograph of a representative specimen including approximately 100 fibers plus short fragments.) Fiber length distribution curves were typical of injection molded materials given in the literature [13,14] and are not included here.

The results in Fig. 14 suggest that the calculated process zone is similar in size to the long fibers for all of the materials. Furthermore, the trend with increasing fiber length is generally good, though there is some scatter. Thus the data are consistent with relationship in Eq 2, but the value of this relationship must be seen as tentative until more data are available. Equation 2 is not correct for cracks growing in the weak direction, where K_Q is reduced to 60 percent or less of the values in Table 4. This is not surprising, since the mode of local crack growth is much smoother, and it is not necessary for the crack to go around the longer fibers, which are almost entirely oriented parallel to the crack.

The model represented by Eq 3 does not necessarily contradict models based on energy considerations such as fiber debonding and pullout friction [15]. The fracture surface work indicated by Eq 3 may be determined by the relationship between K_Q and the critical strain energy release rate, G_c [16]

$$G_c = K_Q^2/E \qquad (4)$$

where E is Young's modulus and Eq 4 is for isotropic materials. Substituting Eq 3 into Eq 4

$$G_c = 2\pi l_f^*(\text{UTS})^2/E \qquad (5)$$

Since the UTS usually varies in a similar fashion to E as reinforcement is added, G_c is expected to increase approximately in proportion to the UTS for moderate increases in the amount of reinforcement. This cannot be extended to the unreinforced case since Eq 3 cannot apply there.

If the notched impact strength behaves in a similar fashion to G_c, then the reduced impact resistance reported for graphite fiber reinforcement [3] may be due solely to the increase in E without a correspondingly high increase in UTS. Factors such as fiber/matrix bond strength, which may increase the UTS, also are expected to increase the toughness. Such an effect has been reported in Ref 14. Finally, the mode of crack growth observed here is not expected to continue as the fiber length becomes much greater, since crack growth then would be likely to involve more fiber failures.

Conclusions

Cracks in the materials tested in this study propagate primarily in a fiber-avoidance mode, with occasional fiber breakage in glass systems. Macroscopically, cracks propagate in a brittle manner to which LEFM appears to apply for initial cracks longer than approximately 4 mm in the toughest case. The value of K_Q varies with location and direction by more than a factor of two in some cases because of fiber orientation. The fracture toughness correlates more closely with the fiber length than with the matrix toughness for this group of materials.

Acknowledgments

This work was supported by the Lord Corp., and materials were supplied by its Keyon Materials facility. Discussions with Dr. A. F. Lewis of the Lord Corp. were helpful.

References

[1] Hall, R. C., Cashulette, C. G., Valentine, J. A., and Chadderdon, G. D. in *Proceedings,* 31st SPI Reinforced Plastics/Composites Institute Conference, Society of the Plastics Industry, Section 19-A, Washington, D.C., 1975.
[2] Theberge, J. E. and Hall, N. T. in *Proceedings,* 24th SPI Reinforced Plastics/Composites Institute Conference, Society of the Plastics Industry, Section 1-B, Washington, D.C., 1969.
[3] Theberge, J. E., Arkles, B., and Robinson, R., *Machine Design,* Vol. 46, 1974, p. 102.
[4] "Torlon, Technical Data and Applications Information," Amoco Chemical Corp. Technical Data Sheet, Commerical Development Department, Chicago, Ill.

[5] Paris, P. C. and Sih, G. C. in *Fracture Toughness Testing and Applications, ASTM STP 381,* American Society for Testing and Materials, 1965, p. 30.

[6] Mandell, J. F. in *Proceedings,* 34th SPI Reinforced Plastics/Composites Institute Conference, Society of the Plastics Industry, Section 20-C, New Orleans, La., 1979.

[7] McGarry, F. J., Mandell, J. F., and Wang, S. S., *Polymer Engineering and Science,* Vol. 16, 1976, p. 609.

[8] Owen, M. J. and Bishop. P. T., *Journal of Composite Materials,* Vol. 7, 1973, p. 146.

[9] Nuismer, R. J. and Whitney, J. M. in *Fracture Mechanics of Composites, ASTM STP 593,* American Society for Testing and Materials, 1975, pp. 117–142.

[10] Kaufman, J. G. in *Review of Developments in Plane Strain Fracture Toughness Testing, ASTM STP 463,* American Society for Testing and Materials, 1970, p. 3–21.

[11] *Modern Plastics Encyclopedia,* J. Agranoff, Ed., McGraw-Hill, New York, 1979, p. 498.

[12] "Fracture Testing of High Strength Sheet Materials," *ASTM Bulletin,* Jan. 1960.

[13] Sawyer, L. C., *Polymer Engineering and Science,* Vol. 19, 1979, p. 377.

[14] Ramsteiner, F. and Teyhsohn, R., *Composites,* Vol. 10, 1979, p. 111.

[15] Kelly, A., *Proceedings, Royal Society of London,* Vol. A319, 1970, p. 95.

[16] Irwin, G. R., *Journal of Applied Mechanics, Transactions,* American Society of Mechanical Engineers, Vol. 24, 1957, p. 361.

R. Y. Kim[1]

On the Off-Axis and Angle-Ply Strength of Composites

REFERENCE: Kim, R. Y., **"On the Off-Axis and Angle-Ply Strength of Composites,"** *Test Methods and Design Allowables for Fibrous Composites. ASTM STP 734.* C. C. Chamis, Ed., American Society for Testing and Materials, 1981, pp. 91–108.

ABSTRACT: This paper is concerned with the static strength of off-axis and angle-ply laminates under uniaxial tension and compression. The experimental data were compared with the tensor polynomial failure criterion, assuming the interaction term based on the Von Mises yield criterion for the isotropic materials. The curing residual stresses are also included in the strength analysis for the angle-ply laminate. The fracture surface was examined using a scanning electron microscope and the results are discussed in conjunction with the failure theory.

KEY WORDS: off-axis, angle ply, strength, failure theory, failure modes, residual stress

Strength tests on off-axis and angle-ply laminates offer useful information for the evaluation of the failure theory and failure mechanism. A number of failure criteria have been proposed to predict the failure of anisotropic composite materials under various combined states of stress. Among these failure theories, the tensor polynomial failure criterion developed by Tsai and Wu accounts for the interaction effects among stress components and offers a significant improvement in operational simplicity over many other failure theories [1].[2] Excellent agreement between the theory and experimental data has been demonstrated under uniaxial tensile loading [2,3]. The theory, however, has not been fully tested in the state of stress induced by uniaxial compressive loading. This appears to be due mainly to the experimental difficulty in compression tests using a flat coupon specimen.

In this paper, the failure theory has been compared with experimental results obtained from the off-axis and angle-ply laminates under applied tensile and compressive loading. A scanning electron microscope (SEM) was

---

[1] Research engineer, University of Dayton Research Institute, Dayton, Ohio 45469.
[2] The italic numbers in brackets refer to the list of references appended to this paper.

employed to examine the fracture surface for selected specimens, and the results were compared with the failure modes implied in the theory. The effect of curing residual stresses on the strength of the angle-ply laminates was considered.

Theoretical Background

Failure Theory

The tensor polynomial failure criterion in stress space is given by Ref 1

$$F_{ij}\sigma_i\sigma_j + F_i\sigma_i = 1 \tag{1}$$

where F_{ij} and F_i are strength parameters associated with the failure criterion and σ_i and σ_j are the stress components in the material axis. When the problem is limited to orthotropic materials under the plane stress state, Eq 1 assumes the following form

$$F_{11}\sigma_1^2 + F_{22}\sigma_2^2 + 2F_{12}\sigma_1\sigma_2 + F_{66}\sigma_6^2 + F_1\sigma_1 + F_2\sigma_2 + F_6\sigma_6 = 1 \tag{2}$$

The strength parameters are determined from material strength and are given by

$$F_{11} = 1/X_L X_L', \qquad F_1 = \frac{1}{X_L} - \frac{1}{X_L'}$$

$$F_{22} = \frac{1}{X_T X_T'}, \qquad F_2 = \frac{1}{X_T} - \frac{1}{X_T'} \tag{3}$$

$$F_{66} = \frac{1}{X_{LT} X_{LT}'}, \qquad F_6 = \frac{1}{X_{LT}} - \frac{1}{X_{LT}'}$$

where

X_L, X_L' = tensile and compressive strength in longitudinal direction,
X_T, X_T' = tensile and compressive strength in transverse direction, and
X_{LT}, X_{LT}' = positive and negative shear strength in longitudinal direction.

To ensure that the failure surface will intercept each stress axis, the magnitude of the F_{12} term is constrained by the following inequality

$$F_{11}F_{22} - F_{12}^2 \geq 0 \tag{4}$$

In view of the difficulty in experimental determination of F_{12}, Tsai recently introduced a dimensionless parameter

$$\overset{*}{F}_{12} = \frac{F_{12}}{\sqrt{F_{11}F_{22}}} \tag{5}$$

He assigned negative $\frac{1}{2}$ to the value of $\overset{*}{F}_{12}$ based on a generalization of the Von Mises yield criterion for isotropic materials. Details can be found in Ref 4.

Calculation of Curing Residual Stresses

Curing residual stresses are inevitably induced in multidirectional composite laminates that are normally cured at elevated temperature. The curing stresses in the angle-ply laminates were calculated from the following relation given by Hahn and Pagano [5]

$$\sigma_i^R = Q_{ij} A_{jk}^{-1} N_K^T - Q_{ij} e_j^T \tag{6}$$

where

Q_{ij} = reduced stiffness at final temperature of interest,
A_{jk} = laminate stiffness,
N_K^T = equivalent stress resultants due to thermal strains, and
e_j^T = thermal strains measured from stress-free state.

Experimental Procedure

The material system used in this study was AS/3501-5A graphite/epoxy supplied in prepreg form by Hercules. Off-axis and angle-ply specimens having various angles were prepared from the panels that were fabricated in an autoclave according to the manufacturer's recommended cure cycle. These specimens were not subjected to postcure. All tension specimens were straight-sided coupons 1.3 to 1.9 cm wide and 7.5 to 12.7 cm long in gage section. In compression tests, dogbone-shaped specimens 1.3 cm wide and 7.5 cm long in gage section were used with the aid of side supports to prevent specimen buckling. Figure 1 shows a sketch of a compression test specimen with side support. A preliminary test was conducted for a few speci-

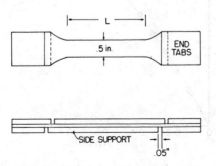

FIG. 1—*Geometry of compression specimen with side support (1 in. = 2.54 cm).*

mens that were instrumented with a back-to-back strain gage; a typical result is shown in Fig. 2. With this technique, specimen buckling appeared to be minimal up to failure.

To prevent any moisture absorption from the laboratory environment, all specimens were stored in a dry cabinet until testing. Each specimen was tested using a closed-loop electrohydraulic MTS testing machine with a set of Instron friction-type jaw grips. Load control in tension and stroke control in compression were employed with the rate corresponding to 16.7 to 33.4 m/ms of strain rate.

Data Reduction

Following is the list of the mean strength values of unidirectional laminates that were determined from 20 replicas

$$X_L = 1448 \text{ MPa} \qquad X_L{}' = 1448 \text{ MPa}$$

$$X_T = 51.7 \text{ MPa} \qquad X_T{}' = 206.9 \text{ MPa}$$

$$X_{LT} = X_{LT}{}' = 93.1 \text{ MPa}$$

Because of experimental difficulty in the determination of compressive strength in the fiber direction, we assumed that compressive strength was equal to tensile strength in the longitudinal direction for this material system. Shear strength was obtained using ± 45-deg angle-ply laminates under

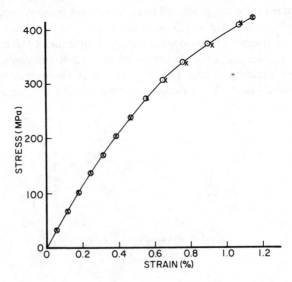

FIG. 2—*Stress-strain curves obtained by back-to-back gages for 15-deg off-axis specimen under compression.*

uniaxial tension. These material strengths allow determination of all the strength parameters, F_{ij} and F_i. The values are

$$F_{11} = 4.771 \times 10^{-7} \,(\text{MPa})^{-2}, \qquad F_{22} = 9.349 \times 10^{-5} \,(\text{MPa})^{-2}$$

$$F_{66} = 1.154 \times 10^{-6} \,(\text{MPa})^{-2}, \qquad F_1 = F_6 = 0$$

$$F_2 = 1.450 \times 10^{-2} \,(\text{MPa})^{-1}$$

$$F_{12} = -3.339 \times 10^{-6} \,(\text{MPa})^{-2} \quad \text{for} \quad \overset{*}{F}_{12} = -\tfrac{1}{2}$$

The thermoelastic properties of unidirectional laminates used in this study are

$$E_L = 137.97 \,\text{GPa}, E_T = 8.96 \,\text{GPa}, G_{LT} = 7.10 \,\text{GPa}$$

$$\nu_{LT} = 0.3, \alpha_L{}^T = -0.3 \,\mu\text{m/m/K}, \text{ and } \alpha_T{}^T = 28.1 \,\mu\text{m/m/K}$$

where $\alpha_L{}^T$ and $\alpha_T{}^T$ are linear thermal expansion coefficients of the unidirectional laminate in the fiber direction and perpendicular to the fiber direction, respectively. The curing residual stresses in angle-ply laminates were calculated by Eq 6 with the stress-free temperature of 121°C instead of the cure temperature of 177°C [6].

Results and Discussion

Typical failures which occurred in the off-axis and angle-ply specimens are shown in Figs. 3 and 4. All failures within the gage section indicate that perturbation in the stress field owing to the end constraint caused by the gripping device appears not to be critical to the strength of the off-axis laminate for the specimen size used in this test. A single failure surface parallel to the fiber direction has been observed for all angles of the off-axis specimens under both tension and compression. The ±5-deg specimen under tension revealed two distinct failure surfaces: one is roughly perpendicular to the loading axis and the other is along the fiber direction as shown in Fig. 4a. Similar failures have been observed for all the ±10-deg and ±15-deg specimens under tension. Unlike the off-axis laminate, the failure surface is different in tension and compression for angles less than ±15 deg. A detailed discussion regarding the failure modes involved will be given later.

Figure 5 shows the experimental data and analytical prediction for the off-axis laminate. The open circles represent the mean values of five to six specimens for each angle. The solid line represents the analytical prediction with $-\tfrac{1}{2}$ for the value of $\overset{*}{F}_{12}$. The broken line also represents an analytical prediction with $\overset{*}{F}_{12} = 0$; that is, no interaction occurs between stress components σ_1 and σ_2. The theory predicts almost identical strength for the two values of $\overset{*}{F}_{12}$.

This does not imply that the interaction term plays no significant role in composite laminate failure. Some results show that the failure theory con-

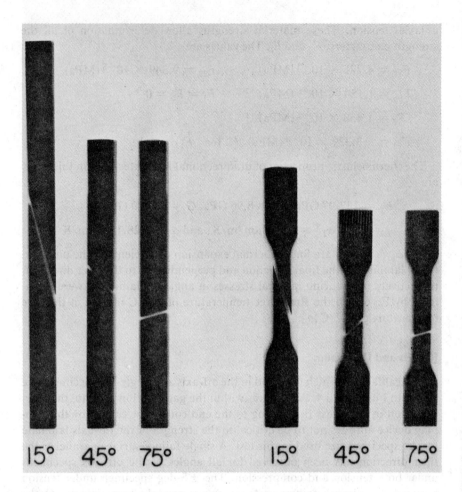

                    (a) Tension   (b) Compression
              FIG. 3—Typical failure of off-axis specimens.

sidered with interaction among the stress components provides the most
accurate prediction of the observed failure behavior [7]. The influence of the
interaction term appears to be dependent upon the magnitude of the stress
component with respect to the lamina strength. This fact suggests that the
off-axis test under either tension or compression is not an effective way to
determine the F_{12}-value for this material system. The agreement between experimental data and prediction of the tensor polynomial failure theory is
excellent in both tension and compression. Both theory and experiment show
greater strength in compression than in tension. This phenomenon can be
easily explained by considering that the transverse strength in compression is
four times greater than that in tension for this material system.

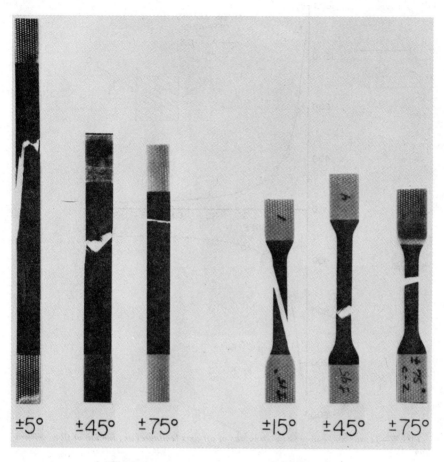

±5°   ±45°   ±75°   ±15°   ±45°   ±75°

(a) Tension   *(b) Compression*
FIG. 4—*Typical failure of angle-ply specimens.*

Figure 6 shows the experimental data of the angle-ply specimens compared with the failure theory for $\overset{*}{F}_{12} = -\frac{1}{2}$ and 0. The solid and broken lines are theoretical prediction for $\overset{*}{F}_{12} = -\frac{1}{2}$ and 0 respectively. The agreement between theory and experiment is very reasonable except for angles smaller than ±15 deg. The discrepancy between theory and experiment in this region is not understood at present and has to be clarified by further work. It is worthwhile noting that both theory and experimental data indicate greater strength in tension for angles up to the neighborhood of 40 deg, and thereafter smaller strength in tension. To understand this behavior, the state of stress in the lamina within the angle-ply laminate was examined. When uniaxial tension

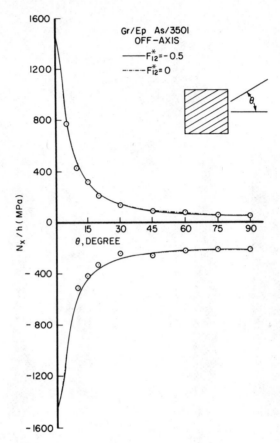

FIG. 5—*Tensile and compressive strengths of off-axis laminates as function of fiber orientation.*

is applied to the specimen, σ_1 is tensile for all angles while σ_2 changes from compression to tension at approximately 40 deg. Conversely, σ_2 changes from tension to compression under uniaxial compression. The change of σ_2 from compression to tension, or the converse, appears to be solely responsible for the observed behavior since the change of σ_1 does not affect the strength based on the assumption $X_L = X_L{}'$.

Unlike the off-axis case, the curing residual stresses are presented in the angle-ply laminate. The residual stresses are plotted as function of ply angle and presented in Fig. 7. The dotted line in Fig. 6 represents the prediction made with curing residual stresses present. The effect of the curing residual stresses appears to be minimal for this material system. This is due largely to the fact that the magnitude of the residual stresses is very small compared with the mechanical stress components at failure, in most cases. Some effect of the curing stresses has been shown around ±45 deg in tension. At the

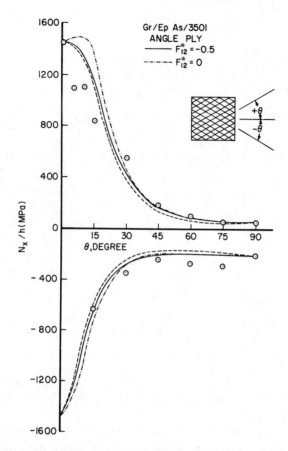

FIG. 6—*Tensile and compressive strength of angle-ply laminates as function of ply angles.*

±45 deg, the magnitude of the transverse residual stress, σ_2^R was found to be approximately 40 percent of X_T, which cannot be neglected. However, the theory considering residual stresses is more conservative and less favorable to the experimental data as shown in Fig. 6. The failure theory for the angle-ply laminate predicts that the stress at the first-ply failure occurred within the laminate. That is, after the layer with either positive or negative orientation has failed, the remainder of the layer, although still intact, cannot carry the existing load. Thus, failure of the entire laminate occurs immediately after the initial failure of either layer. In reality, it is possible that the entire failure of angle ply may not be followed immediately by failure of either layer in some cases. This behavior has been observed by Rotem and Hashin on the testing of a ±45-deg laminate of E-glass/epoxy [8].

The theoretical prediction for $\overset{*}{F}_{12} = 0$ shows that the angle ply strength up to approximately ±10 deg is greater than that in the 0-deg laminate. This

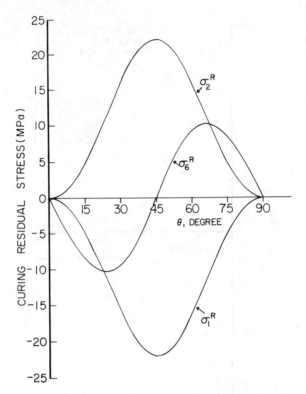

FIG. 7—*Curing residual stresses of angle-ply laminates as function of ply angles.*

contradicts the observed experimental data. Unlike the off-axis case, the interaction among stress components appears to play a significant role in the failure of angle-ply laminates in tension when the angle is small. The effect of the interaction term in compression is smaller than in tension owing to the different state of stress.

The form of the failure theory implies that failure depends upon a function of all stress values in the orthotropic lamina. Therefore, we would like to examine all three stress components to make a judgment on the cause of failure. The stress components, σ_1, σ_2, and σ_6, at failure were normalized with respect to the corresponding strength of the orthotropic lamina. The normalized stresses were plotted as functions of angle, as shown in Figs. 8 and 9. The solid lines were obtained from theory; the circles, triangles, and squares from experiment. The dotted line represents a limiting value, 1, where the magnitude of stress reaches its ultimate strength. A cursory examination of Figs. 8 and 9 reveals that failure can be dependent upon all three stress components rather than on any single component. Failure occurred in some cases before any of the stress components reached their limiting value, while in

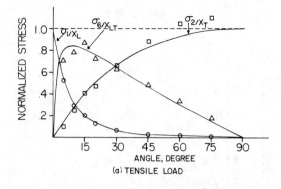

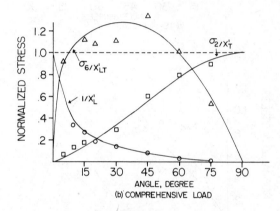

FIG. 8—*Normalized stress components of off-axis laminates as function of fiber orientation at failure.*

other cases the specimens still withstood the existing load after one of stress components had exceeded the limiting value.

In order to identify the fracture modes implied in the failure theory, the fracture surfaces of selected specimens which indicated various types of failure modes were examined using a scanning electron microscope. SEM photographs of all the specimens examined are presented in Figs. 10 through 14. Figure 10 shows the fracture surface of the 75-deg off-axis specimen under tension. The matrix clevage, lack of matrix residues on the fiber surface, and less laceration are the evidence of fracture created mainly by transverse tension. Some fiber breaks which were observed in most of the cases are considered to have been caused by tearing during the final separation of the fracture surface. Figure 11 was obtained from the fracture surface of the 15-deg off-axis specimen under compression. The extensive laceration and matrix residue are created by longitudinal shear stress. Two distinct

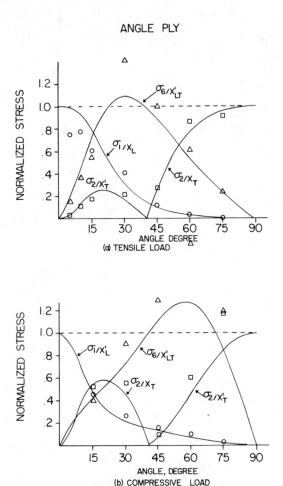

FIG. 9—*Normalized stress components of angle-ply laminates as function of ply angles.*

fracture surfaces have been observed for the ±5-deg angle-ply specimen under tension shown in Fig. 4a. The primary fracture surface appears to be the surface perpendicular to the specimen axis, since the normalized value of σ_1/X_L is far greater than that of σ_6/X_{LT}. The SEM photograph of this fracture surface is shown in Fig. 12 and indicates mainly fiber breaks associated with interfacial debonding, which caused fiber pullout.

The fracture surface of the 30-deg off-axis specimen under tension is presented in Fig. 13 and reveals mainly matrix clevage and laceration caused by a combination of transverse tension and longitudinal shear. Figure 14 shows the fracture surface of the ±15-deg angle-ply specimen under compression. The positively and negatively oriented layer can be identified in the left photo-

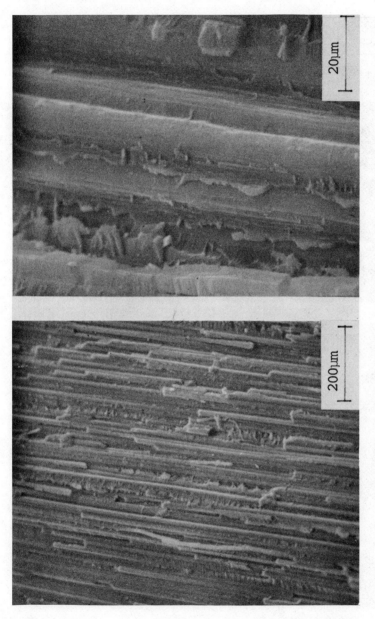

FIG. 10—*SEM photographs of fracture surface of 75-deg off-axis specimen under tension.*

FIG. 11—SEM photographs of fracture surface of 15-deg off-axis specimen under compression.

FIG. 12—SEM photographs of fracture surface of ±5-deg angle-ply specimen under tension.

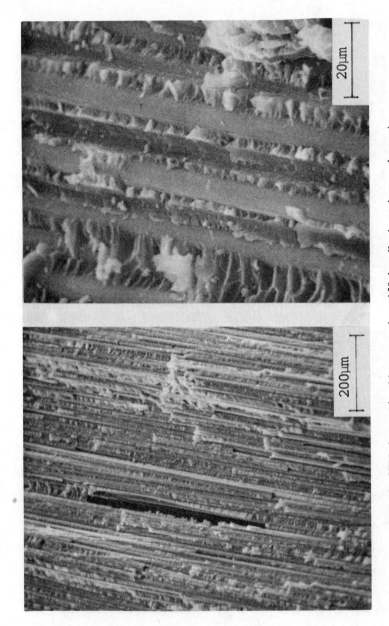

FIG. 13—SEM photographs of fracture surface of 30-deg off-axis specimen under tension.

FIG. 14—*SEM photographs of fracture surface of ± 15-deg angle-ply specimen under compression.*

graph of Fig. 14. The layer showing broken fiber ends is considered to have failed immediately after the other layer. Thus the layer showing the fiber surface is primarily responsible for failure. Whole tow breaks (two places), laceration, and cleavage indicate that all three stress components appear to play a significant role in the failure of this specimen. It is significant to note that the normalized stress value is smaller than 0.6 for all three stress components at final failure, as shown in Fig. 9b. The interpretation of the SEM photograph was based on the work by Sinclair and Chamis [9]. All the SEM observations appear to support the results shown in Figs. 8 and 9.

Conclusions

An experimental investigation has been conducted on the strength and failure modes of off-axis and angle-ply laminates under tension and compression. Experimental data in most cases agree reasonably well with the tensor polynomial failure theory, assuming a value of the interaction term based on the Von Mises yield criterion for isotropic materials. Further work needs to be done to explain the difference between experiment and theory for small angles of the angle-ply laminate under tension. For the angle-ply laminate of the material system used in this study, the theoretical prediction, including the curing residual stresses, was found to be very close to the case without the residual stresses present. The SEM examination of the fracture surface reveals similar failure modes implied in the failure theory.

Acknowledgments

The author wishes to acknowledge Dr. S. W. Tsai of the Air Force Materials Laboratory for his helpful suggestions on this study, and Mr. R. Esterline of the University of Dayton Research Institute for preparation and testing of the specimens. This work was sponsored by the Nonmetallic Materials Division, Air Force Materials Laboratory, under Contract No. F33615-78-C-5102.

References

[1] Tsai, S. W. and Wu, E. M., Journal of Composite Materials, Vol. 9, Jan. 1971, p. 58.
[2] Tsai, S. W., "Strength Characterization of Composite Materials," NASA CR-224, National Aeronautics and Space Administration, April 1965.
[3] Pipes, R. B. and Cole, B. W., Journal Composite Materials, Vol. 7, April 1973, p. 246.
[4] Tsai, S. W. and Hahn, H. T. in Introduction to Composite Materials, to be published, Chapter 7.
[5] Hahn, H. T. and Pagano, N. J., Journal of Composite Materials, Vol. 9, Jan. 1975, P. 91.
[6] Pagano, N. J. and Hahn, H. T. in Composite Materials: Testing and Design (Fourth Conference), ASTM STP 617, American Society for Testing and Materials, 1977, p. 317.
[7] Owen, M. J. and Griffiths, J. R., Journal of Materials Science, Vol. 13, 1978, p. 1521.
[8] Rotem, A. and Hashin, Z., Journal of Composite Materials, Vol. 9, April 1975, p. 107.
[9] Sinclair, J. H. and Chamis, C. C., "Mechanical Behavior and Fracture Characteristics of Off-Axis Fiber Composite I—Experimental Investigation," NASA Technical Paper 1081, National Aeronautics and Space Administration, Dec. 1977.

I. M. Daniel[1]

Biaxial Testing of Graphite/Epoxy Laminates with Cracks

REFERENCE: Daniel, I. M., **"Biaxial Testing of Graphite/Epoxy Laminates with Cracks,"** *Test Methods and Design Allowables for Fibrous Composites, ASTM STP 734*, C. C. Chamis, Ed., American Society for Testing and Materials, 1981, pp. 109–128.

ABSTRACT: An experimental investigation was conducted to study the behavior under biaxial tensile loading of graphite/epoxy plates with cracks and to determine the influence of crack length on failure. The specimens were 40 by 40-cm (16 by 16 in.) $[0/\pm45/90]_s$ laminates with cracks oriented at 30 deg to one side of the specimen. Four crack lengths, 2.54, 1.91, 1.27, and 0.64 cm (1.00, 0.75, 0.50, and 0.25 in.), were investigated. Deformations and strains were measured using strain gages, birefringent coatings, and moiré grids. The specimens were loaded in biaxial tension in a specially designed loading frame. Results were presented in terms of the ratio of the normal to the crack stress and the unnotched tensile strength of the laminate. This ratio was found to be approximately 21 percent below that for uniaxially loaded plates with transverse cracks, indicating an appreciable contribution to failure of the shear stress. This percentage of strength ratio reduction remained nearly the same for all crack sizes. Experimental results were compared satisfactorily with those predicted on the basis of a tensor polynomial failure criterion for the lamina and a progressive degradation model.

KEY WORDS: test methods, biaxial testing of composites, fracture of composites, graphite/epoxy, composite laminates with cracks, failure modes, birefringent coatings, moiré grids, effect of crack length, failure prediction, crack-opening displacement, crack-shearing displacement, composite materials

The behavior of composites with stress concentrations is of great interest in design because of the resulting strength reduction and the damage growth around these stress concentrations in service. Stress distributions in composite plates around stress concentrations, such as holes or cracks, have been treated analytically using linear anisotropic elasticity [1,2][2] and finite-element methods [3–6]. The latter can be used to account for material inhomogeneity, nonlinearity, and inelasticity. Experimental methods, using strain gages, photoelastic coatings, and moiré, have proven very useful in verifying theo-

---

[1] Science advisor, Materials Technology Division, IIT Research Institute, Chicago, Ill. 60616.
[2] The italic numbers in brackets refer to the list of references appended to this paper.

retical solutions in the linear range and supplementing them in the nonlinear range [6-15]. Experimental methods are especially useful in studying failure modes.

The problem of failure and strength of laminated composites with stress concentrations has been dealt with using two major approaches. One approach is based on concepts of linear elastic fracture mechanics carried over from isotropic materials, while the other approach takes into consideration the actual stress distributions near the discontinuity. The first approach was used by Waddoups et al [16], who assumed the existence of two fictitious Griffith-type cracks on the boundary of the hole, and by Cruse [17], who modeled the circular hole with a straight crack having an equivalent stress distribution near its tip. Following the second approach, Whitney and Nuismer [18] proposed simplified stress fracture criteria based on the actual stress distribution near the notch. According to the average stress criterion proposed, failure occurs when the average stress over an assumed characteristic dimension from the boundary of the notch equals the tensile strength of the unnotched material. Comparison with results from uniaxial tensile tests showed satisfactory agreement between predicted and measured strengths for a narrow range of values of the characteristic dimension.

In a similar approach proposed by Wu [19] and applied by Lo and Wu [20], lamina failure criteria are used and a characteristic dimension (volume) is postulated. Failure is said to occur if the average state of stress/strain on the boundary of this characteristic volume falls on the failure envelope of the lamina. To analyze an angle-ply laminate, this requires the study of progressive degradation of the various plies in the vicinity of the notch up to complete failure of the laminate.

Most of the analytical and experimental work above is limited to uniaxially loaded laminates. Very little work has been reported on the behavior of such laminates with stress concentrations under biaxial states of stress. The inhomogeneity of the material, the nonlinearity of response near failure, and the complex interaction of failure modes near notches make it very difficult to predict biaxial behavior on the basis of uniaxial response. An experimental approach dealing directly with biaxial loading of composite plates with stress concentrations is therefore very important.

The objective of this investigation was to study experimentally the deformation and failure under biaxial tensile loading of graphite/epoxy plates containing through-the-thickness cracks of various sizes and to determine the influence of crack length on failure [12]. This study was limited to a quasi-isotropic laminate. The approach used was to load square plate specimens in biaxial tension by means of a specially designed fixture, measure deformations by means of experimental strain analysis techniques, and determine strain distributions, failure modes, and strength reduction ratios. Experimental results were compared with analytical predictions.

Experimental Procedure

The material used was SP-286T300 graphite/epoxy (3M Co.). The basic unidirectional lamina was fully characterized with the determination of moduli, Poisson's ratios, strengths, and ultimate strains. Results from all characterization tests are tabulated in Table 1.

Uniaxial tensile properties were also determined of the unnotched quasi-isotropic $[0/\pm45/90]_s$ eight-ply laminate used in subsequent biaxial specimens with cracks. Properties were determined in the 0- and 90-deg directions. The strain response was found to be linear up to approximately 240 MPa (35 ksi), corresponding to an axial strain of approximately 0.004. Results from these tests were as follows:

$$\text{Modulus: } E_{xx} = 55 \text{ GPa } (8.0 \times 10^6 \text{ psi})$$
$$\text{Poisson's ratio: } \nu_{xy} = 0.30$$
$$\text{Tensile strength: } S_{xxT} = 483 \text{ MPa } (70 \text{ ksi})$$
$$\text{Ultimate strain: } \epsilon^u_{xxT} = 0.010$$

The notched biaxial specimens were 40 by 40-cm (16 by 16 in.) $[0/\pm45/90]_s$ plates with the 0-deg fibers at 30-deg with the sides of the plate. They were tabbed with 5-ply crossply glass/epoxy tabs with the outer fibers parallel to one side of the plate, or at 30-deg with the outer graphite plies. These tabs had a circular cutout at the center of 20.3-cm (8 in.) diameter. Cracks with a 0.08-mm (0.003 in.) tip radius were machined ultrasonically in the center of these specimens. The crack was oriented normally to the outer fibers of the laminate, that is, it was inclined at 30 and 60 deg with the sides of the specimen (Fig. 1). Four crack lengths were investigated: 2.54, 1.91, 1.27, and

TABLE 1—*Properties of unidirectional graphite/epoxy SP-286T300.*

Property	Value
Ply thickness	0.130 mm (0.0051 in.)
Longitudinal modulus, E_{11}	151 GPa (21.9×10^6 psi)
Transverse modulus, E_{22}	10.6 GPa (1.53×10^6 psi)
Shear modulus, G_{12}	6.6 GPa (0.96×10^6 psi)
Major Poisson's ratio, ν_{12}	0.31
Minor Poisson's ratio, ν_{21}	0.014
Longitudinal tensile strength, S_{11T}	1401 MPa (203 ksi)
Ultimate longitudinal tensile strain, ϵ^u_{11T}	0.00925
Longitudinal compressive strength, S_{11C}	1132 MPa (164 ksi)
Ultimate longitudinal compressive strain, ϵ^u_{11C}	0.01040
Transverse tensile strength, S_{22T}	54 MPa (7.8 ksi)
Ultimate transverse tensile strain, ϵ^u_{22T}	0.00541
Transverse compressive strength, S_{22C}	211 MPa (30.6 ksi)
Ultimate transverse compressive strain, ϵ^u_{22C}	0.02565
In-plane shear strength, S_{12}	72 MPa (10.5 ksi)
Ultimate shear strain, ϵ^u_{12}	0.0110

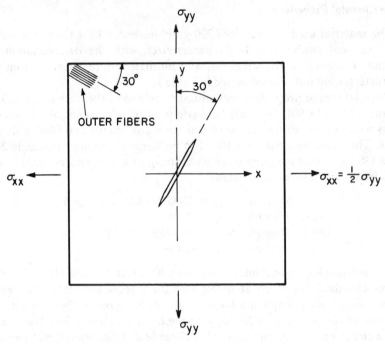

FIG. 1—*Biaxial Loading of [0/±45/90]$_s$ graphite/epoxy specimens with cracks.*

0.64 cm (1.00, 0.75, 0.50, and 0.25 in.). Two specimens were tested for each crack length.

Deformations and strains were measured using strain gages, birefringent coatings, and moiré grids. Miniature single gages and rosettes were mounted near the crack tips along lines perpendicular to the crack axis. Additional two- and three-gage rosettes were mounted along the horizontal (x) and vertical (y) axes of loading away from the crack. A typical gage layout is shown in Fig. 2. A closeup of the area around a crack with strain gages is shown in Fig. 3. In most cases a birefringent coating 0.25 mm (0.01 in.) thick was used around the crack. Also in most cases moiré grids consisting of arrays of 400 lines per centimetre (1000 lines per inch) parallel and normal to the crack were applied to the specimen.

Four 0.95-cm-diameter (0.375 in.) holes were provided on each side of the tabbed specimen for bolting individual pairs of metal grips approximately 5 cm (2 in.) wide and 10 cm (4 in.) long. Loading was introduced by means of four whiffle-tree grip linkages designed to ensure that four equal loads were applied to each side of the specimen. Load was applied by means of two pairs of hydraulic jacks attached to the four sides of a reaction frame. The load was transmitted from the hydraulic cylinders to the grip linkages through cylindrical rods going through the bore of these cylinders. The rods were instru-

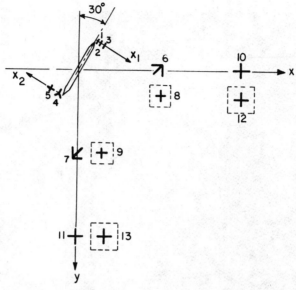

Gage No.	Location mm (in)	Type
1	x_1 = 0.76 (0.03)	EA-06-015DJ-120
2	x_1 = 2.54 (0.10)	EA-06-031DE-120
3	x_1 = 5.08 (0.20)	EA-06-031DE-120
4	x_2 = 1.78 (0.07)	WA-06-030WR-120
5	x_2 = 7.11 (0.28)	WA-06-030WT-120
6	x = 38.1 (1.50)	EA-06-062RB-120
7	y = 38.1 (1.50)	EA-06-062RB-120
8*	x = 38.1 (1.50)	EA-06-062TT-120
9*	y = 38.1 (1.50)	EA-06-062TT-120
10,12*	x = 76.2 (3.00)	EA-06-125TM-120
11,13*	y = 76.2 (3.00)	EA-06-125TM-120

*On opposite side of plate

FIG. 2—*Typical gage layout for biaxially loaded specimens with cracks.*

mented with strain gages and calibrated in a testing machine to establish the
exact relationship between loads and strain-gage signals. These strain-gage
readings were used subsequently both for recording the exact loads applied
to the specimen and as feedback signals for controlling the pressures by
means of the servohydraulic system used. A special fixture was used to help
align the specimen in the loading frame.

The specimens were loaded in biaxial tension with the stress at 30 deg to
the crack twice as large as the stress at 60 deg to the crack. The effective stress
and the exact biaxiality ratio in the test section of the specimen were controlled and determined from the far-field strains. For the quasi-isotropic
laminate in question the ratio of far-field strains is equal to

$$\frac{\epsilon_{yy}}{\epsilon_{xx}} = \frac{1 - k\nu}{k - \nu}$$

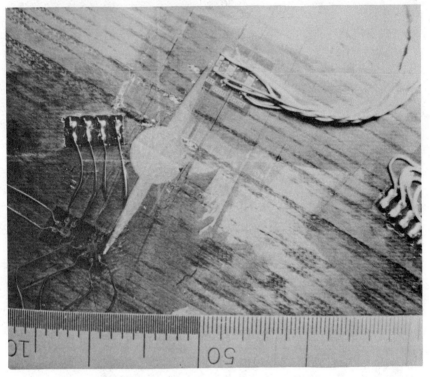

FIG. 3—*Closeup of gage layout near 2.54-cm (1.00 in.) crack (smallest division shown is 0.01 in.).*

where $k = \sigma_{xx}/\sigma_{yy}$. The effective stress σ_{yy} is determined from the expression

$$\sigma_{yy} = \frac{E\epsilon_{yy}}{1 - k\nu}$$

Values of the modulus and Poisson's ratio of the laminate in the direction of the applied loads were determined experimentally.

For the desired biaxiality $k = 0.5$, the required strain ratio is

$$\frac{\epsilon_{yy}}{\epsilon_{xx}} = 3.91$$

The ratio of applied loads was adjusted until the desired strain ratio was obtained. Thereafter, the load ratio was kept constant for the rest of the test.

The loads were applied in increments, the strain gages were recorded at every increment, and photoelastic and moiré fringes were photographed. The loading frame with the specimen and the associated strain recording instrumentation is shown in Fig. 4.

Results and Discussion

Strains in the vicinity of the crack tip and along the horizontal and vertical axes are plotted in Figs. 5 and 6 for a specimen with a 2.54-cm (1.00 in.) crack. In Fig. 5 the strains near the crack tip are lower initially than those farther away. As the load is increased, however, the damage (or crack) grows in the direction normal to the crack and these strains increase at a rapid rate and overtake the strains farther from the crack. The strain near the crack tip (Gage 1 in Fig. 5) at a distance of 0.76 mm (0.03 in.) from the tip is linear up to an applied vertical stress of approximately 76 MPa (11 ksi). The three gages near the crack tip, Gages 1, 2, and 3 in Fig. 5 at distances 0.76, 2.54, and 5.10 mm (0.03, 0.10, and 0.20 in.), failed when the crack propagated through them at stresses of approximately 173, 207, and 240 MPa (25, 30, and 35 ksi), respectively. In Fig. 6 the strain at a distance of 1.52 mm (0.06 in.) from the crack tip is linear up to a stress of 69 MPa (10 ksi); thereafter it increases rapidly up to the point when the crack propagates through the gage at an approximate stress of 186 MPa (27 ksi). The crack reaches the location of Gages 4 and 5 (Fig. 6) at a distance of 5.10 mm (0.20 in.) at a stress of approximately 228 MPa (33 ksi).

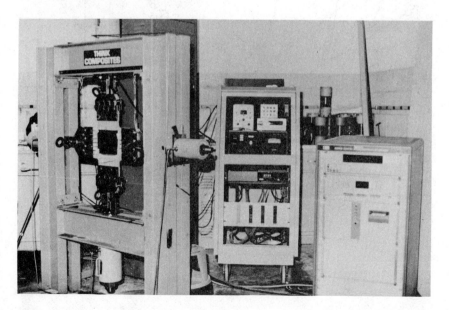

FIG. 4—*Loading frame for biaxial testing of flat laminates and associated strain-recording instrumentation.*

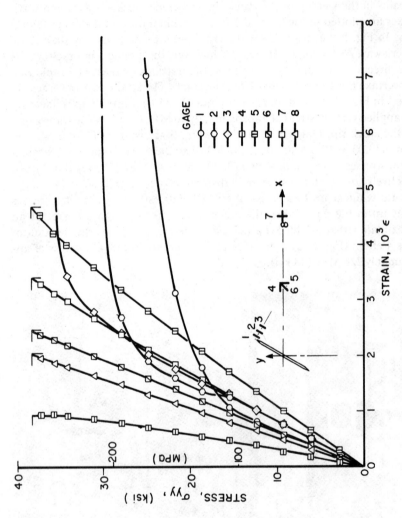

FIG. 5—Strains near crack tip and along horizontal axis of $[0/\pm45/90]_s$ graphite/epoxy specimen with 2.54-cm (1 in.) crack under biaxial loading $\sigma_{yy} = 2\sigma_{xx}$ at 30 deg to crack direction.

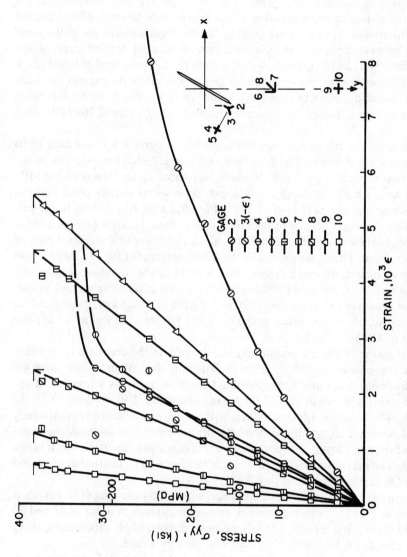

FIG. 6—Strains near crack tip and along vertical axis of $[0/\pm 45/90]_s$ graphite/epoxy specimen with 2.54-cm (1 in.) crack under biaxial loading $\sigma_{yy} = 2\sigma_{xx}$ at 30 deg to crack direction.

Moiré fringe patterns around the crack are shown in Fig. 7 for three load levels. The upper part of the pattern corresponds to displacements normal to the crack direction, the lower part to displacements parallel to it. Relative displacements between any two points can be obtained by the simple expedient of counting fringes. Thus, the crack-opening displacement and the forward sliding or crack shearing displacement were determined and plotted as a function of applied stress (Fig. 8). These displacements are of the same order of magnitude and they become nonlinear at an applied stress of approximately 172 MPa (25 ksi), which is close to the stress level of initial crack propagation as detected by the strain gages. The crack propagates normally to its initial direction or along the outer fibers as illustrated by the moiré fringe patterns. Total failure occurred at an applied stress of 264 MPa (38.3 ksi).

Isochromatic fringe patterns in the coating around a 1.27-cm-long (0.50 in.) crack are shown in Fig. 9 for four load levels. Crack propagation, in the direction normal to the crack, seems to be limited up to a stress of 260 MPa (37.7 ksi). An abrupt jump in crack extension seems to take place between this stress and the next level of 278 MPa (40.3 ksi). In addition to the primary propagation normal to the initial crack, there is crack extension along the original crack direction, probably along the fibers of the central plies of the laminate. This is illustrated by the fringe patterns of Fig. 9. There is also evidence of tertiary crack propagation normal to the initial crack direction but initiating at the tip of the subsurface extended crack. Rapid crack propagation occurs in the stress range 220 to 262 MPa (32 to 38 ksi), corresponding to peak strains of the order of 0.030. Total failure occurred at an applied stress of 313 MPa (45.3 ksi).

A measure of the maximum tangential strain at the crack tip is obtained from the maximum fringe order recorded in the coating. The variation of this fringe order and corresponding tangential strain as a function of applied stress is shown in Fig. 10 for a specimen with a 1.91-cm-long (0.75 in.) crack. The variation appears bilinear with a knee at a strain of approximately 0.011, which is slightly higher than the ultimate strain of an unnotched coupon of the same layup. The computed stress concentration in the linear range is 3.2. The maximum measured strain at failure, which occurred at $\sigma_{yy} = 263$ MPa (38 ksi) in this case, is over 0.025.

A typical failure pattern of a specimen with a 2.54-cm-long (1 in.) crack is shown in Fig. 11. In this case crack extension appears to occur at 45 and 25 deg to the crack direction. The failure mode consists of ply subcracking along fiber directions, local delaminations, and fiber breakage.

Results for all specimens with cracks loaded under a 2:1 biaxial tension are summarized in Table 2. The failure stress along the y-axis as well as the stress components at failure referred to the crack direction are given in this table. The last column represents the ratio of one of these components (the one normal to the crack direction) and the uniaxial strength of the unnotched

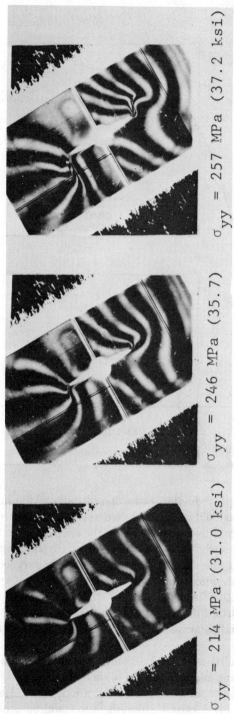

FIG. 7—Moiré fringe patterns around 2.54-cm (1 in.) crack in [0/±45/90]$_s$ graphite/epoxy specimen under biaxial loading $\sigma_{yy} = 2\sigma_{xx}$ at 30 deg to crack direction. (Upper part of pattern corresponds to displacements normal to the crack, lower part to displacements along the crack direction.)

$\sigma_{yy} = 214$ MPa (31.0 ksi) $\sigma_{yy} = 246$ MPa (35.7) $\sigma_{yy} = 257$ MPa (37.2 ksi)

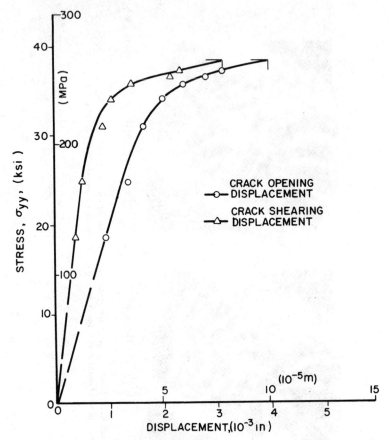

FIG. 8—*Crack-opening and crack-shearing displacements in [0/±45/90]$_s$ graphite/epoxy specimen with 2.54-cm (1 in.) crack under biaxial loading $\sigma_{yy} = 2\sigma_{xx}$.*

laminate. This strength reduction ratio, based only on the normal to the crack component of far-field stress, is plotted as a function of crack length in Fig. 12. As can be seen, this ratio falls below the strength reduction ratio for uniaxially loaded plates with cracks, because of the contribution of the other two stress components, the shear, and the normal stress parallel to the crack direction. The strength ratio for the biaxial loading applied is approximately 21 percent lower than that for the uniaxial case. The contribution of the shear stress is illustrated further in one case where the ratio of shear stress to the normal stress perpendicular to the crack direction was nearly doubled (Fig. 12).

The foregoing results can also be compared with those of unnotched specimens of the same material tested under 2:1 tension. The average stresses at failure for the unnotched material are $S_{yy} = 573$ MPa (83 ksi) and $S_{xx} = 286$ MPa (41.5 ksi). A strength ratio could be expressed as the

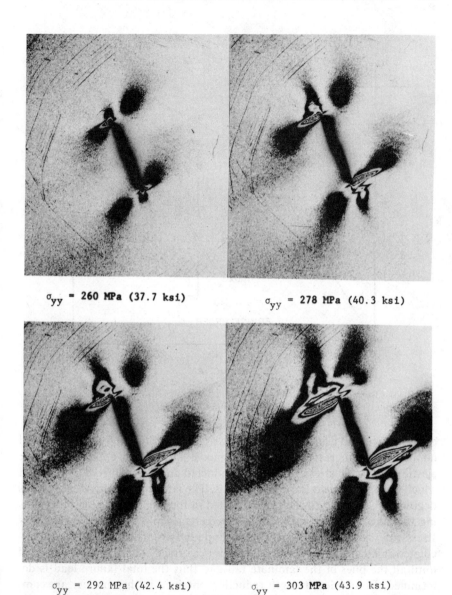

σ_{yy} = 260 MPa (37.7 ksi) σ_{yy} = 278 MPa (40.3 ksi)

σ_{yy} = 292 MPa (42.4 ksi) σ_{yy} = 303 MPa (43.9 ksi)

FIG. 9—*Isochromatic fringe patterns in photoelastic coating around 1.27-cm (0.5 in.) crack in [0/±45/90]$_s$ graphite/epoxy specimen under biaxial loading $\sigma_{yy} = 2\sigma_{xx}$ at 30 deg to crack direction.*

ratio of the maximum applied stresses at failure of the notched and un-notched specimens. This ratio varies between 0.385 and 0.681 for the specimens tested. It is a function of the crack inclination and would decrease as the crack angle from the maximum applied stress increases.

Experimental results were compared with those obtained by an analytical/

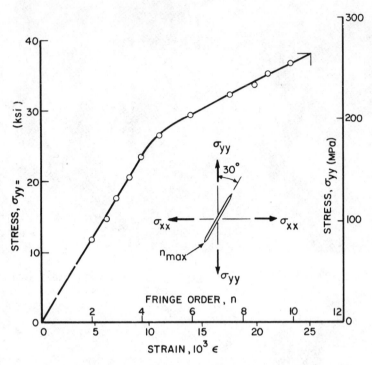

FIG. 10—*Maximum fringe order and tangential strain at crack tip of [0/±45/90]ₛ graphite/ epoxy specimen with 1.91-cm (0.75 in.) crack under biaxial loading* $\sigma_{yy} = 2\sigma_{xx}$.

empirical method using a maximum stress failure criterion for the individual lamina and a progressive degradation model [20]. In this approach a linear elastic anisotropic stress analysis is conducted and the load is determined at which the first ply near the crack tip fails. The laminate undergoes a stiffness degradation. A new stiffness is calculated in the damaged region by assuming that the transverse stiffness of the damaged ply becomes zero. A new stress analysis is conducted until a second ply fails. The new stiffness is calculated again and a new stress analysis is performed. This process is repeated until all the plies of the laminate fail and thus the total failure load is determined. Results for strength reduction obtained in this manner are compared with the experimental results in Fig. 13. The agreement seems to be satisfactory. Similar theoretical results using a tensor polynomial failure criterion have also been obtained [20]. Predicted strength values in the latter case were slightly higher than those in the foregoing.

Summary and Conclusions

An experimental study was conducted of the deformation and failure of [0/±45/90]ₛ graphite/epoxy plates with cracks loaded in biaxial tension,

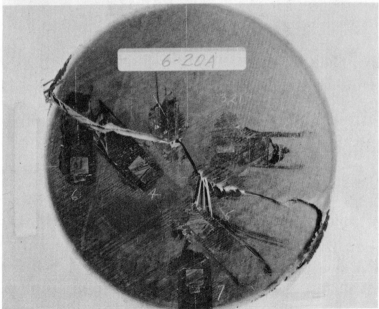

FIG. 11—*Failure pattern in [0/±45/90]ₛ graphite/epoxy specimen with 2.54-cm (1.00 in.) crack under biaxial loading* $\sigma_{yy} = 2\sigma_{xx}$.

TABLE 2—Experimental results for [0/±45/90]s graphite/epoxy laminates with cracks subjected to biaxial loading.

Crack Length, $2a$, cm (in.)	Biaxiality Ratio, σ_{yy}/σ_{xx}	Failure Stresses, MPa (ksi)				Strength Ratio, S_{11}/S_0
		Along y-axis, S_{yy}	Normal to Crack, S_{11}	Parallel to Crack, S_{22}	Shear Along Crack, S_{12}	
2.54 (1.00)	1.98	264 (38.3)	165 (24.0)	232 (33.6)	57 (8.2)	0.343
2.54 (1.00)	2.00	221 (32.0)	138 (20.0)	193 (28.0)	48 (6.9)	0.286
1.91 (0.75)	1.93	263 (38.1)	168 (24.3)	231 (33.5)	55 (8.0)	0.347
1.91 (0.75)	3.57	305 (44.0)	140 (20.2)	249 (36.1)	95 (13.7)	0.289
1.27 (0.50)	2.03	313 (45.3)	194 (28.1)	273 (39.6)	69 (10.0)	0.402
1.27 (0.50)	1.97	343 (49.8)[a]	217 (31.4)[a]	301 (43.7)[a]	73 (10.6)[a]	0.449
0.64 (0.25)	1.98	332 (48.1)[a]	208 (30.1)[a]	291 (42.1)[a]	71 (10.3)[a]	0.430
0.64 (0.25)	1.96	390 (56.5)	247 (35.7)	342 (49.6)	83 (12.0)	0.510

[a] Tab failure.

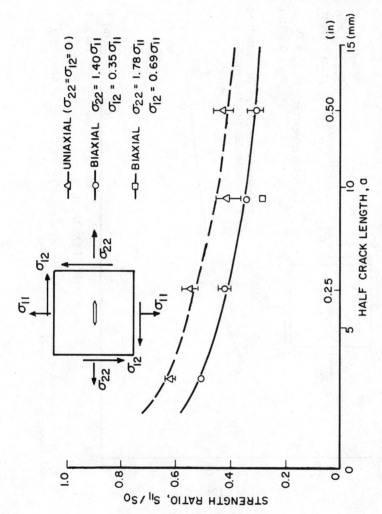

FIG. 12—*Strength ratios as a function of crack length for* $[0/\pm45/90]_s$ *graphite/epoxy plates with cracks under uniaxial and biaxial loading.*

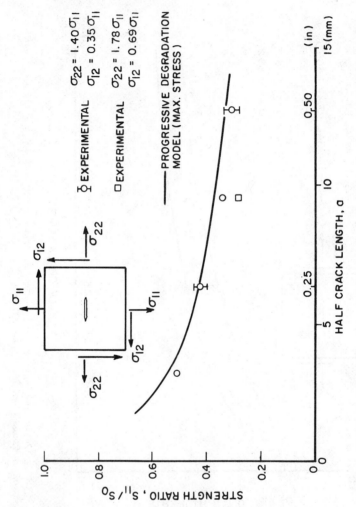

FIG. 13—*Comparison of experimental and theoretical results for strength ratio of* [0/±45/90]ₛ *graphite/epoxy plates with cracks under biaxial loading.*

resulting in a state of stress around the crack with the shear stress equal to 0.35 times the stress normal to the crack direction. Crack lengths investigated were 2.54, 1.91, 1.27, and 0.64 cm (1.00, 0.75, 0.50, and 0.25 in.). Strain gages, birefringent coatings, and moiré grids were used for strain measurement.

Strains near the crack tip are either nonlinear throughout the loading range or they become so at an applied vertical (y-axis) stress of approximately 75 MPa (11 ksi), corresponding to a strain of approximately 0.001. Failure seems to initiate at a point on the crack-tip arc where the tangent is parallel to the highest applied stress. The crack-opening displacement and the crack-shearing (or forward sliding) displacement remain linear up to an applied vertical stress of 172 MPa (25 ksi), which is the level at which initial crack or damage propagation takes place. At this level the crack-opening and crack-shearing displacements begin to increase at a much faster rate. The initial crack extension seems to be at 45 deg to the original crack direction. Thereafter, the damage zone, consisting of subcracks parallel to the fibers of the various plies, delaminations, and fiber breaks, propagates normally to the crack direction in most cases. In one case, crack propagation was observed at 25 deg to the crack direction and in another case damage propagation occurred both parallel and normal to the crack direction. In those cases where the peak strain near the crack tip was recorded up to failure, it exceeded values of the order of 0.03.

The components of far-field stress at failure referred to the crack direction were determined. Results were presented in terms of the ratio of the normal to the crack stress and the unnotched uniaxial tensile strength of the laminate. This ratio was found to be approximately 21 percent below that for uniaxially loaded plates with transverse cracks, indicating an appreciable contribution to failure of the shear stress. This percentage of strength ratio reduction remained nearly the same for all crack sizes.

Experimental results were in good agreement with predictions based on a maximum stress failure criterion for the individual lamina and a progressive degradation model.

Acknowledgment

The work described here was sponsored by the Air Force Materials Laboratory (AFML), Wright-Patterson Air Force Base, Ohio. The author is grateful to Dr. J. M. Whitney of AFML for his encouragement and cooperation, to Dr. T. Liber of Travenol Laboratories and to Messrs. R. LaBedz and T. Niiro of the IIT Research Institute for their assistance.

References

[1] Lekhnitskii, S. G., *Theory of Anisotropic Elastic Body*, Translated by P. Fern, J. J. Brandstatter, Ed., Holden-Day, San Francisco, 1963.

[2] Savin, G. N., *Stress Concentrations Around Holes,* Pergamon Press, New York, London, 1961.

[3] Puppo, A. and Haener, J., Jr., "Application of Micromechanics to Joints and Cutouts," U.S. Army Aviation Materiel Laboratory Report No. 69-25, April 1969.

[4] Levy, A., Armen, H., and Whiteside, J. B., "Elastic and Plastic Interlaminar Shear Deformation in Laminated Composites Under Generalized Plane Stress," Presented at the Air Force Third Conference on Matrix Methods in Structural Mechanics, Wright-Patterson Air Force Base, Ohio, Oct. 1971.

[5] Rybicki, E. F. and Hopper, A. T., "Analytical Investigation of Stress Concentrations Due to Holes in Fiber-Reinforced Plastic Laminated Plates; Three-Dimensional Models," Air Force Materials Laboratory Report AFML-TR-73-100, Wright-Patterson Air Force Base, Ohio, June 1973.

[6] Whiteside, J. B., Daniel, I. M., and Rowlands, R. E., "The Behavior of Advanced Filamentary Composite Plates with Cutouts," Air Force Flight Dynamics Laboratory Report AFFDL-TR-73-48, Wright-Patterson Air Force Base, Ohio, June 1973.

[7] Daniel, I. M. and Rowlands, R. E., *Journal of Composite Materials,* Vol. 5, April 1971, pp. 250-254.

[8] Rowlands, R. E., Daniel, I. M., and Whiteside, J. B., *Experimental Mechanics,* Vol. 13, Jan. 1973, pp. 31-37.

[9] Daniel, I. M., Rowlands, R. E., and Whiteside, J. B. in *Analysis of the Test Methods for High Modulus Fibers and Composites, ASTM STP 521,* American Society for Testing and Materials, 1973, pp. 143-164.

[10] Daniel, I. M., Rowlands, R. E., and Whiteside, J. B., *Experimental Mechanics,* Vol. 14, Jan. 1974, pp. 1-9.

[11] Rowlands, R. E., Daniel, I. M., and Whiteside, J. B. in *Composite Materials: Testing and Design (Third Conference), ASTM STP 546,* American Society for Testing and Materials, 1974, pp. 361-375.

[12] Daniel, I. M., "Biaxial Testing of Graphite/Epoxy Composites Containing Stress Concentrations," Air Force Materials Laboratory Report AFML-TR-76-244, Wright-Patterson Air Force Base, Ohio, Part I, Dec. 1976, and Part II, June 1977.

[13] Daniel, I. M. in *Proceedings,* Second International Conference on Composite Materials, ICCM/2, Toronto, Ont., Canada, 16-20 April, 1978.

[14] Daniel, I. M., *Experimental Mechanics,* Vol. 18, No. 7, July 1978, pp. 246-252.

[15] Daniel, I. M. in *Proceedings,* Sixth International Conference on Experimental Stress Analysis, VDI-Berichte No. 313, Munich, West Germany, 18-22 Sept. 1978, pp. 705-710.

[16] Waddoups, M. E., Eisenmann, J. R., and Kaminski, B. E., *Journal of Composite Materials,* Vol. 5, Oct. 1971, pp. 446-454.

[17] Cruse, T. A., *Journal of Composite Materials,* Vol. 7, 1973, pp. 218-229.

[18] Whitney, J. M. and Nuismer, R. J., *Journal of Composite Materials,* Vol. 8, July 1974, pp. 253-265.

[19] Wu, E. M. in *Proceedings,* Specialists Meeting on Failure Modes of Composite Materials with Organic Matrices and Their Consequences on Design, Munich, West Germany, 6-12 Oct. 1974, (AGARD-CP-163, 1975).

[20] Lo, K. H. and Wu, E. M. in *Advanced Composite Serviceability Program,* Air Force Materials Laboratory, Contract No. F33615-76-C-5344, Seventh Quarterly Report, Wright-Patterson Air Force Base, Ohio, July 1978.

C. T. Herakovich,[1] H. W. Bergner,[1] and D. E. Bowles[1]

A Comparative Study of Composite Shear Specimens Using the Finite-Element Method

REFERENCE: Herakovich, C. T., Bergner, H. W., and Bowles, D. E., **"A Comparative Study of Composite Shear Specimens Using the Finite-Element Method,"** *Test Methods and Design Allowables for Fibrous Composites, ASTM STP 734,* C. C. Chamis, Ed., American Society for Testing and Materials, 1981, pp. 129-151.

ABSTRACT: Four specimens for the in-plane shear behavior of fibrous composites are analyzed for elastic stress distributions. Rigid and elastic fixtures as well as mechanical and thermal loading are considered. Results are presented in the form of stress contours, stress profiles, and stress concentrations for [0], [0/90]$_s$ and [0/±45/90]$_s$ laminates. It is shown that the assumption of rigid-fixture rail shear specimens predicts regions of uniform pure shear, but that the inclusion of elastic fixture effects introduces normal stresses in the test section. These results also indicate that the shear stress in the test section may differ from the average value by as much as 10 percent, depending upon laminate configuration. Results are also presented for slotted coupon, crossbeam, and a double V-notched specimen.

KEY WORDS: composites, shear test, finite elements, elastic, thermal, slotted coupon, crossbeam, notched coupon, rail shear, graphite/epoxy, graphite/polyimide

The accurate determination of shear properties for advanced composite materials, including elastic modulus, nonlinear stress-strain behavior, and shear strength, has proven to be a difficult problem for those concerned with material property characterization. The difficulty has centered around the development of a specimen which would provide a test section with uniform, pure, maximum shear stress. In the "idea" specimen this test section would continue to maintain this stress state during nonlinear material behavior up to and including failure, which would initiate in the test section. Since pure shear is equivalent to biaxial tension and compression along the diagonals of the shear element, the specimen must also be resistant to buckling. Further,

[1]Professor and graduate students, respectively, Department of Engineering Science and Mechanics, Virginia Polytechnic Institute, and State University, Blacksburg, Va. 24061. Coauthor Bergner is now with The Boeing Commercial Airplane Co.

the specimen (or specimens) should be applicable in the cryogenic to cure temperature range, and provide results for in-plane (1-2), as well as transverse (2-3), and out-of-plane (1-3) behavior. Finally, the specimen should be relatively easy to fabricate under high quality control and at reasonable cost.

It is doubtful that any one specimen will meet all of the preceding requirements. However, there is hope that one specimen can be developed for the in-plane behavior and possibly another for the out-of-plane and transverse behavior. This paper is concerned with the evaluation of four specimens considered for in-plane shear behavior.

While the thin-walled tube [1][2] appears to be the most suitable specimen for shear tests, it is very expensive and often difficult to fabricate. (As of this writing, the technology to fabricate graphite/polyimide tubes is not available.) Because of these disadvantages of the tube, considerable efforts have been made to develop flat shear specimens for composites [2-15]. Two major problems arise when using flat specimens: (1) the presence of normal stresses which influence the nonlinear behavior, and (2) stress concentrations which cause early failure outside the test section.

Final acceptance of a test specimen requires experimental verification of the strain distribution in the test section. Strain gages can often provide a reasonably good estimate of these distributions; however, a full field measurement, such as moiré, is more desirable. The fabrication and experimental evaluation of test specimens is a very time-consuming and expensive process. A considerable savings of both time and money can be effected through the use of the finite-element method prior to fabrication and testing. One of the most important advantages of the finite-element method is that an infinite variety of test specimens, loading configurations, and laminates can be investigated using the same computer program. Only minimal changes in input data are required to make substantial changes in the test specimen. This includes changes in specimen and fixture geometry, type of loading, including mechanical or thermal or both, and laminate material and stacking sequence. Another important advantage of the finite-element method is that it provides results over the entire specimen. The results of the finite-element analysis can be used to make more informed decisions as to the specimens to be investigated experimentally.

This paper presents results of linear elastic finite-element investigations of the four specimens for the in-plane shear response of composite laminates shown in Fig. 1. They are (1) slotted coupon, (2) crossbeam, (3) double V-notched coupon, and (4) rail shear. Normalized stress profiles, stress contours, and stress concentrations are presented for a variety of specimen and laminate configurations.

---

[2]The italic numbers in brackets refer to the list of references appended to this paper.

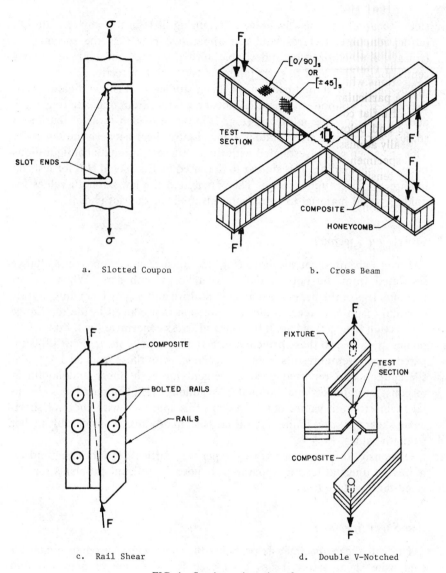

a. Slotted Coupon   b.   Cross Beam

c. Rail Shear   d.   Double V-Notched

FIG. 1—*Specimens investigated.*

Specimens Investigated

Slotted Coupon

A variety of slotted coupon specimens has been used to obtain shear properties of metals and composites. Many of these are reviewed in Ref 6. The slotted coupon has the advantage of requiring little in the way of material, fabrication time, fixtures, and test apparatus. However, the specimen is typically characterized by undesirable normal stresses and high stress concentrations which lead to early failure outside the test section.

The particular slotted coupon specimen studied in this investigation consists of a flat coupon with parallel slots emanating from opposite free edges above and below the midplane (Fig. 1a). When loaded in axial tension, a statically admissible stress state is a line of pure shear between the slot ends. This specimen was investigated with three different laminate configurations under tensile loading. It is similar to that used by Elkin et al [12] for tests on unidirectional graphite/epoxy. They attributed the low strength values for unidirectional material to stress concentrations at the slot ends.

Crossbeam Specimen

The crossbeam shear specimen (Fig. 1b) consists of a composite top flange separated from the bottom flange by a honeycomb core. When the orthogonal legs of the beam are loaded in positive and negative bending, a state of equal magnitude tension and compression is produced in the top flange test section (neglecting core influence and stress concentrations). Stresses on planes at 45 deg to these principal directions are pure shear. Petit [3] compared experimental results from this specimen with those from a $[\pm 45]_s$ tensile coupon and reported good agreement for both linear and nonlinear response of unidirectional laminates. Waddoups [13] compared experimental shear strength values obtained from this specimen with those of short-beam shear tests and found that the crossbeam produced considerably higher strength values.

Obviously, this specimen is very expensive, difficult to fabricate, and requires an unusual testing apparatus. It poses significant problems for elevated-temperature tests.

Rail Shear Specimen

The rail shear specimen is probably the most widely used specimen for composite shear tests. It has been used for a variety of materials and laminate configurations, at room and elevated temperatures. The specimen consists of a flat composite laminate with stiff rails either bonded or bolted to one set of parallel sides forming a long, narrow test section between the rails (Fig. 1c). Tensile or compressive loads are introduced at the rail ends to displace them essentially parallel to one another. A number of variations of the rail shear specimen have been investigated experimentally, analytically, and numerically [2,6,10,11,14,17]. All results presented in this paper are for an aspect ratio (length/width) of 10.0. The influence of aspect ratio on the stress distribution has been discussed in detail in Ref 11.

Double V-Notched Shear Specimen

This specimen (Fig. 1d) consists of a flat laminated composite coupon with symmetric V-notches along the two free edges. The ends of the coupon are gripped by fixtures (bolted, bonded, or clamped) and load is introduced through the application of tensile (or compressive) load to the fixtures as shown. The original specimen used by Iosipescu [16] for metals was notched on all four sides. He conducted an extensive photoelastic study of the specimen and concluded that it produced a region of maximum uniform pure shear stress in the test section. A numerical and experimental investigation of a double V-notched composite specimen was recently reported by Slepetz et al [9].

Mathematical Formulation

Following classical lamination theory, the thermoelastic constitutive equation for a symmetric laminate subjected to in-plane mechanical loading and uniform temperature distribution can be written in the form

$$\{\sigma\} = [a](\{\epsilon\} - \{\epsilon^T\}) \tag{1}$$

where $[a]$ is the matrix of laminate coefficients defined by

$$a_{11} = E_x/\lambda \qquad a_{22} = E_y/\lambda$$

$$a_{12} = E_x \nu_{yx}/\lambda \qquad a_{66} = G_{xy} \tag{2}$$

$$a_{16} = a_{26} = 0 \qquad \lambda = 1 - \nu_{xy}\nu_{yx}$$

and where standard lamination theory notation has been used. The stresses $\{\sigma\}$ are averaged over the laminate thickness, $\{\epsilon\}$ is the matrix of total strains and $\{\epsilon^T\}$ correspond to the free thermal strains. All engineering constants are laminate properties. Results for elastic fixtures were obtained by using laminate analysis to determine the properties of the combined steel/composite laminate in the bonded region of the fixture.

The constitutive equation, Eq 1, is used in conjunction with the finite-element method to determine the unknown nodal displacements u_i and v_i. Back substitution into the strain-displacement and constitutive equations provides the stresses in all elements.

The computer program used in this investigation is an extensively modified version of the code by Desai and Abel [17]. The major modifications to the program include an orthotropic material property capability and a more efficient equation solver. The program uses constant strain, triangular elements, and quadilateral elements composed of four constant-strain triangles. A detailed discussion of the finite-element formulation and the computer code may be found in Ref 6. Programs for plotting finite-element grids, stress contours, and stress profiles have also been written.

Results and Discussion

Table 1 presents a summary of all cases investigated during the course of this study. Because of space limitations, only selected results can be included in this paper. More detailed information, including the finite-element grids for each specimen, numerous additional stress contours for normal and shear stresses, and additional stress profiles and stress concentration factors, may be found in Refs 6 and 14. Slotted coupon and crossbeam results were obtained for typical graphite/epoxy properties, and double V-notch and rail shear results were obtained for both graphite/epoxy and graphite/polyimide. Fixtures were assumed to be steel whenever elastic fixtures were investigated. All results are presented in the form of normalized stresses.

Unidirectional Laminates

Shear tests on unidirectional laminates are obviously very important for understanding the behavior of the lamina, which is the fundamental building block upon which rests laminate analysis. The 10-deg off-axis coupon proposed by Chamis and Sinclair [5] appears very promising for shear modulus and strength determination; however, the biaxial stress state present in this specimen undoubtedly influences the nonlinear shear behavior through stress interaction. Other specimens should be evaluated for the presence (or absence) of normal stresses as well as the magnitude and distribution of shear stress.

Shown in Figs. 2–4 are stress contours for the shear and normal stresses in

TABLE 1—*Summary of cases investigated.*

Specimen	Laminates	Loading/Specimen	Material
Slotted coupon	[0] [90] [±45]$_s$	tension	graphite/epoxy
Crossbeam	[±45]$_s$ [0/90]$_s$	rigid core flexible core	graphite/epoxy
Double V-notch	isotropic [0] [90] [±45]$_s$ [0/90]$_s$ [0/±45/90]$_s$	rigid fixture elastic fixture mechanical loading thermal loading corner doubler rounded notch aspect ratio	steel graphite/epoxy graphite/polyimide
Rail shear	[0] [90] [±45]$_s$ [0/90]$_s$ [0/±45/90]$_s$ [0/±45]$_s$	rigid rail elastic rail tapered rail axial loading diagonal loading thermal loading	graphite/polyimide graphite/epoxy

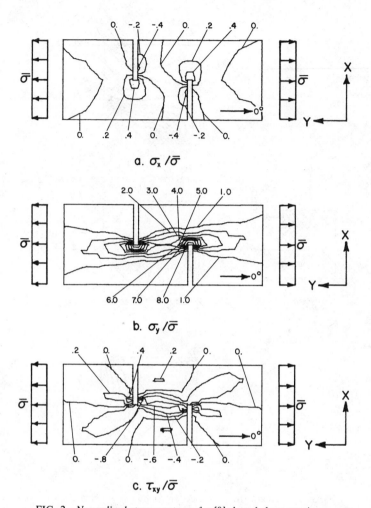

FIG. 2—*Normalized stress contours for [0] slotted shear specimen.*

unidirectional slotted coupon, double V-notched, and rail shear specimens. The stress contours for the slotted coupon in Fig. 2 indicate a region of fairly uniform shear stress with high normal stresses, σ_y, in the test section. Also, high stress concentrations in τ_{xy} and σ_y are indicated at the slot ends.

The results from the rigid-fixture, double V-notched specimen (Fig. 3) indicate that the stress state in the test section is both pure and uniform shear with a normalized maximum shear stress of 1.10 at the center of the specimen. Unlike the slotted coupon, the normal stresses are insignificant in the test section of the double V-notched specimen. The stress concentrations at the corner, C, are not as high as the stress concentrations in the slotted coupon.

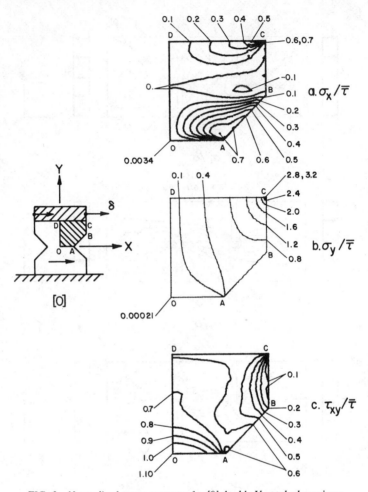

FIG. 3—*Normalized stress contours for [0] double V-notched specimen.*

The rigid-rail shear results in Fig. 4 indicate a region of uniform pure shear in the test section. The normalized shear stress has a value of 1.05 at the center of the specimen. The stress concentrations at the corners are lower than those in the slotted coupon and V-notched coupon, and the region of uniform pure shear extends over a larger portion of the specimen.

The stress contours for normal stresses in Figs. 2–4 are fairly typical of the contours for other laminates, and, because of space limitations, the remainder of the paper will concentrate on shear stresses. The reader is referred to Refs *6* and *14* for additional normal stress contours and stress profiles.

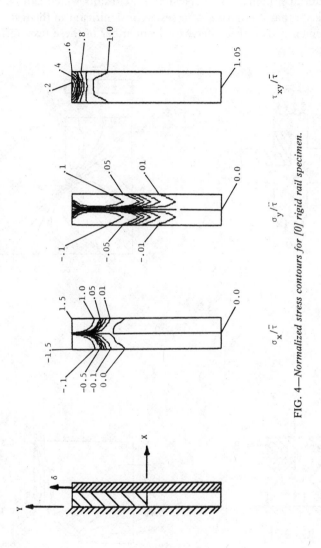

FIG. 4—*Normalized stress contours for [0] rigid rail specimen.*

Crossply Laminates

Figure 5 shows the shear stress distributions in a $[\pm 45]_s$ crossbeam and a $[0/90]_s$ double V-notched specimen. Both specimens give rise to pure shear in the material principal (1-2) coordinates. (Results which can be found in Ref 6 indicate that the normal stresses are insignificant in the test section of both specimens.) The shear behavior from either of these tests differs from

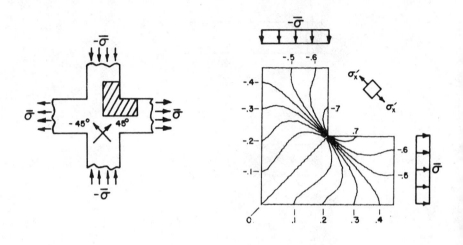

a.   Cross Beam

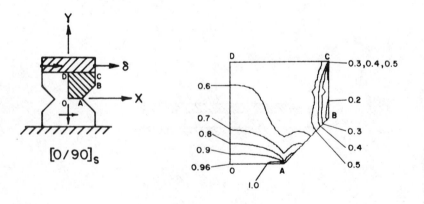

b.   Double V-Notch

FIG. 5—*Normalized shear stresses in crossbeam and rigid fixture double V-notched specimens.*

that on a unidirectional material only to the extent that the residual normal stresses influence the nonlinear behavior through stress interaction. That is, all three tests produce results for the material principal planes. The crossply specimens are less susceptible to stress concentrations than are unidirectional specimens because of the fiber reinforcement in the orthogonal directions. For the crossbeam, the shear stress has a maximum value of 1.13 at the center of the test section. Thus, if the specimen can be made to fail in the test section (possibly through the use of doublers at the corners), this analysis predicts that it would be a shear failure. (The results of this two-dimensional analysis should, of course, be checked further with a three-dimensional analysis to more completely assess the influence of the honeycomb core.) Results for a coupon specimen under biaxial loading have recently been reported by Duggan et al [8] to provide distributions similar to those predicted in this study.

A more uniform shear stress distribution is indicated for the double V-notched specimen with the stress magnitude ranging from 0.96 at the center of the specimen to 1.0 at the notch root. The maximum shear stress is predicted to be 1.02 a small distance in from the notch root. The stress concentration in σ_y at the corner, C, is quite high with a predicted value of 6.0. The stress concentrations in the double V-notched specimen are significantly higher than those in the crossbeam.

As indicated in Fig. 6, the rigid-rail shear specimen is predicted to exhibit a uniform pure shear stress state over more than one half of the specimen. The maximum shear stress occurs at the center of the specimen and is only 1 percent above the average shear stress. Stress concentrations are present at the intersection of the free edge and the rail for one component of normal stress, but they are less than one-half the magnitude of those in the double V-notched specimen.

Quasi-Isotropic Laminates

The stress contours for double V-notched and rigid-rail shear specimens, shown in Figs. 7 and 8, respectively, both indicate regions of uniform pure shear in the test section. The maximum shear stress (1.1) in the double V-notch is very near the notch root, whereas the maximum shear stress in the rail shear (1.03) is at the center of the specimen. Stress concentrations in normal stresses are present at the corners of both specimens, with those in the rail shear being well below those in the double V-notch.

Elastic Rail Shear

Comparisons of stress profiles along the centerline of rigid-rail, axially loaded elastic rail, and diagonally loaded elastic rail specimens are presented for unidirectional, crossply, and quasi-isotropic laminates in Figs. 9–11. The

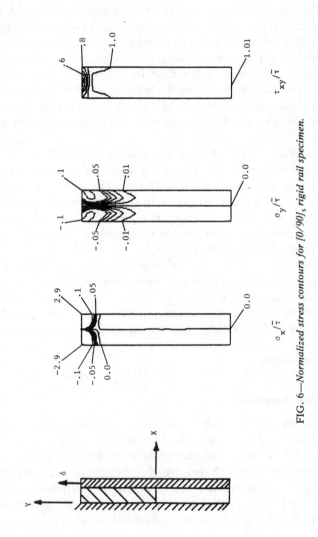

FIG. 6—*Normalized stress contours for [0/90]_s rigid rail specimen.*

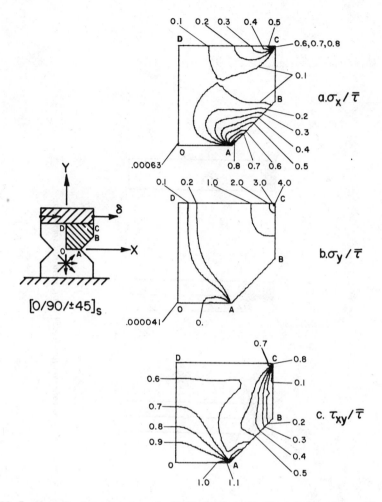

FIG. 7—*Normalized stress contours for [0/90/±45]ₛ double V-notched specimen with rigid fixtures.*

unidirectional and crossply results are similar, indicating regions of uniform shear with the magnitudes at the center approximately equal to 1.0 (Table 2) over almost one half of the specimen for all three loading configurations. The results for the quasi-isotropic laminate are less satisfactory. Both elastic rail cases indicate peak shear stress values near the free edge and a generally nonuniform distribution of decreasing magnitude in the test section. The values of the shear stress at the center of the axially and diagonally loaded quasi-isotropic specimens were 0.94 and 0.91, respectively. Profiles for a [±45]ₛ laminate with elastic rails were similar to those for the quasi-isotropic laminate, but exhibited lower values of shear stress at the center of the

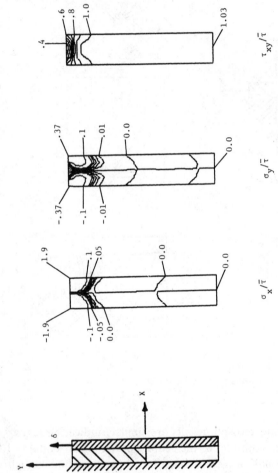

FIG. 8—Normalized stress contours for [0/±45/90]$_s$ rigid rail specimen.

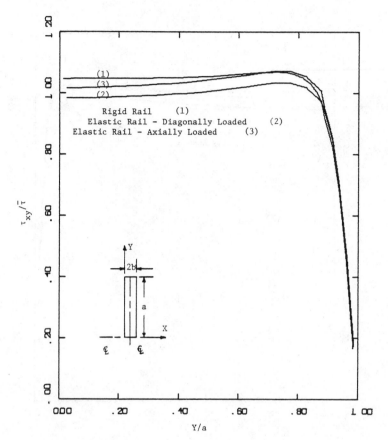

FIG. 9—$\tau_{xy}/\bar{\tau}$ at $X/2b = 0.5$ for [0] rigid and elastic rail specimens.

specimen (Table 2). The inclusion of elastic rails in the analysis also had the effect of increasing the normal stresses in the test section to levels which may be significant ($\simeq 0.4$).

Thermal Loading

Elevated-temperature shear properties are important in many composite applications. The double V-notch and rail shear specimens were studied to determine the effect of the rail constraints on the development of thermal stresses in the specimen. The analysis was linear elastic with constant room-temperature properties. Figure 12 shows the normal and shear stresses in a double V-notch unidirectional specimen which has been subjected to a curing temperature change of $-316°C$ (-600 deg F). All stresses are insignificant in the test section; however, the stress concentrations at the corners

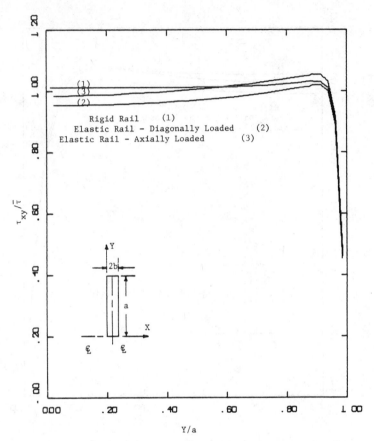

FIG. 10—$\tau_{xy}/\bar{\tau}$ at X/2b $= 0.5$ for $[0/90]_s$ rigid and elastic rail specimens.

(normalized with respect to the ultimate stresses) are severe, indicating failure due to shear and transverse compression.

A thermal analysis of the rail shear specimen (Fig. 13) predicted large σ_y stresses (0.85) in the test section when normalized with respect to the quantity $E_0 \Delta \alpha \Delta T$, where E_0 is the modulus of the rail and $\Delta \alpha$ is the difference in coefficients of thermal expansion between rail and composite along the length of the rail. Stress concentrations in all components of stress were predicted in the specimen corners.

Rounded Notch Effects

The effect of notch rounding was studied for the double V-notch specimen for the case of rigid-rail displacement of [0] and [90] laminates. The results, shown in Figs. 14 and 15, indicate a reduction in stress concentration at the

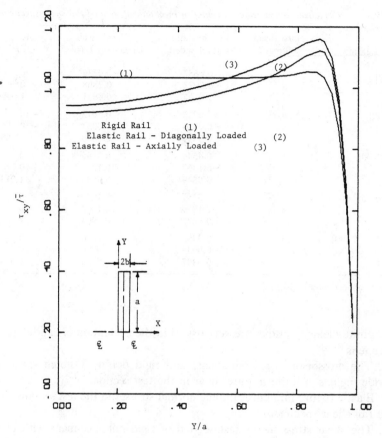

Rigid Rail   (1)
Elastic Rail - Diagonally Loaded   (2)
Elastic Rail - Axially Loaded   (3)

FIG. 11—$\tau_{xy}/\bar{\tau}$ at X/2b = 0.5 for [0/±45/90]$_s$ rigid and elastic rail specimens.

notch root for τ_{xy} and σ_x. The third stress, σ_y, is reduced in the [0] laminate, but is increased significantly in the [90] laminate. This increase in the [90] is due to the high stiffness of this laminate in the y-direction. These results also show that the shear stress is maximum near the notch root of the [90] laminate but attains its maximum value in the center of the [0] laminate. Both figures indicate that the shear stress tends toward zero at the notch root for a rounded notch. As Fig. 15 shows, this is not the case for the sharp V-notch [90] laminate.

Conclusions

Four specimens for in-plane shear behavior of fibrous composites have been analyzed using a linear thermoelastic finite-element analysis. The following general conclusions can be drawn from the results.

TABLE 2—*Normalized stress values at center of mechanically loaded rail shear specimen.*

Laminate	Normalized Stress	Elastic Rail Axially Loaded	Elastic Rail Diagonally Loaded	Rigid Rail
[0]	$\sigma_x/\bar{\tau}$	0.0657	−0.0236	0
	$\sigma_y/\bar{\tau}$	−0.2973	−0.2898	0
	$\tau_{xy}/\bar{\tau}$	1.0164	0.9831	1.0479
[90]	$\sigma_x/\bar{\tau}$	0.6712	0.5012	$\sim 10^{-11}$
	$\sigma_y/\bar{\tau}$	−0.0016	−0.0033	$\sim 10^{-11}$
	$\tau_{xy}/\bar{\tau}$	0.9799	0.9475	1.0089
[0/90]$_s$	$\sigma_x/\bar{\tau}$	0.4544	0.3145	0
	$\sigma_y/\bar{\tau}$	−0.1367	−0.1254	0
	$\tau_{xy}/\bar{\tau}$	0.9849	0.9535	1.0132
[±45]$_s$	$\sigma_x/\bar{\tau}$	0.3157	0.1929	0
	$\sigma_y/\bar{\tau}$	0.1859	0.1003	0
	$\tau_{xy}/\bar{\tau}$	0.9083	0.8903	1.0609
[0/±45/90]$_s$	$\sigma_x/\bar{\tau}$	0.3958	0.2628	0
	$\sigma_y/\bar{\tau}$	−0.0039	−0.0296	0
	$\tau_{xy}/\bar{\tau}$	0.9397	0.9153	1.0333

NOTE: aspect ratio = 10.

1. Finite-element studies are very useful in the evaluation of candidate test specimens.

2. The crossbeam, rigid-rail shear, and rigid double V-notch specimens provide regions of uniform pure shear in the test section.

3. Elastic fixture influences induce normal stresses in the test section of the rigid-rail shear specimen.

4. The shear stress in the test section of rigid-rail specimens with elastic rails may differ from the average value by as much as 10 percent, depending upon laminate configuration.

5. Stress concentrations are present in slotted coupon, crossbeam, rail shear, and double V-notched specimens.

6. Slotted coupon specimens are not recommended for shear tests.

7. Thermal stresses are less severe in double V-notch specimens than in rail shear specimens.

8. The double V-notched specimen should be investigated further, both numerically and experimentally.

Acknowledgment

This work was supported by the NASA-Virginia Tech Composites Program under NASA Grant NGR 47-004-129. Dr. John G. Davis of NASA Langley was the grant monitor, and the authors are grateful to him for his significant contributions during the course of this investigation. Thanks are also due to Ms. Frances Carter for typing the manuscript.

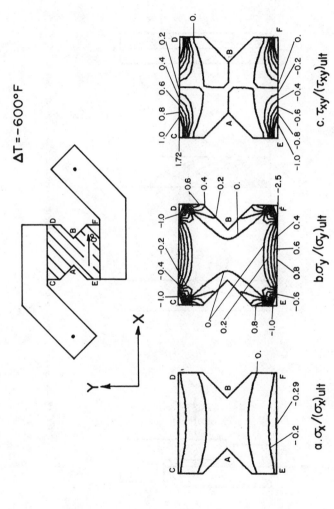

FIG. 12—*Normalized stress contours for thermally loaded [0] double V-notched specimen with elastic fixtures.*

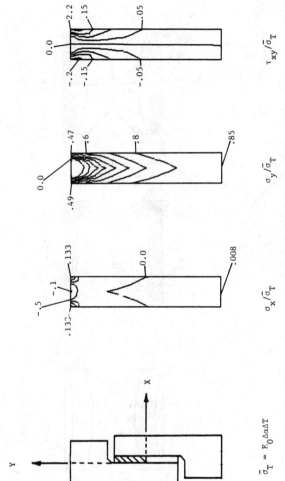

FIG. 13—*Normalized stress contours for [0] thermally loaded elastic rail specimen.*

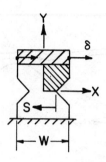

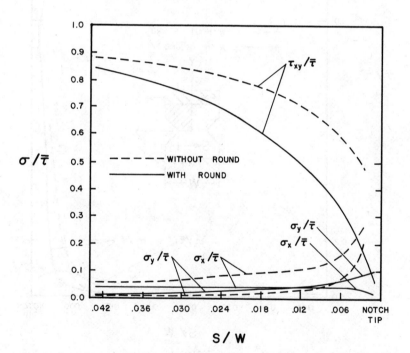

FIG. 14—*Rounded notch effects for the [0] double V-notched specimen along* Y = 0.

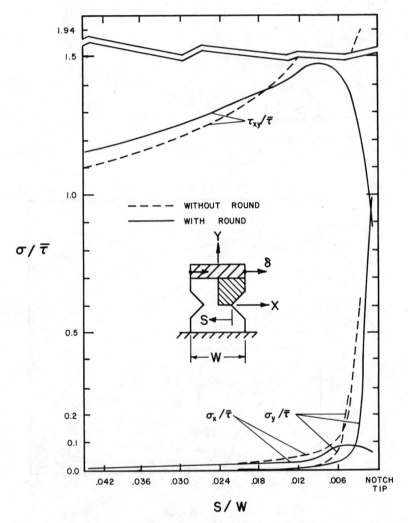

FIG. 15—*Rounded notch effects for the [90] double V-notched specimen along* Y = 0.

References

[1] Chiao, C. C., Moore, R. L., and Chiao, T. T., *Composites,* Vol. 8, No. 3, July 1977, pp. 161–169.
[2] Whitney, J. M., Stensburger, D. L., and Howell, H. B., *Journal of Composite Materials,* Vol. 5, Jan. 1971.
[3] Petit, P. H. in *Composite Materials: Testing and Design, ASTM 460,* American Society for Testing Materials, 1969, pp. 83–93.
[4] Rosen, B. W., *Journal of Composite Materials,* Vol. 6, Oct. 1972.
[5] Chamis, C. C., and Sinclair, J. H., *Experimental Mechanics,* Vol. 17, No. 9, Sept. 1977.
[6] Bergner, H. W., Davis, J. G., and Herakovich, C. T., "Analysis of Shear Test Methods for Composite Laminates," VPI-E-77-14, Virginia Polytechnic Institute and State University, Blacksburg, Va., April 1977; also NASA CR-152704.
[7] Arcan, M., Hashin, Z., and Voloshin, A., *Experimental Mechanics,* Vol. 18, No. 4, April, 1978.
[8] Duggan, M. F., McGrath, J. T., and Murphy, M. A. in *Proceedings,* 19th AIAA/ASME Structures, Structural Dynamics and Materials Conference, American Institute of Aeronautics and Astronautics/American Society of Mechanical Engineers, April 1978, pp. 311–319.
[9] Slepetz, J. M., Zagaeski, T. F., and Novello, R. F., "In-plane Shear Test for Composite Materials," AMMRC TR 78-30, Army Materials and Mechanics Research Center, Watertown, Mass., July 1978.
[10] Weisshaar, T. A. and Garcia, R., "Analysis of Graphite/Polyimide Rail Shear Specimens Subjected to Mechanical and Thermal Loading," NASA CR 3106, National Aeronautics and Space Administration, March 1979.
[11] Garcia, R., Weisshaar, T. A., and McWithey, R. R., *Experimental Mechanics,* Vol. 20, No. 8, Aug. 1980, pp. 273–279.
[12] R. A. Elkin, Fust, G., and Hanley, D. P. in *Composite Materials: Testing and Design, ASTM STP 460,* American Society for Testing and Materials, 1969, pp. 321–335.
[13] Waddoups, M. E. in *Composite Materials Workshop,* S. W. Tsai, J. C. Halpin, and N. J. Pagano, Eds., Technomic Publishing Co., Stamford, Conn. 1968, p. 254.
[14] Herakovich, C. T. and Bowles, D. E., "Thermoelastic Stress Analysis of the Rail Shear Specimen for Composites," VPI-E-79-34, Virginia Polytechnic Institute and State University, Blacksburg, Va., Oct. 1979.
[15] Sims, D. F., *Journal of Composite Materials,* Vol. 7, Jan. 1973, p. 124.
[16] Iosipescu, N., *Journal of Materials,* Vol. 2, No. 3, Sept. 1967.
[17] Desai, C. S. and Abel, J. F., *Introduction to the Finite Element Method,* Van Nostrand Reinhold, New York, 1972.

M. J. Shuart[1]

An Evaluation of the Sandwich Beam Compression Test Method for Composites

REFERENCE: Shuart, M. J., **"An Evaluation of the Sandwich Beam Compression Test Method for Composites,"** *Test Methods and Design Allowables for Fibrous Composites, ASTM STP 734,* C. C. Chamis, Ed., American Society for Testing and Materials, 1981, pp. 152-165.

ABSTRACT: This report evaluates the sandwich beam in four-point bending as a compression test method for advanced composites. To accomplish this evaluation the test method was used to obtain preliminary compressive design data for the temperature range 117 to 589 K (-250 to 600°F) and was analyzed to determine any influence from the honeycomb core on composite mechanical properties. The $[0_8]$, $[90_8]$, $[(\pm 45)_2]_s$, and $[0/\pm 45/90]_s$ laminate orientations were investigated utilizing HTS1/PMR-15 graphite/polyimide composite material.

The experimental and analytical results indicated that the test method can be used to obtain compressive elastic constants for graphite/polyimide composites. Elastic data were obtained for a wide temperature range, although many failures were initiated by either load concentrations or top cover debonds. The analysis indicated that the honeycomb core had negligible effects on composite mechanical properties. The top cover carried essentially all the compressive load in the beam test section, and the stress state in the test section of the top cover was essentially uniform and uniaxial.

KEY WORDS: composites, test methods, compression, mechanical properties, graphite/polyimide, finite-element analysis

Many test methods have been used to obtain compressive design data for composite materials. These methods have used rectangular [1-6],[2] cylindrical [4,7-9], or honeycomb sandwich construction [4,10-13] specimens, and testing of these specimens has been primarily limited to 164 to 450 K (-165 to 350°F) environments. The validity of many of these specimens has been experimentally investigated; however, a minimal number of analytical investigations have been performed.

---

[1]Aerospace technologist, Materials Division, NASA Langley Research Center, Hampton, Va. 23665.

[2]The italic numbers in brackets refer to the list of references appended to this paper.

The sandwich beam specimen in four-point bending has been used to obtain room temperature compressive data for composites [11-13]. The sandwich construction is typical of many composite applications, and the honeycomb core adds stability to the composite test section. Investigators have criticized the beam specimen because of the fabrication cost and the undetermined influence of the honeycomb core on composite mechanical properties.

The objective of this study was to investigate the sandwich beam compressive test method for testing composites. The influence of the honeycomb core was determined and compressive mechanical properties were obtained. A finite-element model was used to predict the stress state in the beam's top cover and the load-carrying capability of the honeycomb core. Compressive ultimate stress, ultimate strain, Young's modulus, and Poisson's ratio were obtained from graphite/polyimide laminates at 117 K ($-250°$F), room temperature, and 589 K (600°F). Specific details of test specimens, procedures, and initial results are available [14,15].

Experimental Program

The experimental program for this study was divided into two activities: (1) testing of sandwich beams and (2) testing of beam constituents. Elastic data from beam constituents were required as input for the finite-element analysis. Four laminate orientations were investigated: $[0_8]$, $[90_8]$, $[(\pm45)_2]_s$, and $[0\pm45/90]_s$. Specimens for both beam and constituent tests were machined from the same composite panel. The panels had fiber volume fractions from 53 to 55 percent and a nominal thickness of 1.5 mm (0.06 in.).

Sandwich Beams

Figure 1 shows a typical composite sandwich beam specimen and beam constituents. The beams were fabricated with composite top covers, metal bottom covers, and metal honeycomb cores. Depending on the test temperature, the bottom cover and the honeycomb core were either aluminum or titanium alloys. The nominal dimensions of the beam specimens, 559 mm (22 in.) long, 25.4 mm (1.0 in.) wide, and 42.8 mm (1.7 in.) thick, were recommended in Ref 16.

Beam specimens were also fabricated with 3.18-mm-thick (0.125 in.) 2024-T3 aluminum alloy top and bottom covers and were tested to obtain Young's modulus and Poisson's ratio data for comparison with documented values. The nominal dimensions of these specimens were the same as those of the composite beams.

The $[0_8]$, $[(\pm45)_2]_s$, and $[0/\pm45/90]_s$ specimens were fabricated with different-density honeycomb core along the length of the beam. Titanium alloy honeycomb beams containing 112- and 178-kg/m^3 (7.0 and 11.1 lb/ft^3)

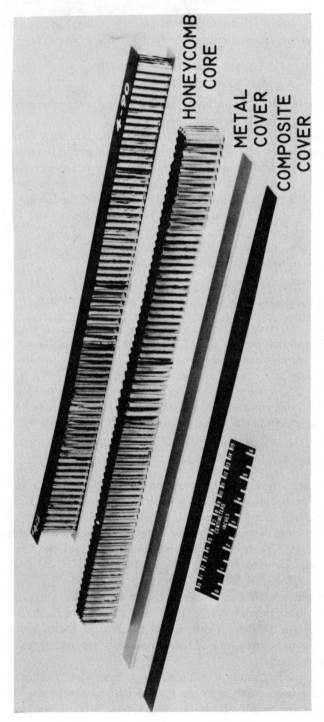

FIG. 1—Sandwich beam specimen and constituents.

core were tested at all test temperatures; aluminum alloy honeycomb beams containing 97.7- and 354-kg/m³ (6.1 and 22.1 lb/ft³) core were tested only at room temperature. The lower-density honeycomb was used in the center section of the beam to minimize any influence from the honeycomb on the composite cover in this region. Due to the transverse (through the beam thickness) shear loading on most of the remainder of the beam, the higher-density honeycomb was used in the outer regions of the beam. The different-density cores were spliced together by bonding or by spot welding.

The [90₈] and aluminum top cover specimens were fabricated with the lower-density aluminum honeycomb or titanium honeycomb along the entire length of the beam since the transverse shear loads for these specimens were predicted to be considerably lower than those for the other specimens. The aluminum top cover specimens were tested primarily to obtain sufficient stress-strain data for calculating elastic properties.

Figure 2 illustrates a beam specimen loaded in four-point bending. The load was applied to the beam through the load platen at two points on the beam's top cover. The beam was supported at two points on the bottom cover. The load and support points were 25.4-mm-diameter (1.0 in.) pins ground flat to distribute the load over a 645-mm² (1.0 in.²) area. Glass/polyimide load pads 1.3 mm (0.05 in.) thick were used under the points of load application to minimize stress concentrations due to loading. To calculate stresses, the load point was assumed to be at the center of the load pads. Strain gages were centrally located on the top cover to obtain strain data.

Beam Constituents

Typical specimens for the beam constituent tests are shown in Fig. 3. The honeycomb used in the test section of the beam was tested in compression at room temperature to obtain both Young's modulus and Poisson's ratio values. These values were unavailable either in the literature or from the manufacturer. Data were obtained for calculating moduli parallel and perpendicular to the ribbon direction and the corresponding Poisson's ratios. The specimen dimensions were approximately 102 by 102 by 38.1 mm (4.0 by 4.0 by 1.5 in.). Linear variable differential transformers (LVDT's) were used to obtain deflection data.

Graphite/polyimide tension specimens were tested at 117 K (−250°F), room temperature, and 589 K (600°F) to obtain Young's modulus and Poisson's ratio data. Tensile elastic properties obtained from these specimens were assumed to be equal to the compressive elastic properties and were used in the analysis. The dimensions of the tensile specimens were nominally 254 by 25.4 by 1.5 mm (10.0 by 1.0 by 0.06 in.). Strain gages were used on these specimens to obtain strain data.

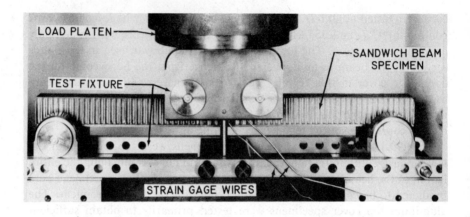

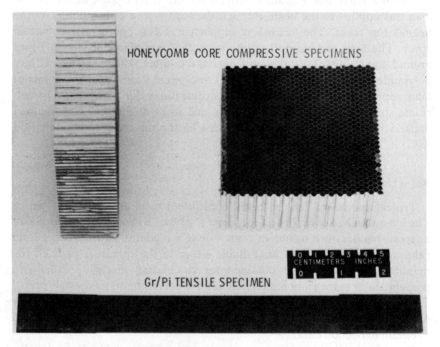

FIG. 3—*Constituent test specimens.*

Experimental Results

Beam Constituent Tests

Honeycomb Compressive Tests—Young's modulus and Poisson's ratio data for unstabilized aluminum and titanium honeycomb are given in Table

1. The modulus values were several orders of magnitude lower and the Poisson's ratio values were significantly higher than the corresponding properties for the composite or metal covers investigated in this study. Because of the different honeycomb geometries, the modulus values of the aluminum honeycomb were higher than those of the titanium honeycomb.

Graphite/Polyimide Tension Tests—Two tension specimens for each laminate orientation were tested at 117 K ($-250°$F), room temperature, and 589 K (600°F). Typical data for a quasi-isotropic laminate are shown in Fig. 4. The Young's modulus and Poisson's ratio data for the $[0_8]$ and $[0/\pm45/90]_s$ laminates were not affected by the test temperature. Temperature did affect the elastic properties of the $[90_8]$ and $[(\pm45)_2]_s$ laminates. The results were expected because the $[0_8]$ and $[0/\pm45/90]_s$ laminates were fiber-dominant and the $[90_8]$ and $[(\pm45)_2]_s$ laminates were matrix-dominant.

TABLE 1—*Honeycomb core average compressive data.*

Honeycomb Description	E_L kPa (psi)	ν_{LW}	E_W kPa (psi)	ν_{WL}
5052 Aluminum alloy density: 97.7 kg/m^3 (6.1 lb/ft^3) cell size: 3.17 mm (0.125 in.)	1765 (256.0)	1.10	1040 (150.8)	0.72
Ti-3Al-2.5V Titanium alloy density: 112 kg/m^3 (7.0 lb/ft^3) cell size: 6.35 mm (0.250 in.)	549.5 (79.7)	0.93	168.2 (24.4)	0.42

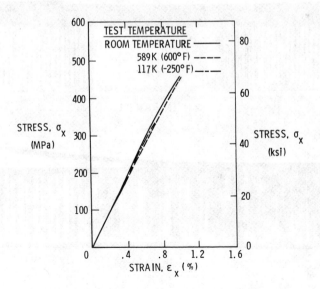

FIG. 4—*Tensile stress-strain behavior for $[0/\pm45/90]_s$ HTSl/PMR-15.*

A complete tabulation of the data and a detailed discussion of the temperature effects are reported in Ref *14*.

Sandwich Beam Tests

A total of 46 graphite/polyimide beam specimens was tested: 12 specimens with aluminum honeycomb at room temperature, 12 specimens with titanium honeycomb at room temperature, 11 specimens with titanium honeycomb at 117 K ($-250°$F), and 11 specimens with titanium honeycomb at 589 K ($600°$F). Generally, three specimens were tested for each laminate orientation.

The three types of failures observed for the composite beam specimens were designated compression, concentrated load, and debond (Fig. 5).

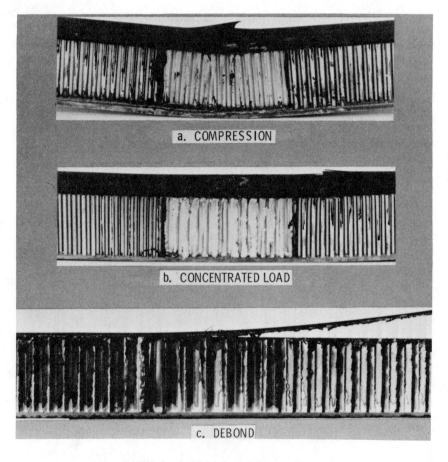

FIG. 5—*Sandwich beam failure modes.*

As shown in Fig. 5a, the compression failure occurred in the beam test section; this was the expected failure for the sandwich beam specimens tested. The $[90_8]$, $[(\pm45)_2]_s$, and $[0/+45/90]_s$ aluminum honeycomb specimens were characterized by compression failure. The $[0_8]$ aluminum honeycomb specimens, the $[90_8]$ and $[(\pm45)_2]_s$ titanium honeycomb room temperature specimens, and all 117 K ($-250°$F) specimens experienced the concentrated load failure shown in Fig. 5b even though thin load pads were used under the load points. Figure 5c illustrates the debond failure that was exhibited by the $[0_8]$ and $[0/\pm45/90]_s$ titanium honeycomb room temperature specimens and all specimens tested at 589 K (600°F). The maximum compressive ultimate stress for laminates that experienced debond failures was not obtainable because the failures were initiated by mechanisms other than material response to compressive load. A summary of the laminate orientations, test temperature, and failure types is given in Table 2.

Young's modulus and Poisson's ratio data were obtained from all composite beam specimens. The elastic properties were a function of test temperature for the matrix-dominant $[90_8]$ and $[(\pm45)_2]_s$ laminates but were not a function of temperature for the fiber-dominant $[0_8]$ laminate. The average values of modulus for $[0/\pm45/90]_s$ laminates tested at 117 K ($-250°$F) and room temperature were equal, whereas the average modulus for these laminates tested at 589 K (600°F) was approximately 15 percent lower. Poisson's ratio data for the $[0/\pm45/90]_s$ laminate were not a function of test temperature. Typical compressive date for a quasi-isotropic laminate are shown in Fig. 6.

A total of six beam specimens having 2024-T3 aluminum top and bottom

TABLE 2—*Summary of composite beam failure modes.*

Laminate Orientation	Test Temperature	Failure Mode
$[0_8]$	117 K ($-250°$F)	concentrated load
	RT-1[a]	concentrated load
	RT-2[b]	debond
	589 K (600°F)	debond
$[90_8]$	117 K ($-250°$F)	concentrated load
	RT-1	compression
	RT-2	concentrated load
	589 K (600°F)	debond
$[(\pm45)_2]$	117 K ($-250°$F)	concentrated load
	RT-1	compression
	RT-2	concentrated load
	589 K (600°F)	debond
$[0/\pm45/90]_s$	117 K ($-250°$F)	concentrated load
	RT-1	compression
	RT-2	debond
	589 K (600°F)	debond

[a]RT-1 = room temperature specimens fabricated with aluminum honeycomb.
[b]RT-2 = room temperature specimens fabricated with titanium honeycomb.

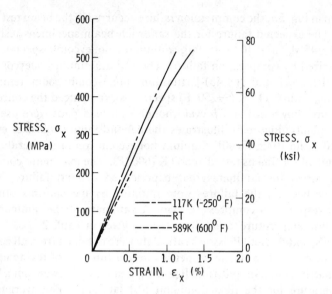

FIG. 6—*Compressive stress-strain behavior for [0/±45/90]ₛ HTSl/PMR-15.*

covers was tested at room temperature: three aluminum honeycomb specimens and three titanium honeycomb specimens. Young's modulus and Poisson's ratio were obtained for all specimens, and the average values were 74.26 GPa (10.77 × 10⁶ psi) and 0.320, respectively. Documented results for Young's modulus and Poisson's ratio are 73.77 GPa (10.7 × 10⁶ psi) and 0.33, respectively, for this alloy [*17*].

Analytical Results

Finite-Element Model

Investigators have often assumed that a uniform uniaxial compressive stress state exists in the beam test section and that all compressive load in the test section is carried by the top cover. To evaluate these assumptions a three-dimensional linear elastic finite-element analysis was used to investigate the stress state in the beam test section and the load-carrying capability of the honeycomb core. These two effects were evaluated as a function of test temperature and the ribbon direction Young's modulus of the core.

The finite-element model used in this study is illustrated in Fig. 7. By taking advantage of quarter symmetry, the portion of the test section to be modeled was significantly reduced. A total of 750 brick elements was used with 1248 nodes; each node had three translational degrees of freedom.

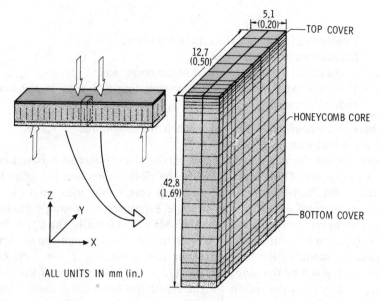

FIG. 7—*Diagram of finite-element model for sandwich beam.*

Brick elements were used to model the composite top cover and the honeycomb core as homogeneous orthotropic materials. Boundary conditions applied to the model were as follows. Along the left surface, which corresponds to the midspan of the beam, displacement in the x-direction equals zero; along the back surface, which corresponds to the mid-width of the beam, displacement in the y-direction equals zero; and at all nodes along the neutral axis, displacement in the z-direction equals zero. A linearly varying displacement field through the thickness of the beam was applied to the model to simulate the moment in the test section. The displacement field corresponded to a 1 percent compressive strain in the top cover and zero strain at the neutral surface.

Stress State in Top Cover

A biaxial stress state was developed in the top cover of the beam test section. Because of the difference in Poisson's ratio between the top cover and the honeycomb core, a transverse (y-direction, see Fig. 7) normal stress, $\bar{\sigma}_y$, existed in the test section of the top cover. The effect of $\bar{\sigma}_y$ on the measured strain response is shown by

$$\epsilon_x{}^0 = \frac{\bar{\sigma}_x}{\bar{E}_x}\left(1 - \bar{\nu}_{xy}\frac{\bar{\sigma}_y}{\bar{\sigma}_x}\right)$$

where

ϵ_x^0 = longitudinal extensional strain in top cover,
$\bar{\sigma}_x$ = longitudinal stress in top cover,
$\bar{E}_x$ = longitudinal Young's modulus of top cover, and
$\bar{\nu}_{xy}$ = Poisson's ratio of top cover corresponding to a load in the longitudinal direction.

The biaxial stress effect is negligible when the magnitude of $\bar{\nu}_{xy}$ $(\bar{\sigma}_y/\bar{\sigma}_x)$ is small compared with unity.

Table 3 gives the magnitude of the biaxial stress effect as a function of Young's modulus of the honeycomb in the ribbon direction for graphite/polyimide and 2024-T3 aluminum alloy top covers. Modulus values in the table were normalized using the average experimental Young's modulus of the aluminum honeycomb [E_L = 1765 MPa (256.0 psi)]. The upper limit of the honeycomb stiffness was chosen to be the through-the-thickness compressive modulus for this particular core geometry, E_z = 1.65 GPa (0.240 Msi), given by the manufacturer [18]. The first column of the table, E_{hc}/E_L = 1.0, gives the results for the aluminum honeycomb beams tested in this study. The subsequent columns give results for analytical cases where the honeycomb modulus was assumed to be one, two, and approximately three orders of magnitude larger than the experimental value. The biaxial stress effect was a function of honeycomb modulus, and this effect was less than 0.02 for values of E_{hc}/E_L up to two orders of magnitude greater than the value for beams tested. Results for titanium honeycomb beams tested at 117 K (−250°F), room temperature, and 589 K (600°F) showed the same effect and were not included in the table. Hence, the effect of $\bar{\sigma}_y$ on the measured strain response was negligible for the beams tested in this study where the ribbon direction Young's modulus of the honeycomb was no greater than 1765 kPa (256.0 psi).

TABLE 3—*Finite-element results for biaxial stress effect in top cover of sandwich beam.*

$$\epsilon_x^0 = \frac{\bar{\sigma}_x}{\bar{E}_x}\left(1 - \bar{\nu}_{xy}\frac{\bar{\sigma}_y}{\bar{\sigma}_x}\right)$$

Top Cover	$-\bar{\nu}_{xy}\dfrac{\bar{\sigma}_y}{\bar{\sigma}_x}$			
	$E_{hc}/E_L = 1.0$	10.0	100.0	937.5
$[0_8]$ gr/pi	0.00001	0.00032	0.00184	0.00520
$[90_8]$ gr/pi	0.00009	0.00065	0.00316	0.01225
$[(\pm45)_2]_s$ gr/pi	0.00025	0.00331	0.01920	0.07630
$[0/\pm45/90]_s$ gr/pi	0.00015	0.00107	0.00575	0.02253
Aluminum	0.00019	0.00062	0.00349	0.01383

NOTE: E_L = 1764 kPa (265.0 psi).

Compressive Load in Test Section

Some of the compressive load in the beam test section was carried by the honeycomb core. Investigators have often neglected this capability of the core. Figure 8 illustrates the load-carrying capability of the honeycomb core as a function of the honeycomb modulus. The ordinate is the ratio of the compressive load in the honeycomb, P_{hc}, to the compressive load in the top cover, P_{tc}, and the abscissa is the stiffness ratio, E_{hc}/E_L. E_{hc}/E_L equal to unity corresponds to the aluminum honeycomb beams tested in this investigation. As shown in the figure, P_{hc} was calculated to be less than 1 percent of P_{tc} for all these test cases. For the $[90_8]$ curve, however, P_{hc} was calculated to be at least 10 percent of P_{tc} for $E_{hc}/E_L \geq 20$. Results for titanium honeycomb beams tested at 117 K ($-250°$F), room temperature, and 589 K ($600°$F) showed the same trend and were not included in the figure. Hence, for the linear elastic analysis of the beams tested, the compressive load supported by the honeycomb core was less than 1 percent of the compressive load carried in the top cover.

Concluding Remarks

The sandwich beam in four-point bending has been experimentally and analytically investigated as a compression test method for advanced composites. Data were obtained for graphite/polyimide beam specimens tested at 117 K ($-250°$F), room temperature, and 589 K ($600°$F), and, as a check case, for 2024-T3 aluminum alloy specimens tested at room temperature.

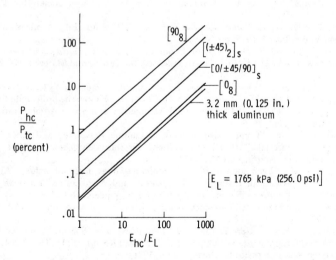

FIG. 8—*Load-carrying capability of honeycomb core as a function of honeycomb stiffness.*

A detailed three-dimensional finite-element model was used to examine the effects of the honeycomb core on measured composite mechanical properties. The results of this investigation led to the following conclusions:

1. The analysis indicated that a near uniaxial compressive stress state existed in the top cover and that essentially all the compressive load was carried by the top cover.

2. This test method can be used to obtain compressive elastic constants over the temperature range 117 to 589 K (-250 to $600°F$).

3. Laminate orientation, test temperature, and type of honeycomb core material were shown to affect the type of beam failure. For specimens with debond failures the maximum achievable compressive stress in the composite may not have been obtained.

The results of this investigation show that the sandwich beam in four-point bending is a viable compression test method for composites subject to the concerns noted herein.

References

[1] Hofer, Kenneth E., Rao, P. N., and Humphreys, V. E., "Development of Engineering Data on the Mechanical and Physical Properties of Advanced Composite Materials," Technical Report AFML-TR-72-205, IIT Research Institute, Chicago, Ill., Part 1, May 1972.

[2] Grimes, G. C., Francis, P. H., Commerford, G. E., and Wolfe, G. K., "An Experimental Investigation of the Stress Levels at Which Significant Damage Occurs in Graphite Fiber Plastic Composites," Technical Report AFML-TR-72-40, Southwest Research Institute, San Antonio, Tex., May 1972.

[3] "Standard Method of Test for Compressive Properties of Rigid Plastics," D 695-69, 1975 Annual Book of ASTM Standards, Part 35, American Society for Testing and Materials, 1975.

[4] Bert, Charles W. in Composite Materials, C. C. Chamis, Ed., Academic Press, New York, Vol. 8, Part 2, 1975, pp. 88-91.

[5] Ryder, J. T. and Black, E. D. in Composite Materials: Testing and Design (Fourth Conference), ASTM STP 617, American Society for Testing and Materials, 1977, pp. 170-189.

[6] Raju, B. Basava, Camarda, Charles J., and Cooper, Paul A., "Elevated-Temperature Application of the IITRI Compression Test Fixture for Graphite/Polyimide Filamentary Composites," NASA TP 1496, National Aeronautics and Space Administration, Washington, D.C., Sept. 1979.

[7] Davis, J. G., Jr., "Compressive Strength of Lamina Reinforced and Fiber Reinforced Composite Materials," Ph.D. Dissertation, Virginia Polytechnic Institute and State University, Blacksburg, Va., May 1973.

[8] Knoell, A. C., "Evaluation of Boron/Aluminum Tubes in Compression," AIAA Paper No. 75-789, American Institute of Aeronautics and Astronautics/American Society of Mechanical Engineers/Society of Automotive Engineers 16th Structures, Structural Dynamics and Materials Conference, American Institute of Aeronautics and Astronautics, May 1975.

[9] Weidner, John C., "New Tensile and Compressive Test Specimens for Unidirectional Reinforced Graphite/Epoxy Composites," Technical Report AFML-TR-70-264, University of Dayton Research Institute, Dayton, Ohio, 1971.

[10] "Standard Method of Test for Edgewise Compressive Strength of Sandwich Constructions," C 364-61 (Reapproved 1970), *1975 Annual Book of ASTM Standards*, Part 25, American Society for Testing and Materials, 1975.

[11] "Advanced Composite Airframe Structures," First Monthly Progress Report for AF Contract F33615-68-C-1301, Grumman Aircraft Corp., Bethpage, N.Y., March 1967.

[12] Waddoups, M. E. in *Composite Materials Workshop*, S. W. Tsai, J. C. Halpin, and N. J. Pagano, Eds., Technomic Publishing Co., Stamford, Conn., 1968, pp. 254-308.

[13] Herakovich, C. T., Davis, J. G., Jr., and Viswanathan, C. N. in *Composite Materials: Testing and Design (Fourth Conference)*, *ASTM STP 617*, American Society for Testing and Materials, 1977, pp. 344-357.

[14] Shuart, Mark J. and Herakovich, Carl T., "An Evaluation of the Sandwich Beam in Four-Point Bending as a Compressive Test Method for Composites," NASA TM 78783, National Aeronautics and Space Administration, Washington, D.C., Sept. 1978.

[15] Davis, John G., Jr., compiler, "Composites for Advanced Space Transportation Systems— (CASTS)," NASA TM 80038, National Aeronautics and Space Administration, Washington, D.C., March 1979, pp. 144-168.

[16] *Advanced Composites Design Guide*, 3rd ed. (third revision), Air Force Flight Dynamics Laboratory, Wright-Patterson Air Force Base, Dayton, Ohio, 1977.

[17] *Metals Handbook*, 8th ed., Taylor Lyman, Ed., American Society for Metals, Metals Park, Ohio, Vol. 1, 1961.

[18] "Mechanical Properties of Hexcel Honeycomb Materials," TSB120, Hexcel Corp., Dublin, Calif., 1976.

N. J. Salamon[1]

Stress Distribution in Sandwich Beams in Uniform Bending

REFERENCE: Salamon, N. J., **"Stress Distribution in Sandwich Beams in Uniform Bending,"** *Test Methods and Design Allowables for Fibrous Composites, ASTM STP 734*, C. C. Chamis, Ed., American Society for Testing and Materials, 1981, pp. 166–177.

ABSTRACT: The three-dimensional elastic stress state in a sandwich beam in uniform bending is given. The analysis is carried out numerically through use of a finite-difference formulation of the exact equations of elasticity. The beam is proportioned similar to one used in the characterization of composite materials and composed of $[\pm 45]_s$ composite face sheets laminated to a typical honeycomb core. It is shown that interlaminar stresses are confined to boundary regions near the free edges and that the in-plane stresses converge to the laminate theory solutions in the interior of the beam despite the fact that the beam is deeper than it is wide. Importantly, the core/face sheet tractions are found to be insignificant.

KEY WORDS: stresses, interlaminar stresses, beams, sandwich structures, tests, characteristics, composite materials, mechanical properties, bending, flexure

The design of an economically and technically suitable test specimen and test method for the determination of the mechanical properties of anisotropic lamina has yet to be agreed upon. And consideration of test methods for layered laminates even further complicates the issue. For all practical purposes the debate can be restricted to two contenders: the uniaxially loaded coupon versus the honeycomb sandwich beam in flexure. The crux of the argument concerns not so much measurement of the elastic moduli, but rather the apparent disagreement in strength measurements; flexure data usually indicate higher strengths and, for that matter, slightly higher moduli. This is generally attributed to advantageous load introduction into the beam. Other topics in contention include, for instance, the difference in shape of the stress-strain curves recorded by each method. But the intent here is not to reiterate test differences, only to recall them to mind.

---

[1] Associate professor, Department of Mechanical Engineering and Mechanics, West Virginia University, Morgantown, W. Va. 26506.

Numerous investigators have reported on composite specimen design and test methods and a few will be pointed out to establish ideas. Waddoups [1][2] reported the results of a group effort to determine primary properties and compared beam test data with tension coupon data. As reasons for inter-method variability, he offered (1) differences in gage-to-gage strain variations (1 percent for the beam versus 5 percent for the coupon), and (2) a beam geometry which approaches a constant-strain specimen. Lantz [2] compiled a large amount of test data to serve as a basis for comparison of properties recorded by the two test methods. Among his several conclusions, the most pertinent to the matter at hand is that "the stiffening or weakening effect of one test over the other varies with ... the orientation of lamination, and the direction of measured strain (axial versus transverse)." A statistical comparison of the two methods reported by Kaminski [3] shows somewhat similar data correlations (less scatter) in ultimate strain measurements, but a poorer correlation (greater scatter) in the computed ultimate stresses for the sandwich beam with either 0-deg or 90-deg face sheets in tension. As a reason for this, he cited the greater complexity introduced by the beam geometry over that of the coupon in stress calculations. Indeed Whitney, Browning, and Mair [4] discussed computational complexities involved in reducing data from flexure tests and particularly pointed out the difficulty in interpreting flex strength; even though their concern was not the sandwich beam, their analysis is applicable. In some of the foregoing, the complexity of the stress state existing in the specimens under test conditions is only hinted at, while in others it is explicitly pointed out. However, the possible influence of interlaminar stresses (thickness direction stresses) is ignored and their inclusion in analyses to settle this issue is certainly desirable.

With a publication by Pipes and Pagano [5], and a host of others since, showing the distribution of interlaminar stresses in finite-width composites in axial extension, it was established that the influence of interlaminar stresses extended inward from the free edges, which run parallel to the extension direction, approximately one laminate thickness (this refers to the entire section thickness). Assuming this interlaminar boundary region also applied to sandwich beams in bending, it raised the concern that due to a depth-to-width ratio in the beam greater than one, interlaminar stresses would diffuse throughout the entire width of the test specimen and thereby utterly confuse the conventionally assumed elementary stress state. To study the implications, Whitney [6] used an approximate analysis based upon the Pipes-Pagano result, and the aforementioned boundary region assumption, to show for a narrow beam a distinctive departure in the in-plane beam stresses from both those predicted by lamination theory and those computed for the tensile coupon. If this approximate analysis is cor-

---

[2] The italic numbers in brackets refer to the list of references appended to this paper.

rect, then serious reservations must be drawn as to the utility of the sandwich beam as a generally valid test specimen.

An efficient analysis for the three-dimensional elastic stress state in a layered composite in pure bending was given by Salamon [7], and in the present paper it is applied to a sandwich beam typical of those used to characterize composite face sheets. After outlining the solution scheme, both interlaminar and in-plane stress distributions acting in particular planes of the beam are presented. In addition, the in-plane strain distribution for the top and bottom surfaces of the beam is given. It is shown that the form of Whitney's solution [6] for multilayered face sheets is essentially correct; however, the assumption as to the extent of influence of interlaminar stresses is not. Furthermore, it is found that the interlaminar stresses at the interfaces between the honeycomb core and face sheets are insignificant.

Problem Formulation

Consider a sandwich beam symmetrically constructed with respect to a midplane about which it is bent by an applied moment only (Fig. 1). Con-

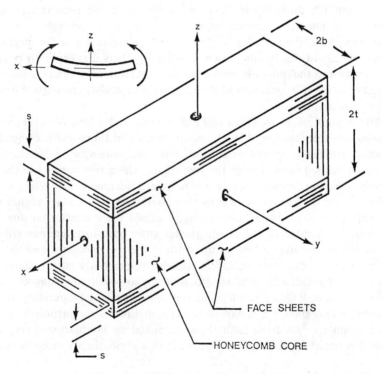

FIG. 1—*Sandwich beam geometry and loading.*

struction is such that bonding is assumed to provide perfect adhesion at all interfaces between the core and face sheets and between individual lamina comprising the face sheets; thus continuity of displacements and tractions is prescribed across the interfaces. The core and individual lamina are treated as homogeneous orthotropic materials subject to the usual linear elastic law [8] which, referred to the x, y, and z coordinate system, is

$$
\begin{bmatrix} \sigma_x \\ \sigma_y \\ \sigma_x \\ \tau_{yz} \\ \tau_{xz} \\ \tau_{xy} \end{bmatrix} = \begin{bmatrix} c_{11} & c_{12} & c_{13} & 0 & 0 & c_{16} \\ & c_{22} & c_{23} & 0 & 0 & c_{26} \\ & & c_{23} & 0 & 0 & c_{36} \\ \text{(symmetric)} & & & c_{44} & c_{45} & 0 \\ & & & & c_{55} & 0 \\ & & & & & c_{66} \end{bmatrix} \begin{bmatrix} \epsilon_x \\ \epsilon_y \\ \epsilon_z \\ \gamma_{yz} \\ \gamma_{xz} \\ \gamma_{xy} \end{bmatrix} \tag{1}
$$

where the c_{ij} are related to the nine orthotropic constants c_{ij}' through the well-known transformation equations [8].

As shown in Fig. 1, let the beam length coordinate be x with uniform bending in the x-z plane such that the resulting stress distribution is independent of x. Then, using the linear strain-displacement relations, the solution takes the form [7]

$$
\begin{aligned}
u &= C_2 xz + U(y, z) \\
v &= C_4 xz + V(y,z) \\
w &= -(C_2/2) x^2 - C_4 xy + W(y, z)
\end{aligned}
\tag{2}
$$

where u, v, and w are the displacements and U, V, and W are the displacement functions associated with x, y, and z, respectively. The constant C_2, which can be viewed as the load input parameter, is the inverse bending radius while C_4 is the twist-coupling constant approximated by

$$
C_4 \sim \frac{D_{16}'}{2D_{11}'} C_2 \tag{3}
$$

where $D_{ij}' = D_{ij}^{-1}$ are the laminate flexural constants [8]. In the laminate theory, Eq 3 is exact; in the present theory, the approximation has been shown to be a good one [7]. It should be noted from Eq 2 that the longitudinal bending strain is linear, that is

$$
\epsilon_x = C_2 z \tag{4}
$$

Hence the problem may be classed as one in uniform bending.

Writing the strains and stresses in terms of the displacements (3) through Eq 1, and substituting into the stress equilibrium equations, yields the field equations

$$c_{66}U,_{yy} + c_{55}U,_{zz} + c_{26}V,_{yy} + c_{45}V,_{zz} + (c_{36} + c_{45})W,_{yz} = 0$$

$$c_{26}U,_{yy} + c_{45}U,_{zz} + c_{22}V,_{yy} + c_{44}V,_{zz} + (c_{23} + c_{44})W,_{yz} = 0$$

$$(c_{36} + c_{45})U,_{yz} + (c_{23} + c_{44})V,_{yz} + c_{44}W,_{yy} + c_{33}W,_{zz}$$

$$= -(c_{13}C_2 + 2c_{36}C_4) \tag{5}$$

where the commas denote partial differentiation with respect to the indicated coordinate subscripts. With the boundary conditions taken as traction-free outer surfaces together with the interface continuity conditions, the field equations (6) are solved by the finite-difference method. Further details of the preceding formulation are presented in Ref 7.

The finite-difference computer program contains two-step extrapolations (three nodes) for accuracy at interfaces and free surfaces and, for economy, is applied only to one quadrant of the beam by employing the symmetry conditions

$$U(y, o) = V(y, o) = W,_z(y, o) = 0$$

$$U(o, z) = V(o, z) = W,_y(o, z) = 0 \tag{6}$$

where the origin of coordinates y and z is taken at the center of the entire cross section. Despite employment of a variable-sized mesh distribution featuring a higher density of nodes at the free edge, the problem required 2160 degrees of freedom for solution.

Results

The results presented are computed for a sandwich beam with a depth-to-width ratio $(2t/2b)$ equal to 1.5 in accordance with typical proportions of a sandwich beam test specimen. The ratio of the face sheet thickness to the half-depth of the beam (s/t) is 0.308, the choice being somewhat restricted by a lack of versatility in the computer code. The beam is composed of a honeycomb core characterized by the material properties

$$E_{xx} = E_{yy} = 276 \text{ MPa } (0.04 \times 10^6 \text{ psi}),$$

$$E_{zz} = 3450 \text{ MPa } (0.5 \times 10^6 \text{ psi}),$$

$$G_{xz} = G_{yz} = 414 \text{ MPa } (0.06 \times 10^6 \text{ psi}), \tag{7}$$

$$G_{xy} = 110 \text{ MPa } (0.016 \times 10^6 \text{ psi}),$$

$$\nu_{xy} = 0.25, \nu_{xz} = \nu_{yz} = 0.02,$$

and face sheets, the same top and bottom, consisting of graphite/epoxy lamina with the properties

$$E_{11} = 152\ 000 \text{ MPa } (22.0 \times 10^6 \text{ psi}),$$

$$E_{22} = E_{33} = 11\ 000 \text{ MPa } (1.6 \times 10^6 \text{ psi}),$$

$$G_{12} = 6900 \text{ MPa } (1.0 \times 10^6 \text{ psi}), \tag{8}$$

$$G_{13} = G_{23} = 4420 \text{ MPa } (0.64 \times 10^6 \text{ psi}), \text{ and}$$

$$\nu_{12} = \nu_{13} = \nu_{23} = 0.25.$$

where the notation is standard [8] and the subscripts 1, 2, 3 denote the material coordinates with 1 the fiber direction. Although unidirectional face sheets were considered and will subsequently be discussed, the [+45, −45, −45, +45] angle-ply units were chosen for the detailed study because this system is prone to the generation of significant interlaminar stresses [7] and it is not uncommon in applications. (Angles orient the 1-axis with respect to the x-axis.)

Selected results are shown in Figs. 2–5, but some explanation is necessary to their understanding. The positive or negative sense of the curves is irrelevant to the present objectives (except for the interlaminar normal stress which displays a sign reversal) and stress or strain distributions should be viewed as absolute values. In order to convey relative orders of magnitude, the following is done:

1. Stresses are normalized with respect to the laminate theory bending stress σ_x^{LT} evaluated at the respective z-coordinate value; specifically, σ_i @ z (or τ_{ij} @ z) is divided by σ_x^{LT} @ z, which, in the figures and hereafter, is denoted by σ.

2. Strains, given only for the top (or bottom) surface, are normalized with respect to the bending strain ϵ_x evaluated at the top (or bottom) surface where it is a constant.

This normalization procedure generalizes the stress results as well in that results involving a ±45-deg interface are virtually identical to those involving a ∓45-deg interface. Furthermore, the following definitions are used in the figures:

±45 INF: ±45-deg or ∓45-deg interface,

±45 INF + : one descretization step away from a ±45-deg or ∓45-deg interface, and

CORE INF + : one descretization step from the core/+45-deg interface inside either face sheet.

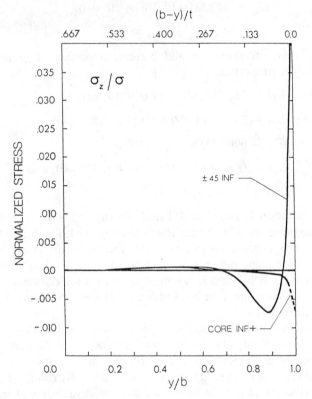

FIG. 2—*Interlaminar normal stresses at face sheet interfaces and the core/face sheet interfaces.*

To simplify things, one need only concentrate upon the upper half of the beam, and much of the following will assume such. The lower horizontal axis in the figures denotes the distance y outward across the beam relative to the half-width b, while the upper axis denotes the distance inward relative to the beam half-depth t.

In Figs. 2 and 3 the interlaminar stresses σ_z/σ and τ_{xz}/σ are shown with σ the laminate theory bending stress as explained in the preceding. Notably, the magnitude of the normal stresses is small, less than 1.0 percent of the bending stress except at the free edge. There the intensity is an order of magnitude less than that for the interlaminar shear τ_{xz}. However, both σ_z and τ_{xz} display the now-familiar singular character, a major cause for concern, at either the ± 45-deg or ∓ 45-deg interface in the face sheets. But one descretization step from the core/face sheet interface inside the $+45$-deg layer (CORE INF $+$), both interlaminar stresses are small; the normal stress so much so that resolution of the rise near the free edge is indefinite (shown dotted). Furthermore, the interlaminar tractions transmitted across the

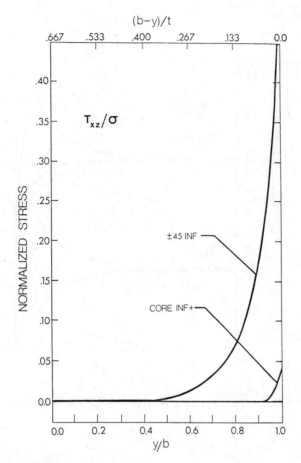

FIG. 3—*Absolute values of the interlaminar shear stresses at face sheet interfaces and the core/face sheet interfaces.*

core/face sheet interface are nil (not shown). Putting magnitudes aside, the other important feature of the interlaminar stress distributions is their general decay, reaching negligible values in the interior of the beam. Clearly they are near zero inward from $y/b = (b - y)/t = 0.4$. Adjacent to the core interface, the shear stress τ_{xz} in the face sheet sharply decreases to zero very near the free edge. The interlaminar shear stresses τ_{yz} are of the same order of magnitude as the normal stresses σ_z and do not contain, due to a free-edge boundary condition, the singular nature displayed by their companions. Hence their importance is secondary and they are not shown.

Face sheet characterization is very dependent upon the distribution of in-plane stresses. In Fig. 4 typical distributions of those stresses which reach appreciable magnitudes are shown. It is evident that all stresses converge

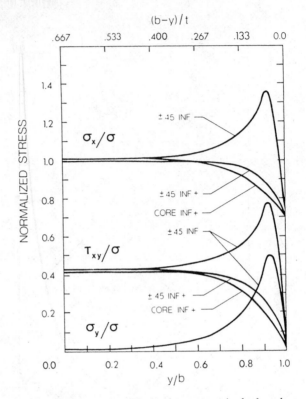

FIG. 4—*Absolute values of the in-plane stresses in the face sheet.*

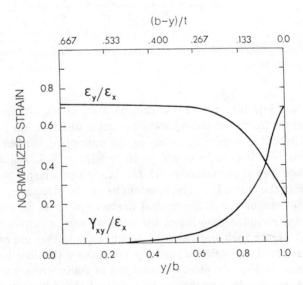

FIG. 5—*Absolute values of the top and bottom surface in-plane strains.*

to constant values in the interior region of the laminate defined by y/b less than 0.4 or $(b - y)/t$ greater than 0.4. These interior constant stresses equal the laminate theory solutions; this is obvious for the plots of σ_x since σ_x/σ approaches unity. Thus significant interlaminar interaction is confined to the free-edge boundary region.

In the free-edge boundary region, the in-plane stress distributions appear in two categories. Those at an interface grow to peak values, then fall off sharply to a common value at the edge, while those adjacent to the interfaces, as if to compensate for the interface peaks, decline moderately to the free-edge values. Likewise, the interlaminar stresses reach higher intensities in an interface. The in-plane stresses display a somewhat gentler slope in the face sheets near the core interface, but within the core itself these stresses decrease everywhere to negligible values, for the most part several orders of magnitude below the face sheet values, and consequently are not shown. Magnitudes for σ_y/σ only reach values appreciably greater than 1 percent at the ± 45-deg or ∓ 45-deg interfaces. On the other hand, τ_{xy}/σ is approximately 42 percent everywhere except in the boundary region. This is to be expected with a ± 45-deg system.

Of interest particularly to experimentalists is the state of strain on the top and bottom surfaces. This is shown in Fig. 5 for the normalized in-plane strains ϵ_y/ϵ_x and γ_{xy}/ϵ_x. In light of Eq 4, the normalized bending strain is unity across the entire laminate width. As is the case for stress distributions, deviations from the laminate theory solutions are limited to the boundary region previously defined.

Conclusions

The objective of this paper is to clarify the influence of interlaminar stresses on the stress state in a sandwich beam in uniform bending. Such a situation typically arises in the four-point sandwich beam flexure test for face sheet properties. The results show that the interlaminar stresses generated in the $[\pm 45]_s$ angle-ply face sheets are restricted to an interlaminar boundary region which is on the order of 40 percent of the beam half-depth (20 percent of the total depth), and hence they do not diffuse throughout the beam width as might have been expected. However, it is more revealing to consider the boundary region as a function not of the beam thickness, but of the face sheet thickness. In this case the boundary region is restricted to a distance from the free edge of 1.3 times the face sheet thickness. This agrees favorably with the extent of the boundary region for laminates (minus a core) in extension [5] and bending [7]. Thus in practical honeycomb beam systems where the face sheet thickness is less than 5 percent of the beam thickness, the boundary region should be less than 7 percent of the total beam depth. Even though computer code inflexibility limited the face sheet thickness in this study to 15.4 percent of the total beam depth, it seems reasonable to conclude that the boundary region will be on the order of the face sheet thickness in soft core systems in pure bending.

It is the softness of the core which appears responsible for confinement of the interlaminar boundary region. This is clearly the reason why the core/ face sheet interface fails to develop appreciable tractions. As seen in Figs. 2 and 3, such tractions appear insignificant. Since the boundary region is confined near the free edge, the in-plane stresses are free to converge to the laminate theory values in the interior of the laminate as seen in Fig. 4. From Fig. 5, the same conclusion is evident for the in-plane strains at the top and bottom surfaces.

The approximate solution given by Whitney [6] as an order-of-magnitude analysis appears to be essentially correct, but his assumption as to the extent of the boundary region is not. Specifically, Whitney's results show the interlaminar boundary region extending inward from the free edge a distance 80 percent of the beam depth, while the present results, as discussed in the foregoing, show only a 20 percent inward extent. However, with a modification of the boundary region to reflect the present results, and perhaps more detailed research to ascertain correct stress magnitudes and to include other beam configurations and material components, his analysis may prove quite satisfactory.

In the process of carrying out this research, single-ply and smeared-out angle-ply unit face sheets were also investigated. But a search for appreciable interlaminar stresses proved futile. In a 0-deg unidirectional face sheet, such stresses are nil. The core/face sheet interface under uniform bending fails to develop appreciable interlaminar tractions in all cases, including, as has been seen, the $[\pm 45]_s$ system. Thus, based upon the influence of interlaminar stresses, the present analysis indicates that the sandwich beam passes as a viable test specimen.

Acknowledgment

This work was supported by the U.S. Air Force Office of Sponsored Research under Grant AFSOR-78-3680.

References

[1] Waddoups, M. E. in *Composite Materials Workshop,* S. W. Tsai, J. C. Halpin, N. J. Pagano, Eds., Technomic Publishing Co., Stamford, Conn., 1968, pp. 254-308.

[2] Lantz, R. B., *Journal of Composite Materials,* Vol. 3, 1969, pp. 642-650.

[3] Kaminski, B. E. in *Analysis of the Test Methods for High Modulus Fibers and Composites, ASTM STP 521,* American Society for Testing and Materials, 1973, pp. 181-191.

[4] Whitney, J. M., Browning, C. E., and Mair, A. in *Composite Materials: Testing and Design (Third Conference), ASTM STP 546,* American Society for Testing and Materials, 1974, pp. 30-45.

[5] Pipes, R. B. and Pagano, N. J., *Journal of Composite Materials,* Vol. 4, 1970, pp. 538–548.
[6] Whitney, J. M. in *Analysis of the Test Methods for High Modulus Fibers and Composites, ASTM STP 521,* American Society for Testing and Materials, 1973, pp. 167–180.
[7] Salamon, N. J., *Fiber Science and Technology,* Vol. 11, 1978, pp. 305–317.
[8] Jones, R. M., *Mechanics of Composite Materials,* Scripta Book Co., Washington, D.C., 1975.

Design Allowables

M. J. Rich[1] *and D. P. Maass*[1]

Developing Design Allowables for Composite Helicopter Structures

REFERENCE: Rich, M. J. and Maass, D. P., **"Developing Design Allowables for Composite Helicopter Structures,"** *Test Methods and Design Allowables for Fibrous Composites, ASTM STP 734,* C. C. Chamis, Ed., American Society for Testing and Materials, 1981, pp. 181–194.

ABSTRACT: The overall procedures for developing composite material static and fatigue design allowables are presented. The allowables are established at the material lamina level accounting for a range of environmental conditions. Design values are derived from an interpolation method, which was confirmed by screening tests. Static test results are presented as a percentage of mean room temperature dry strength and show that the environmental effects are more pronounced for the matrix modes than for the fiber modes of fracture. Kevlar/epoxy is more sensitive to the environmental design conditions than graphite/epoxy. Fatigue test results are presented in a constant-life diagram as a percentage of room temperature wet static tensile strength. The constant-life fatigue diagram shows that the tension-compression and compression-compression range may be the governing aspect for fatigue design. Statistical evaluations are made and indicate normal distribution, and relatively small sample sizes are sufficient to establish reliable design allowables.

KEY WORDS: composites, design allowables, aircraft structures, graphite/epoxy, Kevlar/epoxy, statistical analysis

Nomenclature

$"A"$ Design static strength, 99 percent survivability, 95 percent confidence

D_n Kolmogorov statistic

F_{tu} Ultimate tensile strength, ksi

E Modulus of elasticity, msi

F_{bu} Ultimate bending (flexure) strength, ksi

F_{cu} Ultimate compressive strength, ksi

F_{su} Ultimate in-plane shear strength, ksi

[1]Chief of structures and materials engineer, respectively, Sikorsky Aircraft Division of United Technologies Corp., Stratford, Conn. 06602.

181

F_{siu} Ultimate interlaminar shear strength, ksi
G Shear modulus, msi
N Cycles of stress
R Fatigue ratio, minimum to maximum stress
S Design fatigue strength, ksi
$\overline{S}$ Mean fatigue strength, ksi
$\overline{S}_E, S_E$ Mean and design endurance fatigue strength, respectively, ksi
n Number of data points
σ_V Coefficient of variation, standard deviation to mean strength ratio in percent
ΔM Moisture, percent weight

Abreviations

RTD Room temperature, dry
RTW Room temperature, wet
ETD Elevated temperature, dry
ETW Elevated temperature, wet
K/E Kevlar/epoxy
G/E Graphite/epoxy
SBS Short-beam shear
QA Quality assurance
RH Relative humidity, percent
α, γ Shape curve parameters
ν Major Poisson's ratio

The successful design of lightweight reliable metal aircraft structures is due in a large degree to establishment of accurate material design allowables. The availability of standard design allowables is the result of the combined efforts of government agencies and industry and has resulted in mutually accepted data such as provided in military handbooks [1].[2] However, developing design allowables for composite materials is a far more difficult task. There is no standard material; both fiber type and resin system vary from company to company. The stress analysis requires more property data for composites than for metals, and there does not exist a mutually accepted data bank.

Design guides, such as Ref 2, have helped, but with the rapid changes in materials and technology the data are useful for information only and do not provide an industry-accepted standard. What is required is a document such as Ref 3 which establishes design allowables for the many advanced composite materials currently available. That document contained a wide range

---

[2]The italic numbers in brackets refer to the list of references appended to this paper.

of the then commonly used fiber glass materials and was accepted as a standard.

Thus the burden of obtaining advanced composites design allowables falls on the individual companies. Considerable expense in time and cost are involved, which dictates a careful plan to minimize these efforts. In this paper a procedure for developing design allowables for advanced composite materials is presented. The procedure has been used for design and Federal Aviation Administration (FAA) certification of primary structures. Figure 1 illustrates the overall approach used.

Criteria

Environmental Conditions

The mechanical properties of composite materials are affected by both absorbed moisture and elevated temperatures. It is therefore necessary to establish respresentative quantitative moisture and temperature levels for the design application. The design moisture level is a function of the operational environment and the moisture sensitivity of the particular fiber and especially the resin. To establish the design moisture level a survey of worldwide climate conditions and in-service moisture absorption data [4] was utilized. The moisture absorption characteristic of the fiber/resin system used in Ref 4 is shown schematically in Fig. 2 and results in the definition of an effective relative humidity for combined in-field effects, that is, ambient relative humidity, solar radiation, wind drying, etc. Use of the effective relative

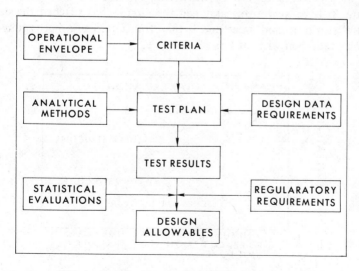

FIG. 1—*Overall approach to develop design allowables.*

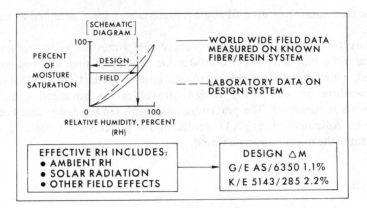

FIG. 2—*Design moisture levels derived from worldwide field data.*

humidity results in design moisture levels of 1.1 and 2.2 percent for AS/6350 graphite/epoxy and 5143/285 Kevlar/epoxy, respectively.

For static strength the design condition is the elevated temperature from the runway storage/takeoff combined with the design moisture level. For fatigue strength the design condition is the specified design moisture level at room temperature since the majority of fatigue cycles are accumulated at ambient flight temperatures. The component is conservatively assumed to be at the design moisture level throughout its life.

A thermal analysis was conducted for the runway storage/takeoff, and the elevated-temperature range, as schematically shown in Fig. 3, was found to be 60 to 77°C (140 to 170°F) depending on location and thickness of laminate. It is conservatively assumed the aircraft has reached the design moisture level in humid areas and is then transferred to a hot region with high solar radiation without loss of moisture.

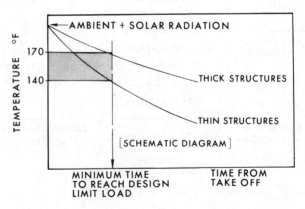

FIG. 3—*Design temperatures determined from runway storage/takeoff conditions.*

Evaluation

Static strength allowables are "*A*" values (99 percent survivability, 95 percent confidence) as determined by sample size and distribution for the design elevated-temperature wet (ETW) condition.

Fatigue strength allowables are mean strength minus three standard deviations. For FAA certification the mean strength is reduced by not less than 20 percent to meet current regulations. The fatigue strength values will be at the design room-temperature wet (RTW) condition.

Both static and fatigue tests will be conducted at room-temperature dry (RTD) and at conditions equal to or exceeding design conditions. Design values will be interpolated as illustrated in Fig. 4. This procedure assures that no extrapolations are required and permits further evaluations should design conditions be altered.

Testing is to be conducted at the lamina level requiring sufficient property data for linear laminate stress analysis. For special tests of specific design configurations, the subcomponent level of testing is to be used with the respresentative laminates.

Test Plan

The scope of testing is shown in Fig. 5 for K/E (woven fabric) and in Fig. 6 for G/E (unidirectional tape). Both static and fatigue strength and the elastic constants are obtained for basic laminate orientations (0 deg G/E, 0/90 and ±45 K/E). The types of test specimens are listed in Table 1. In general, maximum use was made of American Society for Testing and Materials (ASTM) standards with some modifications found to be required.

Compression testing of unidirectional graphite tape was a problem area. Sandwich beams were too expensive for the number of tests required and, in addition, problems were uncovered with laminate-to-core bond under ETW

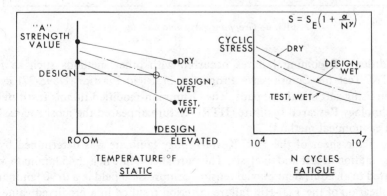

FIG. 4—*Linear interpolation method for static and fatigue strengths.*

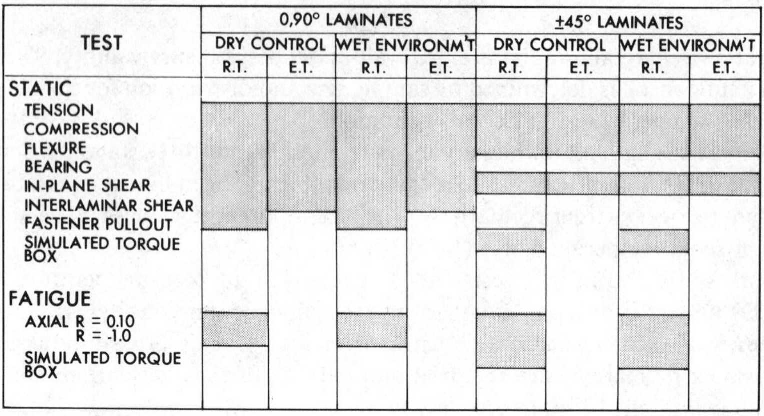

TEST	0,90° LAMINATES				±45° LAMINATES			
	DRY CONTROL		WET ENVIRONM'T		DRY CONTROL		WET ENVIRONM'T	
	R.T.	E.T.	R.T.	E.T.	R.T.	E.T.	R.T.	E.T.
STATIC								
TENSION								
COMPRESSION								
FLEXURE								
BEARING								
IN-PLANE SHEAR								
INTERLAMINAR SHEAR								
FASTENER PULLOUT								
SIMULATED TORQUE BOX								
FATIGUE								
AXIAL R = 0.10								
R = -1.0								
SIMULATED TORQUE BOX								

FIG. 5—*Kevlar/epoxy data base test plan.*

TEST	DRY CONTROL		WET ENVIRONMENTAL	
	R.T.	E.T.	R.T.	E.T.
STATIC				
TENSION				
COMPRESSION				
FLEXURE				
BEARING (0,±45°)				
IN-PLANE SHEAR				
INTERLAMINAR SHEAR				
EDGEWISE SHEAR				
FATIGUE				
AXIAL R = 0.10				
R = -1.0				
TORSION R = -1.0				
FLEXURE				
BEARING (0,±45°)				
COMBINED LOAD				
INTERLAMINAR SHEAR				
EDGEWISE SHEAR				

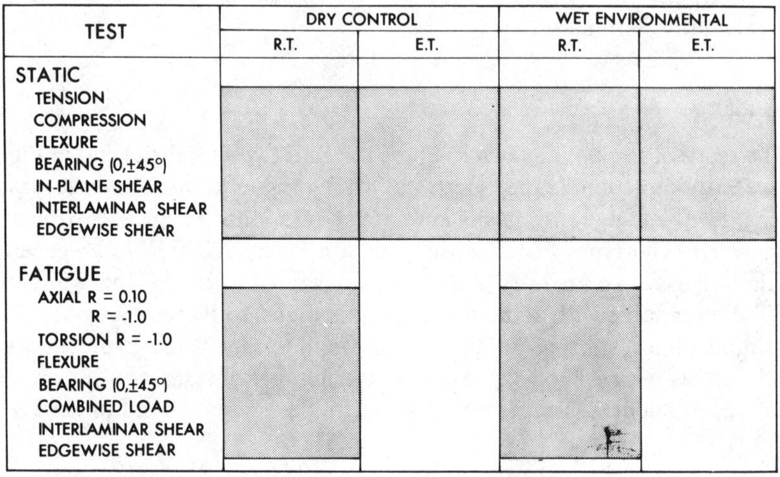

FIG. 6—*Graphite/epoxy data base test plan.*

conditions. Brooming failures occurred with other methods such as the ASTM Test for Compressive Properties of Rigid Plastics (D695-77) even when using soft metal end pads. The Grumman-modified Illinois Institute of Technology Research Institute (IITRI) fixture appeared the most successful and economical method.

In-plane shear of the ±45 Kevlar/epoxy laminate was determined by a combination of test and analysis. The pure shear state in a ±45 laminate was rotated to an equivalent biaxial tension-compression field in a 0/90 laminate. Application of the Tsai-Hill failure criterion resulted in a predicted value of

TABLE 1—*Test specimens.*

Property	Graphite/Epoxy, 0 deg	Kevlar/Epoxy	
		0/90 deg	±45 deg
Static			
F_{tu}, E_t	ASTM 3039	AMS 3903	AMS 3903
F_{cu}, E_c	IITRI[a]	ASTM D 695	ASTM D 695
F_{bu}, F_b	ASTM 790	ASTM 790	ASTM 790
F_{su}, G	ASTM[b]	ASTM[b]	Sikorsky
F_{siu}	ASTM D 2344	ASTM D 2344	...
Fatigue			
axial, $R = 0.1$	ASTM 3039	ASTM 3039	ASTM 3039
$R = -1$	ASTM 3039[c]	ASTM 3039[c]	ASTM 3039[c]

[a]Grumman-modified fixture.
[b]Proposed ASTM procedure.
[c]Sikorsky-modified to prevent instability.

ultimate shear. This predicted value was confirmed by a limited number of torque tube tests, as shown in Table 2. Actual design allowables were derived from the more extensive 0/90 test data.

Finite-element (NASTRAN) studies were made of the short-beam shear (SBS) specimen configuration. Results of a 2D-analysis show that a span-to-depth ratio of 4 is the minimum acceptable for typical G/E systems to maintain the parabolic shear stress distribution assumed in stress calculations (beam theory). The curve in Fig. 7 shows the factor to be applied to correct for span-to-depth ratio and width effect. Tests were conducted for several span-to-depth ratios to confirm the correction factor curve.

TABLE 2—*Comparison of torque tube and stress theory for in-plane shear ±45 K/E.*

Property	Ratio of Stress Theory[a] to Torque Tube Results	
	RTD	82°C (180°F) Dry
F_{su}	1.03	0.81
G	1.17	1.03

$$^a F_{su}(\pm 45) = \frac{F_{tu}(0/90)}{\sqrt{2 + \left(\dfrac{F_{tu}(0/90)}{F_{cu}(0/90)}\right)^2}}$$

$$^a G(\pm 45) = \frac{E(0/90)}{2\left[1 + \nu(0/90)\right]}$$

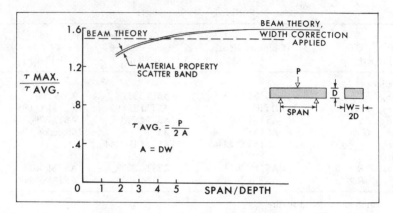

FIG. 7—*Effect of span-to-depth ratio on peak shear stress.*

Test Results

Screening tests were conducted to verify linear interpolation trends. Presented in Fig. 8 are some typical trends found by varying moisture level and temperature for static strength and moisture level for fatigue strength. Within the ranges tested it was found that strength varied linearly with moisture content. For a given moisture level, the strength reduction was almost linear with temperature, with linear interpolation being slightly conservative. The only variation was for K/E compression 0/90 (static) where the wet strength was higher than the dry strength, and this may be due to fiber swelling increasing the compression stability.

In general the data showed more environmental effects for the matrix mode than the fiber mode and K/E was more sensitive than G/E. Figure 9 illustrates the static design allowables as a percentage of the RTD mean

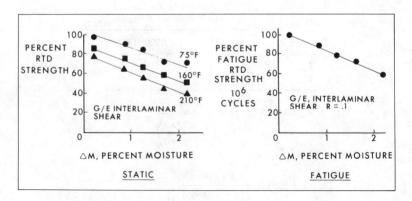

FIG. 8—*Screening tests confirm linear interpolation method.*

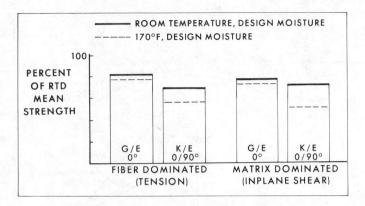

FIG. 9—*Design allowables, percent of RTD mean strength.*

strength and includes the statistical reductions required to obtain "*A*" design values.

For key design properties the sample size was generally about 30. The statistics were checked against some larger samples available from quality-assurance (QA) tests. The end result was that the normal distribution was selected for establishing static design allowables. A more detailed discussion of the statistical aspects is presented in the Appendix.

The fatigue test data generally were obtained by testing at several constant-load levels. The test points were then used to obtain a best fit to a curve shape, and to obtain the standard deviation. For design the curve is then reduced by three standard deviations. However, current FAA regulations (for metals) require not less than a 20 percent reduction for certification, and that value was used if it was greater than 3σ. A further description of the statistical analysis used in fatigue is contained in the Appendix. In general the shape curves were very shallow for all modes of G/E and for K/E (0/90, $R = -1.0$) and steepest for K/E (0/90) axial $R = 0.1$. Design strengths were interpolated to the design moisture content from RTD and RTW data.

The axial fatigue data have been combined in the constant-life diagram presented in Fig. 10. The data for G/E (0 deg) and K/E (0/90) have been plotted after normalization with respect to the static design allowable. Aluminum alloy 2024 is also presented for reference. The constant-life diagram (10^7 cycles illustrated) was created from four data points: static tension, static compression, axial fatigue at $R = 0.1$ (extrapolated to $R = 0$), and axial fatigue at $R = -1.0$. The data points have been linearly combined, which other sources [5-7] indicate may be a conservative measure.

The characteristic shape of the composite constant-life diagram indicates a peak in fatigue capability near the $R = 0$ range. There is a marked reduction in fatigue capability in the tension-compression and compression-compres-

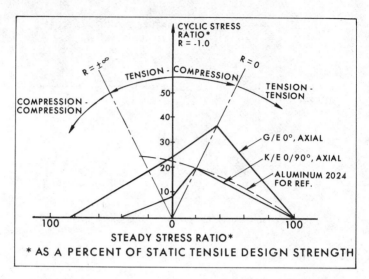

FIG. 10—*Comparison of design fatigue strength at 10^7 cycles.*

sion ranges. The governing fatigue criterion for composites appears to be in a compression range unlike that for metal alloys.

Conclusions

The development of design allowables for composite structures is complicated by the large variety of available fibers and resin systems and the possible combinations of lamina making up a laminate. There is a significant effect of environment, that is, the combination of temperature and moisture that further increases the scope of testing required. The specific conclusions of this paper follow.

1. The lamina level design allowables is the most economical method to establish the basic data for design of composite structures and permits optimization for the laminate.

2. The method of interpolation of design lamina properties, that is for temperature and moisture level, permits the establishing of design allowables subject to the local conditions of the composite structure. The local conditions may well vary throughout the helicopter structure and may be subject to change from future assessments of environmental effects obtained from in-service experience.

3. The use of relatively small sample sizes (in the order of 20) using a simple normal distribution appears to be sufficiently accurate to establish static design allowables. It should be recognized that the accuracy is well within the limits of linear laminate analysis for the composite structure.

4. To establish fatigue strength, Goodman diagrams require testing at various R ratios. The practice of testing at $R = 0.1$ (tension-tension) and extrapolating (as often done for metals) is not applicable to composites. The fatigue results for the composites tested in this paper show a strength reduction for $R = -1.0$ (tension-compression).

5. There is urgent need for an industry and government agency to mutually accept design allowables data bank document for composite materials. Such a document would require sponsorship to enable thorough reviews and agreements to be made on the published data.

APPENDIX

Statistical Analysis for Static Strength

The guidelines of Ref 1 for design static allowables were followed, that is, "A" design values, 99 percent survivability, and 95 percent confidence. The problem is to establish distribution and to account for sample size. For FAA certification it was agreed to use a 30 sample size for key structural properties.

After normalizing the coupon data it was first assumed that a normal distribution would apply and this was tested by means of the Kolmogorov statistic. Section 9.6.1.1 of Ref 1 cites the Kolmogorov test as an acceptable method for determining distribution. For a given sample set, the sample distribution is calculated by r/n, r being the ordered rank and n the total number of data points. The resulting cumulative sample distribution and normal distribution are compared. The Kolmogorov statistic D_N is the maximum difference between the sample and assumed normal distribution. The null hypothesis, that the population is normally distributed, is rejected if D_N exceeds a critical value. A complete description of the method is found in Ref 9. The critical value is a function of sample size and desired confidence level and is tabulated in Ref 10. A level of significance of 0.05 was used in accordance with Ref 1, Section 9.2.6.

The Kolomogorov test was performed on all data sets in the test program and all passed the normal distribution test. Larger-sample tests were obtained from QA tests of bending and interlaminar strengths (111 and 115 data points, respectively). A plot of the bending strength distribution is shown in Fig. 11 and illustrates the good fit to normal distribution.

In addition, a two-parameter Weibull distribution was tested on the larger sample size and no significant difference was found in the Kolmogorov D_N-value or the resulting "A"-values as determined from using a normal distribution. Smaller sample sizes of all other properties were evaluated using the two-parameter Weibull distribution and again no significant difference in "A"-values was found from those determined by the normal distribution method. The conclusion reached was that the simpler normal distribution would be used for design.

The design "A" allowables using a 30-sample size and the larger sample sizes available for G/E bending and interlaminar shear were for all practical purposes the same. However, the question still arises as to how small the sample size can be and still result in valid answers.

A study was conducted to assess the effect of sample size on design allowables. The largest data set available was used. The sample statistics (mean, standard deviation, "A"-value) were calculated for the first n data points of the set, where n was varied

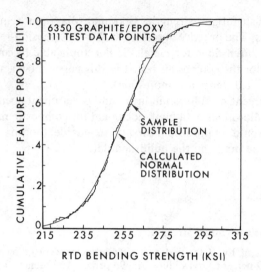

FIG. 11—*Normal distribution fits test data.*

from 2 to the total number of data points in the set. This process was repeated many times for the randomly reordered data set. In this way, the ordering effects of the particular data set were eliminated.

The results of the study are shown in Fig. 12, where the mean value (normalized to the infinite population value) and scatter are plotted as a function of sample size. As expected, the scatter of the "*A*"-value is very high for small sample size due to the influence of the first few points taken. The mean "*A*"-value is very low because of the large statistical reduction required. As the sample size increases, however, the mean "*A*"-value stabilizes and the scatter continues to decrease. Based on this study, it

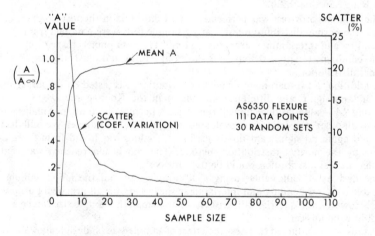

FIG. 12—*Effect of sample size on design "A"-value.*

would appear that a reasonable sample size is in the neighborhood of 20 specimens. It should be noted that this value is about the same as suggested in Ref 5, although the latter study was done for fatigue.

Statistical Analysis for Fatigue Strength

The region of interest for helicopter structures is generally from 10 000 cycles to a very high cycle range. Therefore the statistical reductions are in strength rather than life. The fatigue test data are reduced to a mean shape curve of the form

$$\frac{\overline{S}}{S_E} = 1 + \frac{\alpha}{N} \gamma \tag{1}$$

where

$\overline{S}$ = mean fatigue strength, ksi,
S_E = mean endurance fatigue strength, ksi,
N = millions of cycles, and
α, γ = shape parameters.

This shape form has been used by other companies and was first proposed by Weibull and used by Henry [12] and by Lariviere [13].

For any set of shape parameters the mean endurance strength is calculated from the n fatigue data points (S_i, N_i) by

$$\overline{S}_E = \frac{1}{n} \sum_{i=1}^{n} \frac{S_i}{1 + \dfrac{\alpha}{N_i} \gamma} \tag{2}$$

and the standard deviation and coefficient of variation are computed. Normalized data were pooled to find shape curves common to several data groups and to increase

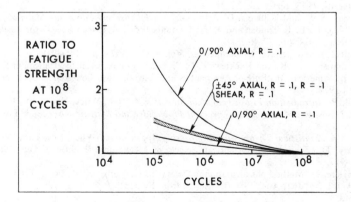

FIG. 13—*Kevlar/epoxy fatigue shape curve.*

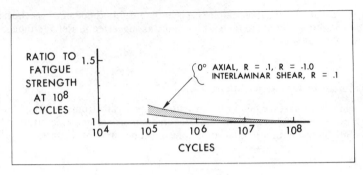

FIG. 14—*Graphite/epoxy fatigue shape curve.*

confidence in the derived shapes. Typical shape curves are shown in Fig. 13 for K/E and in Fig. 14 for G/E. These curves were normalized to strength at 10^8 cycles.

The design fatigue strength is then derived by

$$S = \bar{S} \left(1 - \frac{3\sigma_v}{100} \right) \tag{3}$$

where σ_v is the coefficient of variation, in percent.

References

[1] "Military Standardization Handbook Metallic Materials and Elements for Aerospace Vehicle Structures." MIL-HDBK-5.

[2] "Advanced Composites Design Guide," U.S. Air Force.

[3] "Plastics for Aerospace Vehicles Part 1, Reinforced Plastics." MIL-HDBK-17A.

[4] Pride, R. "Environmental Effects on Composites for Aircraft," presented at CTOL Transport Technology Conference, March 1978.

[5] Ryder, J. T. and Walker, E. K. in *Fatigue of Filamentary Composite Materials, ASTM STP 636,* K. L. Reifsnider and K. N. Lauraitis, Eds., American Society for Testing and Materials, 1977, pp. 3-26.

[6] Ramani, S. V. and Williams, D. P. in *Fatigue of Filamentary Composite Materials, ASTM STP 663,* K. L. Reifsnider and K. N. Lauraitis, Eds., American Society for Testing and Materials, 1974, pp. 27-46.

[7] Schutz, D. and Geharz, J., *Composites,* Vol. 8, Oct. 1977, pp. 245-250.

[8] Kollmansberger, R. and Rackiewicz, J., "Kevlar Composites in the Helicopter Environment," American Institute of Aeronautics and Astronautics, Conference, St. Louis, Mo., April 1979.

[9] Hoel, P., *Introduction to Mathematical Statistics,* 4th ed., Wiley, New York, 1971.

[10] Von Mises, R., *Mathematical Theory of Probability and Statistics,* Academic Press, New York, 1964.

[11] Hald, A., *Statistical Theory with Engineering Applications,* Wiley, New York, 1952.

[12] Henry, D., *Transactions,* American Society of Mechanical Engineers, Aug. 1955, pp. 913-917.

[13] Lariviere, J. "Method of Calculation to Determine Helicopter Blade Life," Technical Note STA/UT No. 165, Aug, 1956 (translated from French).

D. W. Wilson,[1] *R. B. Pipes,*[1] *D. Riegner,*[2] *and J. Webster*[1]

Mechanical Characterization of PMR-15 Graphite/Polyimide Bolted Joints

REFERENCE: Wilson, D. W., Pipes, R. B., Riegner, D., and Webster, J., **"Mechanical Characterization of PMR-15 Graphite-Polyimide Bolted Joints,"** *Test Methods and Design Allowables for Fibrous Composites, ASTM STP 734,* C. C. Chamis, Ed., American Society for Testing and Materials, 1981, pp. 195-207.

ABSTRACT: Design-allowable data characterizing the static strength and viscoelastic performance of PMR-15/Celion 6000 graphite/polyimide composite bolted joints are presented. Two laminate configurations in the $[0_i/\pm45_j/90_k]_{ns}$ family, one-fiber-dominated and one matrix-dominated, were studied. Temperature-dependent strength characteristics were measured over the 21 to 315°C temperature range for several joint geometries. Effects of strain rate on strength and failure mode behavior were assessed at room temperature for rates ranging between 0.002 and 1.00 s^{-1}. Creep behavior was examined for the case of bearing deformations at temperatures of 21 and 177°C. Ultrasonic inspection techniques were used to assure material integrity and to verify post test damage and failure mode.

KEY WORDS: graphite/polyimide, composite material, bolted joint, ultrasonic inspection, creep, strain rate sensitivity

Material anisotropy and inhomogeneity cause mechanically fastened joints to exhibit complex strength and failure mode behavior. Laminate configuration, material system, and joint geometry significantly affect bolted joint efficiency. This complex interaction of material parameters with the bolted joint geometry necessitates the development of design-allowable data bases for each different material system and laminate configuration.

While extensive studies have characterized the strength and failure mode [1-4][3] behavior and environmental sensitivity [5,6] of epoxy matrix

---

[1]Research associate II, director, and graduate student, respectively, Center for Composite Materials, Department of Mechanical and Aerospace Engineering, University of Delaware, Newark, Del. 19711.

[2]Project engineer, General Motors Manufacturing Development, Detroit, Mich.

[3]The italic numbers in brackets refer to the list of references appended to this paper.

composites, similar design-allowable data are not available for the graphite/polyimide system. Current design methods, whether modified strength of materials approaches or sophisticated numerical solutions [7-10], require design data bases. Design allowables must therefore be developed for bolted joints fabricated from graphite/polyimide composites. Of primary importance is a quantitative understanding of bolted joint strength and failure mode behavior as related to temperature, load rate, load duration, joint geometry, and laminate configuration. This paper presents data describing the elevated-temperature (315°C) performance of Celion 6000/PMR-15 graphite/polyimide composite bolted joints. The results include effects of the aforementioned parameters on strength and failure mode. The macromechanical behavior characterized for composite bolted joints in the PMR-15 material system is compared with similar data generated for other materials and then used in the development of analytical strength models.

Procedure

The test specimens were fabricated from a single Celion 6000/PMR-15 panel of the $[45/0_2/-45/0_2/45/0_2/-45/90]_s$ laminate configuration processed by the National Aeronautics and Space Administration (NASA) Langley Research Center to eliminate processing as a test variable. Two laminate configurations were obtained by sectioning the panel into subpanels oriented parallel and perpendicular to the laminate principal direction. Subpanels oriented parallel to the laminate principal direction, defined as Laminate I, result in fiber-dominated laminates with a $[45/0_2/-45/0_2/45/0_2/-45/90]_s$ orientation. The perpendicular subpanels, defined as Laminate II, possess a $[-45/90_2/45/90_2/-45/90_2/45/0]_s$ matrix-controlled orientation. The test specimen geometry and dimensions are illustrated in Fig. 1.

To eliminate machining-induced damage, each test coupon was machined to size with a diamond-impregnated circular saw. Beveled end tabs were applied for load introduction on the 21 and 177°C tests while the 315°C coupons were not tabbed. After fabrication the test coupons were ultrasonically inspected to detect any damage introduced by the fabrication process.

A specially designed clevis fixture (Fig. 2) was used to simulate bolted load reaction through the hole in the coupon while standard friction grips introduced load at the tabbed end of the specimen. Annular disks with a 6.5-mm^2 contact area formed the constraining surfaces to simulate torqued fastener conditions, and high-strength tool-steel dowels were used as the bolt shaft. Various-size shafts were available to insure excellent fit with the machined hole.

The testing procedure was similar to that used for standard static strength characterization. Loading was applied by a model TTC Instron fitted with an

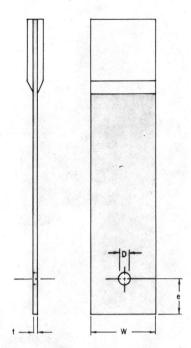

TYPICAL DIMENSIONS

w mm(in.)	e mm(in.)	D mm(in.)	t mm(in.)
38.10 / (1.50)	9.53 (.375)	5.08 (.200)	3.65 (.144)
38.10 / (1.50)	9.53 (.375)	5.08 (.200)	3.65 (.144)
19.05 / (.75)	19.05 (.750)	5.08 (.200)	3.65 (.144)
19.05 / (.75)	19.05 (.750)	5.08 (.200)	3.65 (.144)

FIG. 1—*Specimen geometry.*

Applied Test Systems environmental chamber. Three replicates of each specimen geometry were tested at each of the test temperatures for the two laminate configurations—totaling 72 tests. Failure load was defined as the maximum load attained as indicated by the load/deflection curve.

Strain-rate sensitivity tests were conducted at 21°C in an Instron Model 1322 high-rate servohydraulic test machine. The specimen geometry was held constant while strain rates of 0.002, 0.01, 0.10, and 1.00 s^{-1} were imposed. Load and deflection versus time data were recorded on a dual-trace storage oscilloscope.

Bearing creep response was measured by applying a constant load and monitoring deflection of bearing surface with a specially designed exten-

FIG. 2—*Bolted joint test fixture.*

someter. Constant-load conditions were achieved by operating an Instron Model 1321 servohydraulic machine in load control. The Applied Test Systems environmental chamber was used for temperature control. Specimens of Laminate II with $e/D = 4$ and $W/D = 4$ were loaded to 50 percent of the average static strength, and deflection data were recorded at known time intervals over the 8-h duration of the test. Data points were taken at 5-min intervals for the first 15 min, at 15-min intervals for the next 45 min, and at 1-h intervals thereafter.

Results

Material Integrity

The glass transition temperature (T_g) was determined, by a du Pont Thermal Mechanical Analyzer (TMA), to be $>340°C$, indicating that the material was adequately cured. Microstructural verification of the stacking sequence resulted in the discovery of extensive microcracking in the laminate structure. This cracking, shown in Fig. 3, is the result of thermal residual

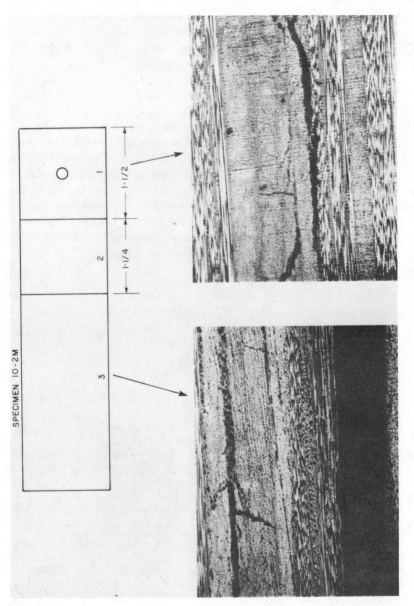

FIG. 3—*Microstructure of PMR-15 laminate.*

stresses introduced into the laminate during processing. Some large delaminations were also detected (Fig. 3) in coupons fabricated from random areas of the panel. The exact cause of this phenomenon was not determined. It is known that the delaminations were not caused by machining operations directly as verified in postmachining ultrasonic inspection. The impact of the microcracking on the bolted joint strength results was minimized by censoring the defective specimens from the test program.

Elevated-Temperature Strength

The bolted joint strength of the PMR-15 system was evaluated at 21, 177, and 315°C. The test coupon geometries were designed to provide strength data for bearing, shearout and net tension failure modes for both the fiber and matrix-dominated laminates. Ultimate strength results as a function of temperature are presented in Figs. 4–6 for each of the three failure modes. Laminates I and II exhibited distinctly different bearing failure behavior over the test temperature range as seen in Fig. 4. Ultimate bearing strength for Laminate II decreases linearly with increasing temperature from 640 MPa at 21°C to 400 MPa at 315°C, a degradation in strength of 37 percent.

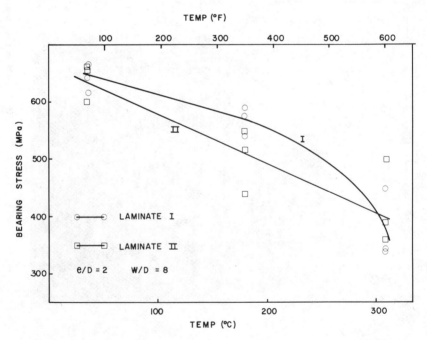

FIG. 4—*Ultimate bearing strength versus temperature for Laminates I and II.*

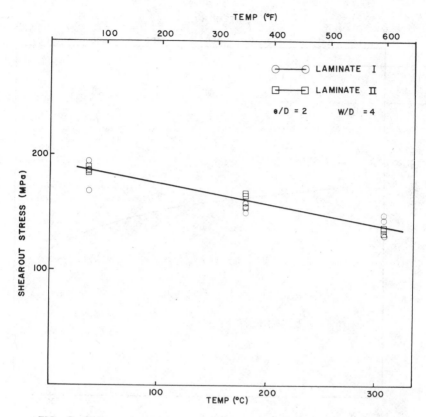

FIG. 5—*Ultimate shearout strength versus temperature for Laminates I and II.*

Laminate I showed a lesser decrease in strength over the 21 to 200°C range and then developed a nonlinear degradation in the high-temperature region for an effective 41 percent reduction in bearing strength over the total excursion.

Shearout strength behavior over the test temperature range was very similar for both laminate types. Ultimate shearout strength decreased linearly from 190 MPa at 21°C to 140 MPa at 315°C for a net reduction of 26 percent in strength (Fig. 5).

Net tension failures were exhibited only by Laminate II, the 90-deg-dominated laminate (Fig. 6). A 31 percent reduction in ultimate net tension strength was observed for the 21 to 315°C temperature excursion. The temperature-dependent strength behavior appeared to be independent of edge distance (e/D), and slightly nonlinear in the higher-temperature region (>200°C).

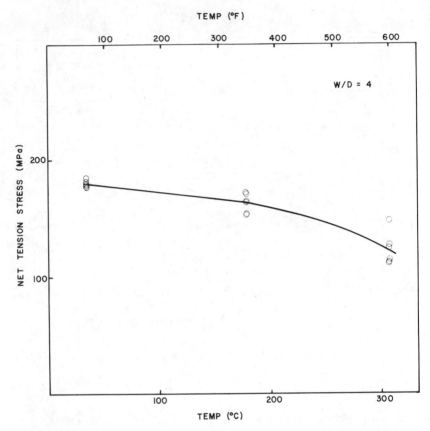

FIG. 6—*Ultimate net tension strength versus temperature for Laminate II.*

Effect of Edge Distance on Strength

The effect of edge distance on bolted joint strength behavior can be visualized by analyzing failure load as a function of the nondimensional parameter, *e/D*. Separate results are presented in Figs. 7-9 for each laminate and specimen width. For Laminate I (Fig. 7) with a *W/D* of 3.71, sharp increases in ultimate failure load were recorded with increasing e/D for the 21 and 177°C test temperatures. A lesser strength increase with *e/D* was observed at 315°C, presumably due to the elevated-temperature effects. A failure-mode shift from shearout to bearing accompanied this increased load-carrying capacity with increasing *e/D* while temperature did not affect failure-mode behavior. Similar increased load-carrying capacity with increasing e/D was observed for Laminate II with a *W/D* of 7.42. The failure mode changed from shearout to bearing with increasing *e/D* while remaining independent of temperature variation (Fig. 8). Laminate II tests where *W/D*

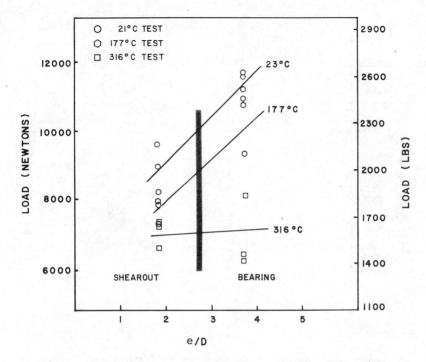

FIG. 7—*Failure load versus* e/D; *0-deg-dominated laminate;* W/D = 3.71.

was 3.71 resulted in only net tension failures. As shown in Fig. 9, failure load increased with increasing *e/D,* but not as sharply as for the shearout and bearing cases.

Strain-Rate Sensitivity

Ultimate load-carrying capacity was determined for strain rates of 0.0022 (static), 0.01, 0.10, and 1.00 s^{-1}. Combined failure modes occurred for all strain rates above 0.0022 s^{-1} which exhibited a bearing failure. Thus the results in Fig. 10 are expressed as the bearing strength nondimensionalized with respect to the "static" bearing strength as a function of strain rate. The results indicate that both laminates exhibit strain-rate sensitivity in the form of increased strength with increasing strain rate. Laminate II, the matrix-controlled laminate, displays a sharp monotonically increasing strength behavior for rates above 5×10^{-3} s^{-1} while remaining relatively insensitive at slower rates. The fiber-controlled laminate exhibited a very slight sensitivity, 3.66×10^{-4} s^{-1}. Similar analysis of the results in terms of net tension and shearout failure modes yielded the same results, since combined failures occurred for all the specimens.

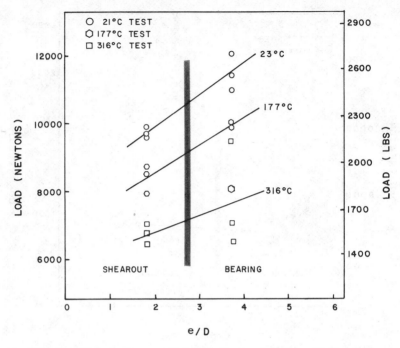

FIG. 8—*Failure load versus* e/D; *90-deg-dominated laminate;* W/D = 7.42.

Bearing Creep Sensitivity

Results from this study are preliminary in nature since the characterization was carried out only for short time duration. At 50 percent of the ultimate bearing load-carrying capacity, bearing creep became significant at the 177°C temperature (see Fig. 11). The total deformation at 177°C after 8 h was equivalent to a 5 percent elongation of the hole diameter.

Discussion

Several of the results from this work have been helpful in gaining a basic understanding of composite bolted joint failure mechanisms. This insight can be used in the development of analytical models to predict composite bolted joint strengths for each of the three basic failure modes.

The existence of two distinct failure mechanisms may be inferred from the different behavior exhibited for bearing strength as a function of temperature for the two laminate types. The relatively linear behavior of the matrix-controlled laminate suggests a failure mechanism similar to local plastic yielding in metals, while the nonlinear developments in the fiber-controlled laminate behavior are indicative of an instability phenomenon. Further experimentation is necessary to distinguish the nature of these bear-

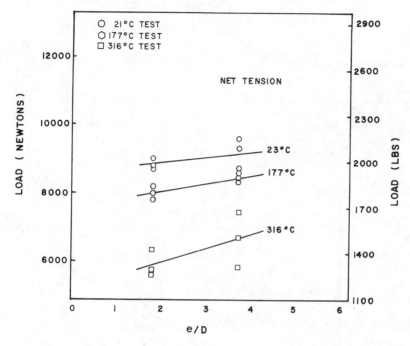

FIG. 9—*Failure load versus* e/D; *90-deg-dominated laminate;* W/D = 3.71.

ing failure mechanisms and relate that to laminate configuration, and joint geometry.

The shearout strength versus temperature results exhibited no difference between the strengths of the two laminate types. The significant implication of this result is that the ±45-deg fibers carry the shearing load in each case. That the shearout strength exhibits the lowest strength dependence on temperature supports this conclusion. Thus the shearout strength is primarily a function of the number of ±45-deg plies in the laminate.

Conclusions

Bearing, shearout, and net tension strength behaviors have been characterized as a function of temperature, geometry, and laminate configuration for the Celion 6000/PMR-15 material system. The results implicate two distinct mechanisms responsible for bearing failure which are related to laminate configuration. It was also found that primarily the ±45-deg plies carry the shearing load in composite bolted joints. The material was seen to exhibit strain-rate sensitivity which varied with laminate configuration. The matrix-controlled laminate showed the most significant strain-rate sensitivity and also displayed a significant bearing creep response at elevated temperature. Though preliminary in nature, the creep results indicate that the creep behavior in PMR-15 composite joints warrants further study.

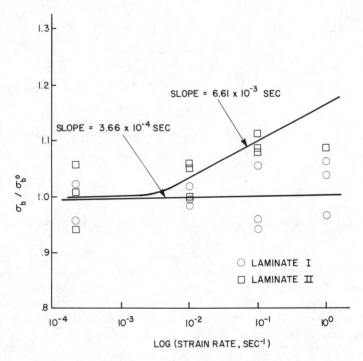

FIG. 10—*Strain-rate sensitivity of ultimate bearing strength.*

Acknowledgment

The research reported herein was supported under NASA grant NSG-1409; Dr. Paul Cooper, project monitor.

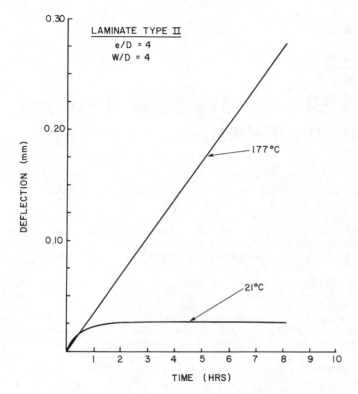

FIG. 11—*Bolted joint creep response.*

References

[1] Ashton, J. E., Burdouf, M. L., and Olson, F. O., "Design Analysis and Testing Advanced Composite F-111 Fuselage," Convair/General Dynamics, SMD-059, Fort Worth, Tex., 20 April 1971.

[2] Hart-Smith, L. J., "Bolted Joints in Graphite-Epoxy Composites," Douglas Aircraft Co., NASA CR-144899, Long Beach, Calif., June 1976.

[3] Padawer, G. E. in *Composite Materials: Testing and Design (Second Conference). ASTM STP 497,* American Society for Testing and Materials, 1972, pp. 396–414.

[4] Van Siclen, R. C. in *Proceedings,* Army Symposium on Solid Mechanics, AMMRC MS 74-8, Bass River, Mass., Sept. 1974.

[5] Kim, R. Y. and Whitney, J. M., *Journal of Composite Materials,* Vol. 10, April 1976, p. 149.

[6] Wilkins, D. J. in *Composite Materials: Testing and Design (Fourth Conference). ASTM STP 617,* American Society for Testing and Materials, 1977, pp. 497–513.

[7] deJong, Theo, *Journal of Composite Materials,* Vol. 11, July 1977, p. 313.

[8] Waszczak, J. P. and Cruse, T. A., "A Synthesis Procedure for Mechanically Fastened Joints in Advanced Composite Materials," AIAA/ASME/SAE 14th Structures, Structural Dynamics, and Materials Conference, Williamsburg, Va., March 1973.

[9] Waszczak, J. P. and Cruse, T. A., *Journal of Composite Materials,* Vol. 5, July 1971, p. 421.

[10] Harris, H. G. and Ojalvo, I. U. in *Proceedings,* Army Symposium on Solid Mechanics, AMMRC MS 74-8, Bass River, Mass., Sept. 1974.

J. A. Suarez[1]

Cost-Effective Mechanical Property Characterization

REFERENCE: Suarez, J. A., **"Cost-Effective Mechanical Property Characterization,"** *Test Methods and Design Allowables for Fibrous Composites, ASTM STP 734,* C. C. Chamis, Ed., American Society for Testing and Materials, 1981, pp. 208–228.

ABSTRACT: This paper describes current mechanical property characterization practices for advanced composites in automotive and primarily stiffness-critical aircraft secondary structure applications. The use of extensive costly test programs to develop statistically valid static and fatigue design allowables is avoided. Use is made of small property-characterization programs that make maximum use of cost-effective coupon-type specimens and that are tailored to the special application at hand.

KEY WORDS: composite materials, secondary structures, aircraft, automotive, hybrids, test methods, cost

Up to now, a very low-risk development attitude has inhibited a more rapid application of advanced composites in the aerospace and automotive industries. The application of advanced composite materials has been preceded by costly design allowables test programs. This attitude has become less tolerable today when the emphasis is no longer on unitary composite structure, but on mixed fiber or hybrid structure, which is fabricated using a large variety of processes (Table 1). Today more than ever, the composites design engineer is designing both the structure and the material. He must generate design values for the various, sometimes exotic material and process combinations available to him. He must do this if he is to design the lowest weight or lowest cost structure or both.

Costly design allowables programs must be avoided in order to achieve large increases in the application of composites, which utilize innovative techniques in processing and fabrication. A possible alternative is the use of small property-characterization programs which are tailored to the special application at hand, and which are sufficient in most automotive and primarily stiffness-critical secondary aircraft structure design. Secondary

---

[1]Group leader, Structural Mechanics Section, Grumman Aerospace Corp., Bethpage, N.Y. 11714.

TABLE 1—*Composite materials and processes.*

• Resin System	• Fiber Glass Reinforcement Form
—polyimide	—tape
—epoxy	—cloth
—polyester	—mat
—acrylic	—knitted
—polypropylene	
—nylon	
• Process	• Special High-Performance
—autoclave cure	Reinforcement
—compression molding	—boron
—vacuum bag cure	—graphite
—injection molding	—Kevlar
—thermoplastic stamping	
—vacuum forming	
• Special Materials/Processes	
—RIM (reaction injection molding)	
—ERM (elastic reservoir molding)	
—VRIM (vacuum reaction injection molding)	
—in mold coating	

structure in aircraft is generally not a safety of flight component. It usually has high safety margins, is designed to minimum gage considerations for appearance, durability, producibility, or all three, and is proof-loaded.

Test Methods

The problem of obtaining meaningful test data in a cost-effective manner is still an unresolved issue. Design engineers object to the use of small-size test specimens, since they bear little or no relationship to the size, thickness, loading, and failure mechanisms actually experienced by the complete structure. For example, design engineers have had serious doubts about the structural validity of both flexural and short-beam horizontal shear tests (Fig. 1), even though these tests are accepted material engineering standards for development and qualification of composites.

Another example is the design engineer's preference for sandwich beam over coupon-type compression tests. Because of premature buckling and end-load introduction failures, compression testing of advanced composites is not yet well standardized. Costly honeycomb sandwich beams (Fig. 2), unsupported end-loaded specimens, and short, fixture-supported, end-load specimens have been used with varying results. Grumman is currently successfully using a modified Illinois Institute of Technology Research Institute (IITRI) fixture (Fig. 3).[2] The compression coupon specimen (Fig. 4) is fiber

[2]Whiteside, J. B. and Wolkowitz, W., "Enviromental Sensitivity of Advanced Composites," Fourth Quarterly Progress Report, Air Force Contract F33615-76-C-5324, Grumman Aerospace Corp., Bethpage, N.Y., Sept. 1977.

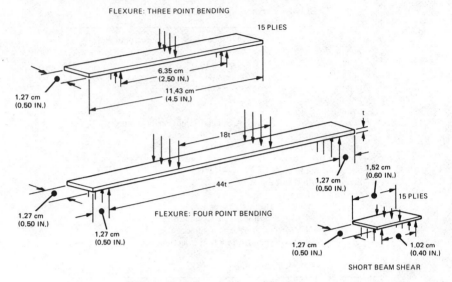

FIG. 1—*Typical materials qualification tests.*

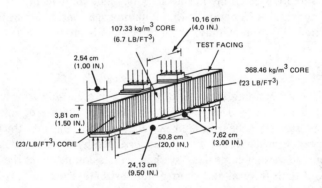

FIG. 2—*Compression sandwich beam specimen and test fixture.*

glass tab ended. Load introduction is accomplished by a gripping arrangement similar to that used in tension testing, wherein load is introduced by shear rather than end-loading.

Our current approach to generating cost-effective design values is to maximize the use of coupon-type specimens for mechanical property characterization. Tension tests are preformed using the standard Grumman tension coupon, with fiber glass end tabs for load introduction by shear (Fig. 5). In-plane shear moduli are obtained from ±45-deg tension coupons (Fig. 5),[3]

---

[3]Rosen, B.W., *Journal of Composite Materials,* Vol. 6, 1972, pp. 552–554.

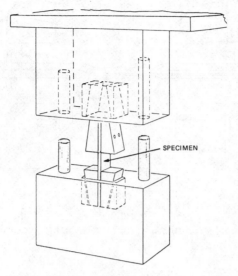

FIG. 3—*Modified IITRI compression test fixture.*

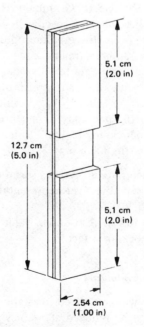

FIG. 4—*Compression coupon.*

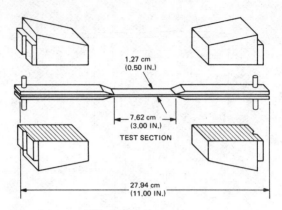

FIG. 5—*Tension coupon.*

and in-plane shear strengths are obtained using the rail shear method (Fig. 6).

Aerospace Application

The trailing edge of the EF-111A horizontal stabilizer (Fig. 7) was selected to demonstrate constrained vacuum-pressure oven-processing techniques for single-cure assemblies. The USAF/Grumman EF-111A is the latest Air Force tactical aircraft to be dedicated specifically to electronic warfare.

When vacuum-pressure oven-cure graphite/epoxy was first considered for the EF-111A horizontal stabilizer trailing edge (Fig. 8), a review showed insufficient mechanical property data to be available. The available industry data looked promising, but did not constitute a sufficient base for mechanical property characterization. As a result, a limited test program was initiated to establish the necessary design properties, and at the same time demonstrate the producibility of the selected fabrication method. The basic program consisted of the following:

1. Establishment of material and process specifications.

2. Selection of a conservative, yet representative, long-term moisture-absorption criterion.

3. Coupon static tests to establish design values with environmental and vacuum-pressure-cure knockdown factors.

Mechanical Property Determination

A limited test program was conducted to establish material acceptance requirements, and to verify the reduced allowables used in the design of the composite trailing edge.

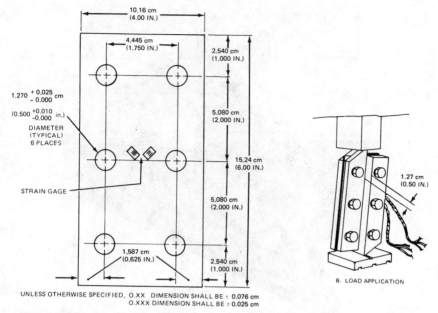

UNLESS OTHERWISE SPECIFIED, O.XX DIMENSION SHALL BE ± 0.076 cm
O.XXX DIMENSION SHALL BE ± 0.025 cm

NOTES:

1. FIBER ORIENTATION TO BE IDENTIFIED ON SPECIMEN.
2. SPECIMEN EDGES TO BE PARALLEL TO ± 0.013 cm

A. SPECIMEN CONFIGURATION

FIG. 6—*Test configuration for in-plane shear tests using rail shear procedure.*

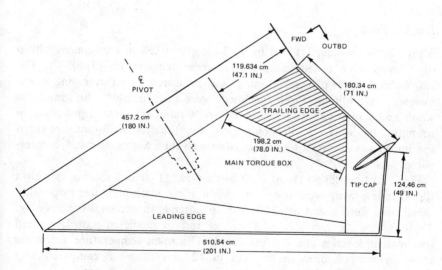

FIG. 7—*EF-111A horizontal stabilizer trailing edge (left hand).*

FIG. 8—*EF-111A graphite/epoxy horizontal stabilizer trailing edge.*

Tension Tests

Five 27.94-cm-long (11 in.) by 1.27-cm-wide (0.50 in.) specimens with a (0_{8T}) layup were tested in tension at room temperature (Table 2). The longitudinal strain of the specimens was measured by an averaging extensometer. In addition, three specimens were instrumented with transverse strain gages in order to generate Poisson's ratio data. No degradation in strength, modulus, or Poisson's ratio for these vacuum-pressure-cured specimens was observed when compared with their autoclave-cured counterparts.

Five 27.94-cm-long (11 in) by 2.54-cm-wide (1.00 in.) specimens with a $(0_4/90_2\pm45_3)$ layup were tested in tension at room temperature; three additional ones were tested at 431 K (317°F) after environmental conditioning. Table 3 lists the test results. Ratios of test to predicted autoclave-cured specimen properties are also presented. The room temperature specimens show an average drop in strength of 12 percent when compared with autoclave-cured specimens.

TABLE 2—*Test tensile strength and modulus for vacuum-pressure-cured 3501-5A/AS-1 graphite/epoxy laminates (0_{8T}).*

Specimen No.	Thickness, in.	Width, in.	Test Temperature, °F	$F^{tu\,a}$ test, ksi	F^{tu} autoclave, ksi	$\dfrac{F^{tu}\text{ test}}{F^{tu}\text{ autoclave}}$	$E^{t\,a}$ test, msi	E^{t} autoclave, msi	$\dfrac{E^{t}\text{ test}}{E^{t}\text{ autoclave}}$	$\nu_{xy\,\text{test}}$	$\nu_{xy\,\text{autoclave}}$	$\dfrac{\nu_{\text{test}}}{\nu_{\text{autoclave}}}$
100-1	0.044	0.500	RT	216.2	211.3	1.023	19.0	18.5	1.027	0.280	0.25	1.120
100-2	0.042	0.488	RT	225.9	211.3	1.069	20.0	18.5	1.081	0.246	0.25	1.984
100-3	0.043	0.497	RT	231.4	211.3	1.095	19.1	18.5	1.032	0.274	0.25	1.096
100-4	0.044	0.497	RT	175.3	211.3	0.830	18.6	18.5	1.005	...	0.25	...
100-5	0.043	0.505	RT	220.2	211.3	1.042	19.0	18.5	1.027	...	0.25	...
Avg				213.8		1.012	19.1		1.032	0.267		1.068

^{a}All test results normalized for an average layer thickness of 0.00525 in.
in. × 2.54 = cm.
(°F + 459.67) × 5/9 = K.
ksi × 6.895 = MPa.
msi × 6.895 = GPa.
RT = room temperature.

TABLE 3—*Test tensile strength and modulus for vacuum-pressure-cured 3501-5A/AS-1 graphite/epoxy laminates ($0_4/90_2/\pm45_3$).*

Specimen No.	Thickness, in.	Width, in.	Test Temperature, °F	F^{tu}_{test}, ksi	$F^{tu}_{autoclave}$, ksi	$\dfrac{F^{tu}_{test}}{F^{tu}_{autoclave}}$	E^{ta}_{test}, msi	$E^{t}_{autoclave}$, msi	$\dfrac{E^{t}_{test}}{E^{t}_{autoclave}}$	$\nu_{xy\,test}$	$\nu_{xy\,autoclave}$	$\dfrac{\nu_{test}}{\nu_{autoclave}}$
102-1	0.065	1.005	RT	83.9	94.4	0.889	9.48	8.43	1.125	0.437	0.39	1.121
102-2	0.065	1.011	RT	81.5	94.4	0.863	8.87	8.43	1.052	0.427	0.39	1.095
102-3	0.065	1.011	RT	80.5	94.4	0.853	8.79	8.43	1.043	0.409	0.39	1.049
102-4	0.066	1.012	RT	86.0	94.4	0.911	9.39	8.43	1.114	...	0.39	...
102-5	0.066	1.010	RT	83.6	94.4	0.886	9.39	8.43	1.114	...	0.39	...
Avg				83.1		0.880	9.18		1.089	0.424		1.087
103-1	0.065	1.005	317	81.4[b]	84.3	0.966	7.62[e]	7.92	0.962	0.417	0.41	1.017
103-2	0.066	1.004	317	94.1[c]	84.3	1.116	7.52[e]	7.92	0.949	0.422	0.41	1.029
103-3	0.066	1.008	317	84.4[d]	84.3	1.001	7.76[e]	7.92	0.980	0.436	0.41	1.063
Avg				86.6		1.028	7.63		0.964	0.425		1.037

[a] All test results normalized for an average layer thickness of 0.00525 in.
[b] Wet conditioned (1.25% moisture content).
[c] Wet conditioned (1.22% moisture content).
[d] Wet conditioned (1.93% moisture content).
[e] Using extensometer readings.

in. $\times$ 2.54 = cm.
(°F + 459.67) $\times$ 5/9 = K.
ksi $\times$ 6.895 = MPa.
msi $\times$ 6.895 = GPa.
RT = room temperature.

Compression Tests

Five 10.16-cm-long (4.0 in.) by 2.54-cm-wide (1.0 in.) specimens cut from Panel No. 202 with a 0_{15T} laminate orientation were tested in compression at room temperature (Table 4). The coupon-type specimens were instrumented with back-to-back strain gages. The specimens failed at approximately 52 percent of the autoclave compression strength value of 1516.9 MPa (220.0 ksi). The room temperature strength reduction is acceptable since the trailing edge of the EF-111A horizontal stabilizer is a stiffness-critical component. Furthermore, the critical design condition for the trailing edge is a supersonic maneuver that produces maximum loading on the trailing edge due to the aft center-of-pressure location. The associated aerodynamic heating produces a temperature of 431 K (317°F).

Five additional specimens 12.7 cm (5.0 in.) long by 2.54 cm (1.0 in.) wide, cut from Panel 204 with a (0_{15T}) laminate orientation, were tested in compression at room temperature. The longer tabs were designed to facilitate testing. Results obtained from the specimens, which were also instrumented with back-to-back strain gages, are presented in Table 4. Results are similar to those obtained from Panel 202.

Another set of five specimens was conditioned at 333 ± 12 K (140 ± 5 deg F) and 95 ± 3 percent relative humidity for approximately seven days (1.94 percent moisture content) and tested at 431 K (317°F). The specimens failed in shear instability.

In a similar manner, five 10.16-cm-long (4 in.) by 2.54-cm-wide (1.0 in.) specimens with a $(0_4/90_2/\pm45_3)$ laminate orientation were tested in compression at room temperature (Table 5). A second set of five specimens was conditioned at 333 ± 12 K (140 ± 5 deg F) and 95 ± 3 percent relative humidity for approximately seven days (2.15 percent moisture content) and tested at 431 K (317°F). The room temperature specimens failed at approximately 48 percent of the autoclave-cured compression strength value. The results also show a 9 percent average reduction in Young's modulus. Note that the wet, 431 K (317°F), average compression value of 237.9 MPa (34.5 ksi) is approximately 45 percent of the estimated dry autoclave-cure value.

In-Plane (Rail) Shear Tests

Five 15.24-cm-long (6-in.) by 10.16-cm-wide (4-in.) specimens with a $(0_4/90_2/\pm45_3)$ laminate orientation were tested in rail shear at room temperature (Table 6). The specimens were instrumented with rosette gages. One of the specimens had backed-up rosette gages. No shear strength degradation was observed—only a 9 percent drop in shear modulus. One of the specimens failed prematurely (bearing failure in holes) and was therefore not included in the average.

TABLE 4—*Test compressive strength and modulus for vacuum-pressure-cured 3501-5A/AS-1 unidirectional graphite/epoxy laminates.*

Specimen No.	Thickness, in.	Width, in.	Test Temperature °F	F^{cu} test,[a] ksi	F^{cu} autoclave, ksi	$\dfrac{F^{cu} \text{ test}}{F^{cu} \text{ autoclave}}$	E^{c} test,[a] msi	E^{c} autoclave, msi	$\dfrac{E^{c} \text{ test}}{E^{c} \text{ autoclave}}$
202-1	0.075	1.000	RT	114.7	220.0	0.521	16.52	18.50	0.893
202-2	0.075	1.000	RT	107.3	220.0	0.488	16.29	18.50	0.881
202-3	0.076	1.001	RT	117.9	220.0	0.536	16.58	18.50	0.896
202-4	0.074	1.004	RT	121.6	220.0	0.553	16.73	18.50	0.904
202-5	0.072	.999	RT	107.2	220.0	0.487	16.38	18.50	0.885
Avg				113.7		0.517	16.50		0.892
204-1	0.071	1.002	RT	117.1	220.0	0.532	16.08	18.50	0.869
204-2	0.073	1.002	RT	120.2	220.0	0.546	16.50	18.50	0.892
204-3	0.069	1.000	RT	113.6	220.0	0.516	15.81	18.50	0.855
204-4	0.077	1.002	RT	113.1	220.0	0.514	16.50	18.50	0.892
204-5	0.066	1.002	RT	110.2	220.0	0.501	15.17	18.50	0.820
Avg				114.8		0.522	16.01		0.865
111-1	0.044	0.992	317	$NT^{b, d}$	50.0[c]	...	...	...	...
111-2	0.044	1.003	317	45.6[b]	50.0[c]	0.912	...	...	...
111-3	0.043	1.002	317	48.2[b]	50.0[c]	0.964	...	...	...
111-4	0.044	1.003	317	43.7[b]	50.0[c]	0.874	...	...	...
111-5	0.043	1.003	317	50.4[b]	50.0[c]	1.008	...	...	...
Avg				47.0		0.940			...

[a] All tests results normalized for an average layer thickness of 0.00525 in.
[b] Wet conditioned (1.94% moisture content)—specimens failed in shear instability.
[c] Estimated for 80% saturated (1.6% moisture content) autoclave-cured graphite/epoxy at 317°F.
[d] NT—invalid test.
[e] in. × 2.54 = cm.
(°F + 459.67) × 5/9 = K.
ksi × 6.895 = MPa.
msi × 6.895 = GPa.
RT = room temperature.

TABLE 5—Test compressive strength and modulus for vacuum-pressure-cured 3501-5A/AS-1 graphite/epoxy laminates $(0_4/90_2/\pm 45_3)$.

Specimen No.	Thickness, in.	Width, in.	Test Temperature °F	$F^{cu,a}$ test, ksi	F^{cu} autoclave, ksi	$\dfrac{F^{cu} \text{ test}}{F^{cu} \text{ autoclave}}$	$E^{c,a}$ test, msi	E^c autoclave, msi	$\dfrac{E^c \text{ test}}{E^c \text{ autoclave}}$
112-1	0.064	0.994	RT	44.7	91.3	0.490	7.86	8.43	0.932
112-2	0.063	0.986	RT	42.7	91.3	0.468	7.87	8.43	0.934
112-3	0.061	0.999	RT	46.4	91.3	0.508	7.56	8.43	0.897
112-4	0.062	1.000	RT	40.9	91.3	0.448	7.45	8.43	0.884
112-5	0.065	1.000	RT	46.0	91.3	0.504	7.74	8.43	0.918
Avg				44.1		0.483	7.70		0.913
113-1	0.066	0.998	317	39.6[b]	31.6[c]	1.253	...	...	...
113-2	0.067	1.001	317	30.9[b]	31.6[c]	0.978	...	...	...
113-3	0.067	0.992	317	35.2[b]	31.6[c]	1.114	...	...	...
113-4	0.067	1.001	317	30.1[b]	31.6[c]	0.953	...	...	...
113-5	0.067	1.002	317	36.8[b]	31.6[c]	1.165	...	...	...
Avg				34.5		1.092			

[a] All test results normalized for an average layer thickness of .00525 in.
[b] Wet conditioned (2.15% moisture content).
[c] Estimated for 80% saturated (1.6% moisture content) autoclave-cured graphite/epoxy at 317°F.
[d] in. × 2.54 = cm.
(°F + 459.67) × 5/9 = K.
ksi × 6.895 = MPa.
msi × 6.895 = GPa.
RT = room temperature.

TABLE 6—Test in-plane rail shear strength and modulus for vacuum-pressure-cured 3501-5A/AS-1 graphite/epoxy laminates $(0_4/90_2/\pm45_3)$.

Specimen No.	Thickness, in.	Length, in.	Test Temperature, °F	F^{su}_{test}, ksi	$F^{su}_{autoclave}$, ksi	$\dfrac{F^{su}_{test}}{F^{su}_{autoclave}}$	G_{test}, msi	$G_{autoclave}$, msi	$\dfrac{G_{test}}{G_{autoclave}}$
120-1	0.066	6.0	RT	33.8	33.8	1.000	2.35	2.75	0.855
120-2	0.065	6.0	RT	29.8	33.8	0.882	2.34	2.75	0.851
120-3	0.065	6.0	RT	33.8	33.8	1.000	2.52	2.75	0.916
120-4	0.062	6.0	RT	36.8	33.8	1.089	2.79	2.75	1.015
120-5[b]	0.064	6.0	RT	...	33.8	...	...	2.75	...
Avg				33.6		0.994	2.50		0.909

[a] All test results normalized for an average layer thickness of 0.00525 in.
[b] No test—bearing failure in holes.

in. × 2.54 = cm.
ksi × 6.895 = MPa.
msi × 6.895 = GPa.
RT = room temperature.

±45-deg Tension Tests

Five 27.94-cm-long (11 in.) by 2.54-cm-wide (1.0 in.) specimens with a ($\pm45_{4T}$) laminate orientation were tested in tension at room temperature. One specimen was instrumented with backed-up rosette gages; the remaining ones were instrumented with a single rosette gage. Test results are given in Table 7. The main purpose of these tests was to generate basic layer shear stress-strain data for vacuum-pressure-cured laminates. The 20 percent drop in ultimate shear strength for this relatively narrow specimen is not unreasonable because of the very significant edge effects in this ±45-deg laminate. The initial shear moduli obtained fell in the range of 4.716 to 5.723 GPa (0.684 to 0.830 msi), which agrees well with autoclave-cured specimen data.

Design Values

From the mechanical property tests the preliminary design values used in the design of the trailing-edge component were established. Table 8 gives these design values as ratios of dry strength at temperature of autoclave-cured laminates.

These static design values were obtained at a fraction of the cost of a program that would generate statistically valid design allowables.

The cost benefits of a reduced allowables test program must of course be weighed against the risks associated with a small property-characterization program with regard to possible failure-mode changes and possible increased scatter in particular failure modes. In addition, obtaining design values for a new material by applying factors to the properties of a well-characterized material assumes the failure modes and variability of the new material to be similar to those of the old.

Automotive Application

Grumman is currently applying advanced composites technology to the new De Lorean DMC-12 sports car. The De Lorean is a brushed stainless steel and reinforced-plastic vehicle. Some 6000 to 8000 are scheduled for production in 1980, and 20 000 to 30 000 the following year. The designs of the sports car's roof spider (Fig. 9) and gullwing doors (Fig. 10) were performed under developmental programs sponsored by the new car company. The roof spider was designed as an adhesive bonded assembly of outer and inner panels. The windshield frame is graphite/fiberglass vinyl ester and the remaining structure is fiberglass vinyl ester produced by the elastic reservoir molding process (ERM). The gullwing doors were designed in a similar fashion (Figure 10). Grumman has started fabrication of two prototype doors.

TABLE 7—Test tensile strength and modulus for vacuum-pressure-cured 3501-5A/AS-1 graphite/epoxy laminates ($\pm 45_{4T}$).

Specimen No.	Thickness, in.	Width, in.	Test Temperature °F	F^{tu} test,[a] ksi	F^{tu} autoclave, ksi	$\dfrac{F^{tu} \text{ test}}{F^{tu} \text{ autoclave}}$	E^{t} test,[a] msi	E^{t} autoclave, msi	$\dfrac{E^{t} \text{ test}}{E^{t} \text{ autoclave}}$	ν_{xy} test	ν_{xy} autoclave	$\dfrac{\nu \text{ test}}{\nu \text{ autoclave}}$
130-1	0.045	1.005	RT	18.8	23.8	0.79	2.44	2.40	1.02	0.760	0.778	0.977
130-2	0.045	1.006	RT	19.4	23.8	0.82	2.68	2.40	1.12	0.841	0.778	1.081
130-3	0.046	1.003	RT	18.9	23.8	0.79	2.53	2.40	1.05	0.751	0.778	0.965
130-4	0.045	1.003	RT	19.1	23.8	0.80	2.53	2.40	1.05	0.811	0.778	1.042
130-5	0.046	1.010	RT	18.9	23.8	0.79	2.56	2.40	1.07	0.780	0.778	1.003
Avg				19.0		0.80	2.55		1.06	0.789		1.014

[a] All test results normalized for an average layer thickness of 0.00525 in.

in. × 2.54 = cm.
ksi × 6.895 = MPa.
msi × 6.895 = GPa.
RT = room temperature.

TABLE 8—*Preliminary design values for vacuum-pressure-cured graphite/epoxy laminates.*[b]

Property	Ratio of VPC to Autoclave-Cured Dry Laminate at RT[f]	Ratio of VPC Wet Laminate to Autoclave-Cured Dry Laminate at 431 K (317°F)[a]
F_t^d/F_t	0.88	0.81
F_c^d/F_c	0.50	0.49
F_s^d/F_s	0.90	0.49[c]
E_t^d/E_t	1.00	0.91
E_c^d/E_c	0.87	0.77[c]
G^d/G	0.91	0.74[c]

[a]Moisture absorption is 1 to 1.4%.
[b]Per-ply thickness of vacuum-pressure-cured laminate is 0.0133 cm (0.00525 in.).
[c]Estimated.
[d]VPC[e] laminate property.
[e]VPC = vacuum-pressure-cured.
[f]RT = room temperature.

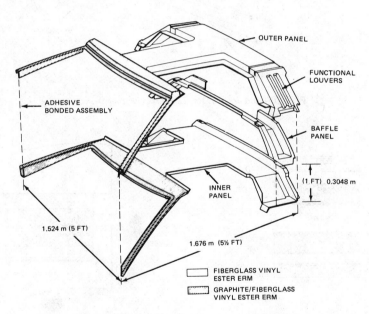

FIG. 9—*DeLorean DMC-12 sports car roof spider.*

Elastic reservoir molding, one of the newest reinforced-plastic processes (Fig. 11), is basically taking a flexible open-cell urethane foam and impregnating it with a thermoset resin, applying a dry reinforcement, and placing the sandwich into a heated matched mold tool at approximately 7.03 kg/cm² (100 psi) to form the shape of the part. After the resin has been cured, which is related to the curing mechanism and the heat used, the part is then complete. The main advantages of ERM are low weight [with no rein-

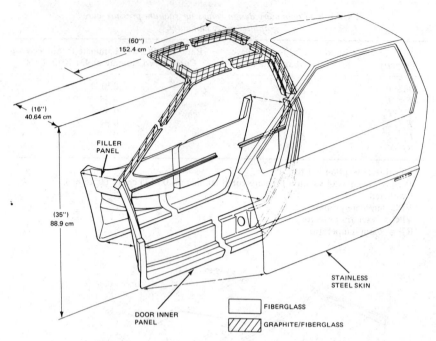

(60")
152.4 cm

(16")
40.64 cm

FILLER
PANEL

(35")
88.9 cm

STAINLESS
STEEL SKIN

DOOR INNER
PANEL

FIBERGLASS

GRAPHITE/FIBERGLASS

FIG. 10—*DeLorean DMC-12 sports car door structure.*

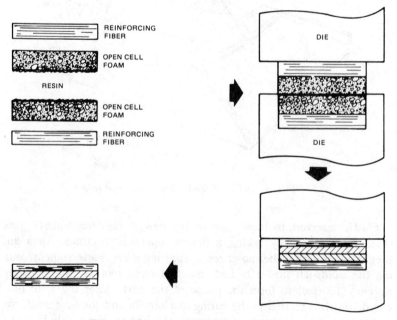

REINFORCING
FIBER

OPEN CELL
FOAM

RESIN

OPEN CELL
FOAM

REINFORCING
FIBER

DIE

DIE

FIG. 11—*Elastic reservoir molding (ERM).*

forcement 1.3426 g/cm^3 (0.0485 lb/in.^{3}) and with 18 percent graphite rein-forcement 1.5779 g/cm^3 (0.057 lb/in.^{3}), higher impact resistance, low-cost tooling, and the ability to make tailored hybrid composites. It is this latter ability, which affords the designer a wide choice of composite reinforcement and resin system combinations, that results in high costs for mechanical property characterization. Therefore, a cost-conscious approach to mechan-ical property characterization was undertaken. Tradeoff studies using dif-ferent resin systems, forms of fiber glass, and special high-performance reinforcement (Table 1) were performed using flexural and short-beam hori-zontal shear specimens.

Flexure and Short-Beam Horizontal Shear Tests

The flexural stiffness of the ERM horizontal shear specimens required that the specimen's span length (Fig. 1) be reduced from the $5t$ used for graphite/epoxy to $3t$, where t is the thickness of the specimen. The normal shear modulus, G_z, was obtained from the load-deflection curves. The flex-ural modulus was obtained from the flexure tests corrected for shear defor-mation. An average of three specimens was tested for each reinforce-ment/resin/cure combination. Table 9 gives the average of each set of tests for some of the more promising combinations tested.

Tension and Compression Tests

A limited number of tension (Fig. 5) and compression coupons (Fig. 4) were tested (Table 9).

Lap Shear Tests

The assembly of the roof spider requires bonding subassemblies together and to the underbody structure. For the critical car rollover design load con-dition, bond stresses were determined at various locations in the roof spider structure. Although the adhesive chosen for bonding, Goodyear Pliogrip 6000, is used throughout the automotive industry, design mechanical prop-erty data pertinent to the materials chosen for the roof spider were not available. A simple overlap tension shear test program was performed. Stan-dard 2.54-cm-wide (1 in.), 1.83 and 3.66-kg/m^2 (6 and 12 oz/ft^2) glass mat specimens with 0.0762-cm-thick (0.030 in.) bond line and 2.54 and 5.08-cm (1 and 2 in.) overlaps were tested. To evaluate the additional strength that might be derived by preventing peeling action, specimens containing fasteners at the ends of the bond overlap were also tested. The test data are presented in Table 10. The resulting design values, based on bond location and adjustments for length, as applicable, together with maximum antici-pated stresses, are given in Table 11.

TABLE 9—Test panel data.

Panel No.	Material	Tension		Compression		Flexure		Shear	
		F_t, ksi	E_t, msi	F_c, ksi	E_c, msi	F_b, ksi	E_b, msi	F_s, ksi	G_z, ksi
	6 oz/sq ft glass mat design	9.76	1.18	9.76	1.18	29.3	2.0	1.25	50.0
51	epoxy/anhydride	12.6	1.20	12.2	1.25	38.0	2.21	1.25	50.7
60	vinyl ester (mod-temp)	12.3	1.02	19.5	1.13	36.6	1.78	1.25	48.0
64	vinyl ester (hi-temp)	15.0	1.23	24.7	1.29	41.8	2.09	2.21	51.0
	6-oz/sq ft glass mat + 2 plies graphite A-pillar design	11.9	5.95	11.9	5.95	23.4	10.5	1.73	69.2
52	epoxy/pitch plain weave	19.5	5.10	12.9	4.72	34.0	9.42	1.87	...
55	epoxy/pitch nuff	...	...	11.0	5.10	31.3	10.5	...	...
57	epoxy/fortafil 5 nuff	...	...	20.9	5.80	59.0	10.7	1.97	49.4
59	VE (mod-temp)/Fortafil 5 nuff	32.2	5.66	13.3	...	41.6	11.1	2.28	76.4
63	VE (hi-temp)/fortafil 5 nuff	29.3	5.75	21.7	5.58	68.7	11.2	2.91	78.6
66	VE (mod-temp)/pitch 4H satin weave	20.8	4.50	7.7	4.43	16.8	8.2	1.74	82.1

ksi = 6.895 MPa.
msi = 6.895 GPa.
oz/sq ft = 0.305 kg/m².

TABLE 10—*Adhesive bonded joint strength test data.*

	Overlap Length		
	1 in.	2 in.	2 in. with Rivets
ERM with 6 oz/ft² glass mat	1257 psi[a]	776 psi	737 psi
ERM with 12 oz/ft² glass mat	1860 psi	1205 psi	1295 psi

[a] Average of three test points.
psi = 6895 Pa.
oz/ft² = 0.305 kg/m².

TABLE 11—*Summary of maximum bond stresses.*

	Maximum Applied Stress	Bond Shear Design Value[a]
B-frame header	390 psi	1000 psi
AFT hinge fitting	700 psi	1490 psi
B-pillar/underbody attachment	770 psi	1125 psi

[a] 80% of average of test data.
psi = 6895 Pa.

Design Values Verification

Because of the developmental nature of the ERM process, a manufacturing feasibility verification component for the roof spider (Fig. 12) was fabricated and tested. The three-point bending beam results correlated well with predictions for stiffness, failure level, and failure mode, based on the coupon-generated design values. The successful prediction of the component's test results engendered additional confidence in the fabrication process, and in the low-cost coupon-type specimens for generating design values.

Conclusions

1. Abreviated coupon testing for static design values for secondary structures can be successfully applied to expand utilization of composite structures.

2. Rapid processing of composites for high production rates or minimum manufacturing costs can result in reduction of compression strength. Reductions in tension and in-plane shear strength and axial stiffness are moderate.

Acknowledgments

The work described under Aerospace Application was performed for the Air Force Materials Laboratory (AFML) under Contract F33615-78-C-5234.

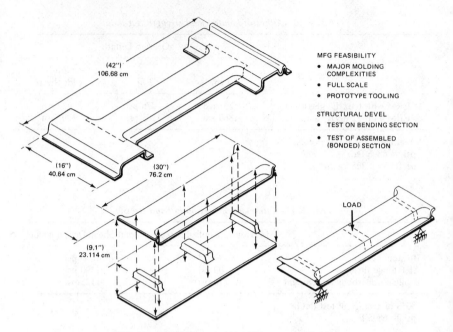

FIG. 12—*DMC-12 roof spider T-beam test element.*

Mr. H. S. Reinert Jr., AFML/LTN, is the Air Force project engineer. The work described under Automotive Application was performed for the De Lorean Motor Co.

L. F. Tenn[1]

Statistical Analysis of Fibrous Composite Strength Data

REFERENCE: Tenn, L. F., "**Statistical Analysis of Fibrous Composite Strength Data,**" *Test Methods and Design Allowables for Fibrous Composites, ASTM STP 734,* C. C. Chamis, Ed., American Society for Testing and Materials, 1981, pp. 229-244.

ABSTRACT: Using standard statistical procedures, distributional forms for fibrous composite strength data are examined for goodness of fit. Six examples of strength data for epoxy reinforced with Kevlar 49, fiber glass, and Thornel 300 graphite fabrics, as well as unidirectional T300/5208 graphite/epoxy composites, were examined. The data were evaluated using the chi-square and Kolmogorov-Smirnov goodness-of-fit statistical tests and the correlation coefficient from linear regression analysis. The distributional forms tested were normal, log normal, 2-parameter Weibull, and 3-parameter Weibull. At the 5 percent significance level, none of these was rejected.

KEY WORDS: statistics, Kolmogorov-Smirnov, chi-squared, normal, log normal, Weibull, composites, strength, design allowables, graphite/epoxy, fiber glass materials, composite materials

In the design of aerospace structures, strength values which are acceptable to the Federal Aviation Administration (FAA) and to other government agencies are required for structural analysis. The strengths of many metallic alloys are published in MIL-HDBK-5 [1].[2] Allowable strengths of fibrous composites have not been defined by these agencies, nor are they likely to be in the immediate future. Structures which are classed as redundant, fail-safe, or secondary are usually designed with B-basis design values. At least 90 percent of the population of values are expected to equal or exceed the B-basis mechanical property allowable with a confidence of 95 percent. A-basis values are exceeded by 99 percent of the population with a 95 percent confidence and are used for single-load-path structural components, whose failure would be catastrophic.

---

[1]Design specialist, Lockheed-California Co., Burbank, Calif. 91520.

[2]The italic numbers in brackets refer to the list of references appended to this paper.

Present Study

This study has two objectives: to determine, where possible, the A- and B-basis allowables for several fibrous composite materials as governed by Lockheed specifications, and to examine the fit of several distribution functions to available data. Normal distribution has traditionally been used to describe the ultimate and yield strengths of metallic materials, but several workers in fibrous composites have proposed Weibull distributions. For completeness, the distribution of the logarithm of the strength data is also included.

Data

Room temperature dry tensile strengths of fiber-reinforced epoxy resins were available for study. These are described in Table 1, Items a through e. Item f, Table 1, is the wet compressive strength of a structural fiber glass fabric/epoxy material.

Discussion of Data

Many statistical techniques are available in the literature for making up a representative sample; however, in real life, flawed samples are usually found. Samples used for the computation of A- and B-basis allowables should be for one material only, should represent many lots from several sup-

TABLE 1 —*Summary of fiber-reinforced epoxy materials.*

Sample	Reinforcement	Epoxy	No. of Specimens	Load Direction [6]	Comments
a	Style 181 Kevlar 49	5209	24	warp	(1)
b	T300 woven fabric	5208	28	fill	(2)
c	T300 woven fabric	5208	30	warp	(2)
d	T300 woven fabric	5208	45	warp	(3)
e	T300 unidirectional tape	5208	127	L	(4)
f	Style 181 fiber glass fabric	general purpose	171	warp	(5)

(1) Batch acceptance testing of four lots, three per lot by the material vendor and three per lot by Lockheed Quality Assurance.

(2) All tests from the same lot of material to investigate within batch variability.

(3) Batch acceptance testing of 12 lots. Three lots had six tests each; the remainder had three tests each.

(4) Batch acceptance testing of 28 batches, 11 batches of 0.19-mm (7.5 mil) tape and the balance 0.127-mm (5 mil) tape.

(5) Batch acceptance testing of 29 lots, five tests each by Lockheed Quality Assurance and one test by vendor. Tests are wet compressive strengths.

(6) Warp is longitudinal direction and fill is transverse direction in cloth terminology. L refers to fiber direction for unidirectional laminates.

pliers, and the data should not be biased in favor of any one supplier, lot, or test laboratory. Sufficient data should be available to determine the form of the distribution and sample statistics. In real life, resin formulations are usually proprietary and single-source items; the same applies to graphite fibers and Kevlar 49 fibers.

Sample *a* is from four material lots; three tests from each lot were by the material vendor and three by the Lockheed Quality Assurance Laboratory. There are insufficient tests for the reasonable prediction of even the B-basis allowable, but the strength distribution is similar to all the other data.

Samples *b* and *c* were examined for distribution within one lot. It is of interest to determine if the variation of strength between lots and within lots is significant for fiber-reinforced materials.

Sample *e* is from batch acceptance tests for material being used for the L-1011 composite fin and aileron structures being developed under contract to National Aeronautics and Space Administration (NASA). Most of the values are from the material vendor, the remainder from tests by Lockheed Quality Assurance or Avco Corp.

Sample *f* is from batch acceptance testing of wet compression strength for 29 lots of fiber glass laminates. The material is used for honeycomb face sheets as well as laminates.

Statistical Analysis

The data previously described were statistically analyzed. The first objective was to determine which probability distribution form best fits the six samples, and the second was to determine A- and B-basis design allowables. Where data were insufficient, an S-basis strength is proposed.

The probability distributions examined were:

1. Two-parameter normal distribution.
2. Two-parameter log normal distribution.
3. Two-parameter Weibull distribution.
4. Three-parameter Weibull distribution.

Sample Statistics

The sample means and standard deviations were used as estimates of the population means and standard deviations. Definitions and equations for these calculations may be found in many statistics texts; for example, Ref 2.

Weibull Distribution

The probability density function for the Weibull distributions are [3]

$$f(x) = [k(x-c)^{k-1}/\theta^k] \exp [(c-x)/\theta]^k, \theta, k \geq 0, x \geq c \geq 0 \quad (1)$$

where

x = random variate or specimen measurement,
θ = scale parameter,
k = shape parameter, and
c = location parameter.

When $c = 0$ the equation reduces to the two-parameter Weibull distribution. The maximum-likelihood parameters for the Weibull distribution were calculated with a FORTRAN computer program provided by H. C. Harter of Wright-Patterson Air Force Base. This program is described in Ref. 3.

Goodness of Fit

Several methods were used to determine the goodness of fit of the several distributions to the data. These were the chi-squared test [2], the Kolmogorov-Smirnov test (K-S) [4], and the correlation coefficient from regression. The preceding two tests were carried out at the 5 percent significance level.

Results of Statistical Tests

Chi-Squared Tests

To a significance level of 5 percent, none of the proposed distributions was rejected by the chi-squared test. The number of intervals was rounded up from $1.5 + 3.322 \times [\log 10 (\text{sample size})]$. The results are tabulated in Table 2.

Kolmogorov-Smirnov Tests

This test is reputed to be more powerful than the chi-squared test for the rejection of distributions. Tables of critical values for D_n may be found in the

TABLE 2—*Results of the chi-Squared Tests.*

			Chi-Squared Statistic			
				Weibull	Critical Chi-Squared	
Sample	Normal	Log Normal	2-Parameter	3-Parameter	5 %	25 %
a	3.0	3.0	3.0	3.0	7.8	4.1
b	5.0	7.6	5.9	5.9	7.8	4.1
c	1.6	4.4	2.4	2.4	7.8	4.1
d	2.6	4.7	3.4	7.9	7.8	4.1
e	3.1	4.0	5.4	4.5	11.1	6.6
f	6.6	6.3	5.0	4.1	11.1	6.6

literature [4]. The 5 percent significance level critical values are shown as a curve in Fig. 1. These values are for normally distributed samples with the mean and variance unknown. We have been unable to locate values for Weibull distribution with scale and shape parameters unknown, but the values for normal distribution should be approximately correct. From Fig. 1 it is concluded that, generally, all four distributions are acceptable descriptions of fiber-reinforced epoxy strengths.

Regression

Probability plotting of the strength data is used to visually illustrate the fit of various distributions. These are shown in Figs. 2 through 9. Visually, these are fairly straight lines and the three samples with 45 or more specimens are judged to be linear. In this paper the plotting points are taken as $(i - 0.5)/n$ where i is the rank and n is the number of specimens in the sample.

Probability plotting of normal distributions uses normal deviates for a linear scale. The normal deviates, or the number of standard deviations by which an observed value exceeds the mean, is approximated by the following equation from Ref 5.

Normal deviate =

$$\left[\pm T - \left(\frac{2.515517 + 0.802853T + 0.010328T^2}{1 + 1.432788T + 0.18927\,T^2 + 0.001308T^3} \right) \right] \qquad (2)$$

where

$$T = \left[\ln\left(\frac{n}{i - 0.5} \right)^2 \right]^{1/2}$$

Using Eq 2, the error in the deviate is less than 0.00045.

For Weibull probability plotting the following parameters are used. The ordinate is

$$Xp = \ln\left[\ln\left(\frac{1}{1 - P} \right) \right] \qquad (3)$$

where $P = (i - 0.5)/n$ and the abscissa is ln (strength) for the two-parameter Weibull or ln (strength $- c$) for the three-parameter Weibull distribution. The rationale for a straight-line relationship may be found in Ref 6. These plots were produced by a standard statistical computer program (Statistical Package for the Social Sciences) [7], which provided sample statistics of the data. The straight-line fit of the data provides estimates of

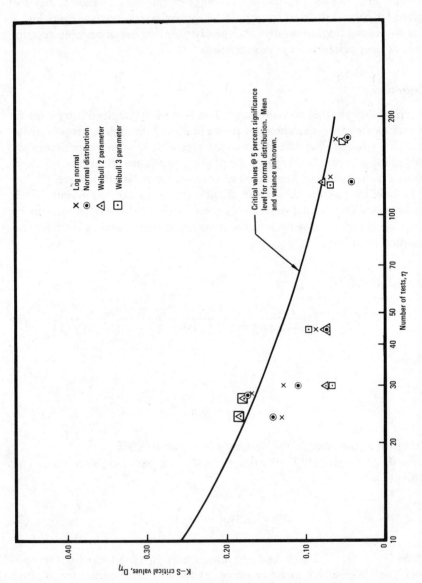

FIG. 1—*Kolmogorov-Smirnov test—fiber-reinforced composites.*

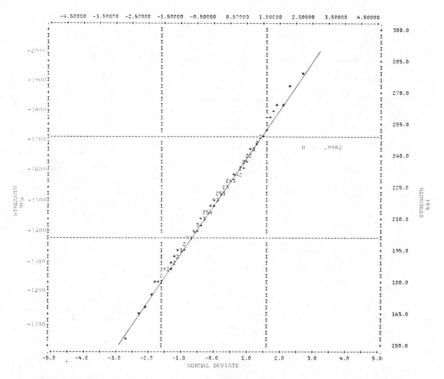

FIG. 2—*Probability plotting of distributional assumption. Material: unidirectional graphite T300/5208 epoxy; distribution: normal; n = 127; Avg = 217.2; s = 22.492.*

the Weibull parameters. The slope of the line estimates the shape parameters, and the strength at probability 0.625 estimates the scale parameter.

Allowables

The quantity and characteristics of a population required for the computation of design allowables for aerospace vehicles are defined in Chapter 9 of Ref 1. A brief summary of the requirements has already been given under "Discussion of Data." In this presentation, A and B allowables are calculated as follows.

Normal Distribution

The following equations are used

$$A = \overline{X} - k_A s \qquad (4)$$

$$B = \overline{X} - k_B s \qquad (5)$$

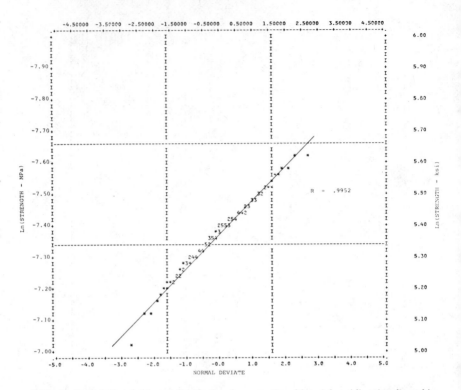

FIG. 3—*Probability plotting of distributional assumption. Material: unidirectional graphite T300/5208 epoxy; distribution: log normal;* n = 127; Avg = $e^{5.3753}$; s = $e^{0.1050}$.

where

$\overline{X}$ = sample mean based on n observations,

s = standard deviation,

k_A = one-sided tolerance-limit factor corresponding to a proportion at least 0.99 of a normal distribution and a confidence coefficient of 0.95, and

k_B = one-sided tolerance-limit factor corresponding to a proportion at least 0.90 of a normal distribution and a confidence coefficient of 0.95.

Tables of k_A and k_B may be found in Ref *1*.

Log Normal Distribution

The natural logarithms of the data are taken and the equations for normal distribution are used to compute the natural logarithm of the A and B allowables.

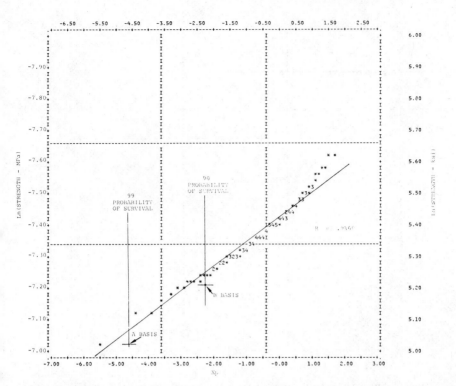

FIG. 4—*Probability plotting of distributional assumption. Material: unidirectional graphite T300/5208 epoxy; distribution: 2-parameter Weibull;* n = 127; θ = [1567.1 MPa (227.3 ksi)]; k = 10.2034.

Weibull Distributions

The A and B allowables for the Weibull distributions utilize the linear relationship between the Weibull probability function given by eq 3 and ln (strength). From Ref 2, the 0.95 lower confidence interval estimate for a single value on a regression line is below the line by

$$t_{0.975} Sy \left[1 + \frac{1}{n} + \frac{(X_p{}^1 - \overline{X}_p)^2}{\Sigma(X_p - \overline{X}_p)^2} \right]^{1/2} \qquad (6)$$

where

$t_{0.975}$ = from tables of the percentiles of the t-distribution,
n = sample size, or number of specimens
Sy = standard error,
X_p = defined by Eq 3,

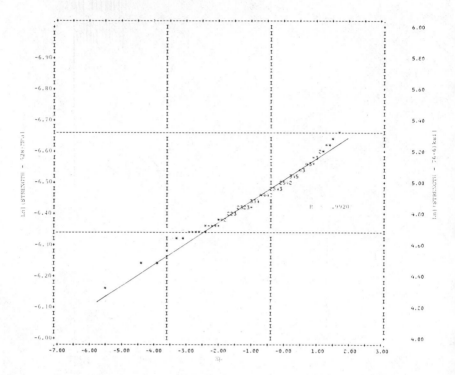

FIG. 5—*Probability plotting of distributional assumption. Material: unidirectional graphite T300/5208 epoxy; distribution: 3-parameter Weibull;* $n = 127$; $c = [527.8 MPa (76.5638 ksi)]$; $\theta = [1036.3 MPa (150.315 ksi)]$; $k = 6.764$.

$\overline{X_p}$ = mean X_p of population,

$X_p{}^1 = X_p$ at the allowable probability,

$X_p{}^1 = -2.2504$ for B-basis (0.90 probability), and

$X_p{}^1 = -4.6001$ for A-basis (0.99 probability).

Results

The chi-squared test results given in Table 2 failed to reject any of the distributions for all samples at the 5 percent significance level. If the chi-squared statistic is less than the critical value, the fit of the data to the distribution function cannot be rejected. Sample *d*, Table 1, was marginal for the three-parameter Weibull distribution. Similarly, the Kolmogorov-Smirnov test failed to reject any of the distributions, although the samples with less than 30 observations were marginal. Log normal distributions provided the highest allowables and the worst linear fits. A- and B-basis allowables are given in Table 3 for Samples *e* and *f* for all three distributions. These are for T300/5208 unidirectional tape and 181-style fiber glass/epoxy.

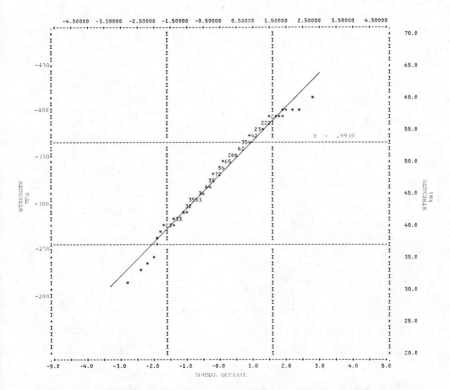

FIG. 6—*Probability plotting of distributional assumption. Material: 181-style fiber glass/epoxy; distribution: normal;* n = 171; Avg = 48.2; s = 5.557.

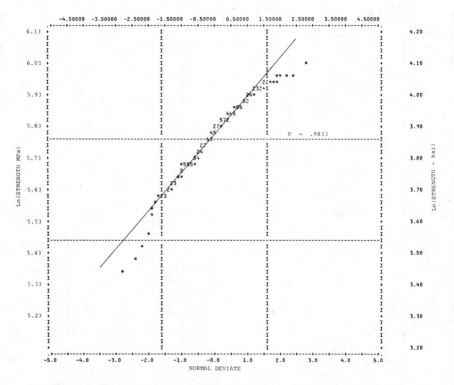

FIG. 7—*Probability plotting of distributional assumption. Material: 181-style fiber glass/epoxy; distribution: log normal; n = 171; Avg = $e^{3.869}$; s = $e^{0.1195}$.*

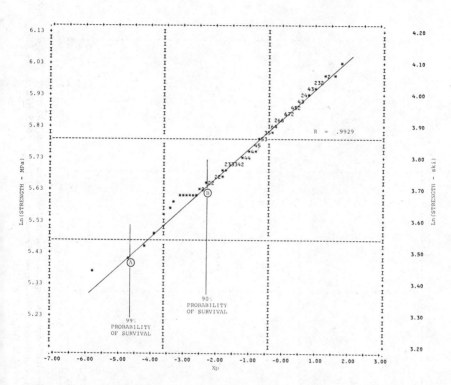

FIG. 8—*Probability plotting of distributional assumption. Material: 181-style fiber glass/epoxy; distribution: 2-parameter Weibull:* n = 171; θ = [349.2 MPa (50.65 ksi)]; k = 10.072.

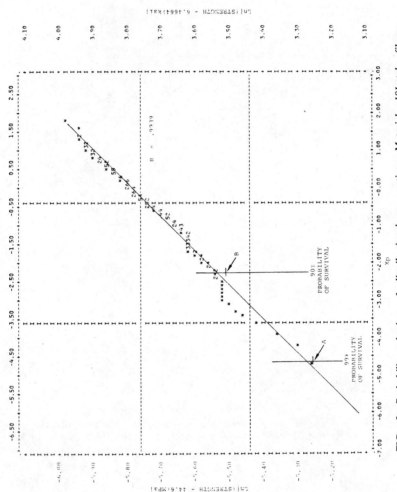

FIG. 9—*Probability plotting of distributional assumption. Material: 181-style fiber glass/epoxy; distribution: 3-parameter Weibull; n = 171; c = [44.5 MPa (6.4664 ksi)]; θ = [304.4 MPA (44.15 ksi)]; k = 8.7454.*

TABLE 3—Summary of allowables, MPa (ksi).

Sample (see Table 1)	a	b	c	d	e	f
Sample size	24	28	30	45	127	171
Sample mean	598.0 (86.69)	487.0 (70.70)	512.0 (74.26)	553.0 (80.21)	1497.0 (217.18)	333.0 (48.23)
Coefficient of variation	0.055	0.100	0.076	0.098	0.104	0.115
Material properties	F_{tu}			F_{tu}	F_{tu}	F_{cu}
A-Basis allowables						
normal	...	...	...	...	1088.0 (157.8)	233.0 (33.8)
log normal	...	...	...	...	1129.0 (163.7)	242.0 (35.1)
2P Weibull	...	...	...	...	1043.0 (151.2)	221.0 (32.1)
3P Weibull	...	...	...	...	1078.0 (156.3)	225.0 (32.6)
B-Basis allowables						
normal	523.0 (75.9)	...	...	462.0 (67.0)	1265.0 (183.5)	276.0 (40.1)
log normal	527.0 (76.4)	...	...	465.0 (67.5)	1273.0 (184.6)	277.0 (40.2)
2P Weibull	527.0 (76.5)	...	...	466.0 (67.6)	1262.0 (183.1)	276.0 (40.0)
3P Weibull	529.0 (76.7)	...	...	467.0 (67.7)	1272.0 (184.5)	277.0 (40.2)
S-Basis (tentative)	483.0 (70.0)	...	...	393.0 (57.0)	...	...

B- and S-basis allowables are also shown for Samples *a* and *d*, which are 181-style Kevlar/5209 epoxy and T300 graphite fabric/5208 epoxy, respectively. Samples *b* and *c* are all from one plate and are therefore not appropriate for the computation of allowables. Samples *a* and *d* are of insufficient size to meet the criteria of Ref *1* for A-allowables.

References

[*1*] "Metallic Materials and Elements for Aerospace Vehicle Structures," MIL-HDBK-5C Department of Defense, Washington, D.C., 15 Sept. 1976.
[*2*] Natrella, M. G., *Experimental Statistics* (with corrections) National Bureau of Standards Handbook 91, U. S. Government Printing Office, Washington, D.C., Oct. 1966.
[*3*] Harter, H. L. and Moore, A. H., *Technometrics,* Vol. 7, No. 4, Nov. 1965.
[*4*] Lillifors, H. W., *Journal of the American Statistical Association,* Vol. 62, 1967, pp. 399–402.
[*5*] *Handbook of Mathematical Functions* (with corrections), M. Abramowitz and I. A. Stegun, Eds., National Bureau of Standards Applied Mathematics Series 55, U. S. Government Printing Office, Washington, D.C., Nov. 1964.
[6] King, J. R. *Probability Charts for Decision Making,* Industrial Press Inc., New York, 1971.
[*7*] Nie, N. H. et al., *SPSS, Statistical Package for the Social Sciences,* 2nd ed., McGraw-Hill, New York, 1975.

G. P. Sendeckyj[1]

Fitting Models to Composite Materials Fatigue Data*

REFERENCE: Sendeckyj, G. P., **"Fitting Models to Composite Materials Fatigue Data,"** *Test Methods and Design Allowables for Fibrous Composites, ASTM STP 734,* C. C. Chamis, Ed., American Society for Testing and Materials, 1981, pp. 245-260.

ABSTRACT: A new procedure for fitting fatigue models (consisting of a deterministic equation defining the shape of the S-N curve and a probabilistic description of the data scatter) is presented. The procedure consists of (1) transforming the fatigue data into equivalent static strength data by using the deterministic equation in the fatigue model with unknown parameters, (2) obtaining the maximum-likelihood estimates for the parameters of the (two-parameter Weibull) distribution describing the equivalent static strength data, and (3) repeating Steps 1 and 2 until the largest maximum-likelihood estimate of the shape parameter is obtained. The procedure can handle runouts and tab failures through progressive censoring. Finally, the procedure is well defined when small fatigue data samples are available.

KEY WORDS: composite materials, fatigue data, fatigue models, S-N curves, two-parameter Weibull distribution, maximum-likelihood estimators, censoring

Fatigue experiments are normally conducted to achieve specific objectives. The problem faced by the researcher is to design the experiment to achieve the objectives at a minimum expenditure of resources and effort. This can be done by using efficient fatigue data analysis and interpretation procedures. Herein, a new procedure is presented for analyzing and interpreting fatigue data in experiments designed to characterize the S-N (applied cyclic stress versus fatigue life/residual strength) behavior of composite materials. The procedure is based on the following three assumptions:

1. The S-N behavior can be described by a deterministic equation. The equation can be based on either theoretical considerations or experimental observations of the fatigue damage accumulation process.

2. The static strengths are uniquely related to the fatigue lives and residual strengths at runout (termination of cyclic testing). The specific relationship

---

* This paper is based on in-house work performed at the Flight Dynamics Laboratory under the Air Force Wright Aeronautical Laboratories Solid Mechanics Project funded by the Mechanics Directorate of the Air Force Office of Scientific Research.

[1] Aerospace engineer, Structures and Dynamics Division, Flight Dynamics Laboratory, Air Force Wright Aeronautical Laboratories, Wright-Patterson Air Force Base, Ohio 45433.

assumed herein is that the strongest specimen has either to longest fatigue life or the highest residual strength at runout. It should be noted that this assumption may not hold if competing failure modes are observed during the fatigue tests.

3. The static strength data can be described by a two-parameter Weibull distribution. This assumption is based on the observation that the two-parameter Weibull distribution provides a good fit of the static strength data for composite materials and is attractive because of the simplifications inherent in using the Weibull distribution.

The description of the S-N behavior, based on the foregoing three assumptions, will be referred to as the *fatigue model*. Different fatigue models will arise from using different deterministic equations to describe the S-N behavior. The deterministic equation depends on parameters, referred to as *fatigue model parameters*, describing particular features of the S-N behavior. The fatigue model defines a unique curve for each static strength value. The curve passing through the scale parameter of the Weibull distribution, describing the static strength distribution, will be referred to as the S-N *curve*. This definition is made for the sake of convenience in graphical presentation of the results.

The data analysis procedure, described in the present paper, simultaneously determines the parameters of the deterministic equation (fatigue model parameters) and the parameters of the underlying Weibull distribution describing the static strength. This is done by (1) assuming values of the fatigue model parameters and transforming the fatigue data into equivalent static strength data, (2) fitting a two-parameter Weibull distribution to the equivalent static strength data by the maximum-likelihood method, (3) repeating Steps 1 and 2 until an optimality criterion is satisfied, and (4) applying a goodness-of-fit test to determine whether the particular fatigue model fits the data. The optimality criterion adopted herein is that the best choice of the fatigue model parameters is the one that maximizes the maximum-likelihood estimate for the shape parameter of the two-parameter Weibull distribution for the equivalent static strength data.

A detailed derivation and justification of the procedure is given in the next section. This is followed by a number of examples illustrating its application and a discussion of some of its salient features. Finally, recommendations for additional work are made.

Theoretical Development

Nature of Fatigue Data

The complete set of data in a simple fatigue experiment to characterize the S-N behavior of a composite material will fall into the following five groups:

1. *Static strength data:* Data in this group consist of either tensile or compressive static strength at a loading rate comparable to the cyclic loading rate encountered during fatigue testing. If the composite material is not strain-rate sensitive, static strength data at conventional static test loading rates may be used. Otherwise, high-loading-rate data must be used. The static strength data will be considered herein to be fatigue failure data during the first cycle ($N = 1$) at applied cyclic stress (σ_a) equal to the static strength (σ_s).

2. *Fatigue failure data:* Data in this group consist of the number of cycles to failure (N) at some applied cyclic stress (σ_a)

3. *Residual strength data:* Data in this group consist of the static residual strength (σ_r) after a predetermined number of cycles (n) at some applied cyclic stress (σ_a). The residual strength data for a strain-rate sensitive material must be at a loading rate corresponding to the cyclic loading rate.

4. *Runouts:* Data in this group consist of the number of cycles (n) to run-out (termination of cyclic testing prior to fatigue failure) at some applied cyclic stress (σ_a), with no corresponding residual strength data. In the analysis to follow, these data will be treated as being *censored;* that is, it will be assumed that either the fatigue life is actually longer than n or the residual strength at n cycles is higher than σ_a.

5. *Tab failures:* Data in this group consist of data from groups 1 through 4 for which the specimen failure mode is suspect due to the fact that the specimen has failed at a tab or some obvious defect, making the data suspect. Data in this group will be treated by censoring. For a static or residual test tab failure, the data will be interpreted by the statement that the actual static or residual strength of the specimen is higher than the measured value. For a fatigue tab failure, the data will be treated as a runout.

More general types of fatigue data can occur in more complex experiments, but these will not be considered herein.

Wearout Model

Before discussing the procedure for fitting fatigue models to experimental data, let us consider the structure and statistical implications of the wearout model proposed by Halpin et al [1].[2] The wearout model is a direct application of metal crack growth concepts to composite materials. In its formulation, a dominant crack is assumed to be present. As cyclic loading is applied, the crack grows until the specimen can no longer support the applied cyclic load, resulting in a fatigue failure. If the cyclic test is terminated prior to fatigue failure, the residual strength is related to the crack length through a fracture mechanics calculation. The form of the resulting deterministic equation relating the residual strength, applied cyclic load, cycles, and initial static strength depends on the particular form of crack growth law assumed.

---

[2] The italic numbers in brackets refer to the list of references appended to this paper.

Because of objections to the dominant crack assumption, the model was recast by a number of authors [2-6] in terms of residual strength degradation.

The particular form of the wearout model, developed by the author and adopted herein, is based on the following deterministic equation

$$\sigma_e = \sigma_a[(\sigma_r/\sigma_a)^{1/S} + (n-1)C]^S \tag{1}$$

where σ_e, σ_a, σ_r, and n are the *equivalent static strength*, maximum applied cyclic stress, residual stength, and number of cycles, respectively. The parameter S is the absolute value of the asymptotic slope at long life on a log-log plot of the $S-N$ curve. The parameter C is a measure of the extent of the "flat" region on the $S-N$ curve at high applied cyclic stress levels. The formulation of the wearout model is completed by assuming that the equivalent static strength is two-parameter Weibull distributed; that is, the probability that the static strength is higher than σ_e is given by

$$P(\sigma_e) = \exp[-(\sigma_e/\beta)^\alpha] \tag{2}$$

where α and β are the Weibull shape and scale parameters, respectively. Here, $P(x)$ is the reliability function related to the cumulative distribution function $F(x)$ by

$$P(x) = 1 - F(x)$$

It should be noted that the use of the reliability and cumulative distribution functions instead of the probability density function tends to simplify the derivations because of their simpler change of variable behavior.

Let us first examine the structure of Eq 1. For fatigue failure, $\sigma_r = \sigma_a$, $n = N$, and Eq 1 reduces to

$$\sigma_e = \sigma_a(1 - C + CN)^S \tag{3}$$

which is the fatigue failure criterion describing the shape of the S-N curve. For $C = 1$, Eq 3 reduces to the classical *power law fatigue failure criterion*. For $C > 1$, Eq 3 gives an S-N curve that steepens at low cycles on a log-log plot as illustrated in Fig. 1. This behavior has not been observed to occur for composite materials. For $C < 1$, Eq 3 gives an S-N curve that flattens out at low cycles on a log-log plot as shown in Fig. 1. As C tends to zero, the size of the flattened region increases. This type of behavior is common for composite materials. Finally, rewriting Eq 1 as

$$\sigma_r = \sigma_a[(\sigma_e/\sigma_a)^{1/S} - C(n-1)]^S \tag{4}$$

we see that the residual strength is a monotonically decreasing function of the number of cycles. The rate of decrease of the residual strength depends on the values of C and S. For $C = 1$, Eq 4 gives the residual strength for the power law fatigue failure criterion.

Let us now examine the statistical implications of the wearout model. The

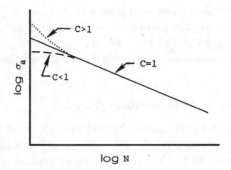

FIG. 1—*Shape of the* S-N *curve as function of parameter* C *in the wearout model (Eq 3).*

distribution of fatigue lives at a given maximum applied cyclic stress is obtained by substituting Eq 3 into Eq 2 and rearranging terms to give

$$P(N) = \exp\{-[(N - A)/\beta_f]^{\alpha_f}\} \tag{5}$$

where

$$\alpha_f = S\,\alpha \tag{6}$$

$$\beta_f = (\beta/\sigma_a)^{1/S}/C \tag{7}$$

$$A = -(1-C)/C \tag{8}$$

From Eq 5 it can be seen that the fatigue lives at a specific maximum applied cyclic stress are three-parameter Weibull distributed, with shape, scale, and location parameters given by Eqs 6 through 8, respectively. The negative location parameter, A, is a direct consequence of the fact that there is a finite probability that some specimens may fail during the first cycle since their static strength may be below the maximum applied cyclic stress. A negative location parameter has been encountered by some authors [7] when fitting a three-parameter Weibull distribution to fatigue data.

The negative location parameter in Eq 5 is a point of controversy since it implies negative lives which are physically impossible. This difficulty can be eliminated by using conditional probabilities in deriving the expression for fatigue life from the reliability function for the equivalent static strength distribution. The reliability function such that $\sigma_e \geq \sigma_a$ is given by

$$P(\sigma_e|\sigma_e > \sigma_a) = \exp[-(\sigma_e/\beta)^\alpha + (\sigma_a/\beta)^\alpha] \tag{9}$$

and Eq 5 is replaced by

$$P(N|\sigma_e > \sigma_a) = \exp\{-[(N-A)/\beta_f]^{\alpha_f} + (\sigma_\alpha/\beta)^\alpha\} \tag{10}$$

which does not suffer from the difficulties associated with Eq 5. For $N = 1$, Eq 10 reduces to $P(N = 1) = 1$ as required. Note that Eq 10 does not define a

Weibull distribution. As can be seen from a comparison of Eqs 5 and 10, the form of the fatigue life distribution depends on whether conditional probabilities are used in its derivation and not on the fatigue model. Similar statements can be made about the residual strength distribution.

Heuristic Derivation of the Data-Fitting Procedure

In order to explain the reasoning behind the S-N curve-fitting approach proposed herein, consider a heuristic fatigue experiment in which replicate specimens were tested at two distinct maximum applied cyclic stress levels until fatigue failure. The fictitious fatigue data are shown as open circles in Fig. 2, which is a log-log plot of the data. To emphasize the fictitious nature of the data, the scales have been left off the axes in the figure. It is necessary to fit a fatigue model to the data. Since data at only two maximum applied cyclic stress levels are available, it is natural to assume that the S-N curve can be described by Eq 3 with $C = 1$, that is, by a power law fatigue failure criterion. Note that, if the scatter in the fatigue data at the two maximum applied cyclic stress levels is different, a more complex fatigue model could be used.

The problem at hand is to determine the value of the slope parameter (S in Eq 3 with $C = 1$) that gives the best fit of the data. This could be done by using regression analysis in log-log space [8]. Since regression analysis is based on the implicit assumption that the logarithms of the data are normally distributed, there is some question about the applicability of this approach to data governed by a Weibull distribution. Hence, an alternative method is required.

A hint at the alternative method can be obtained by using Eq 3 with $C = 1$

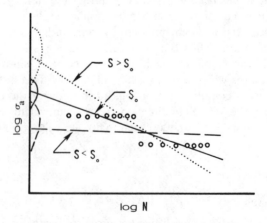

FIG. 2—*Effect of the slope parameter in the power law fatigue criterion on the scatter in the equivalent static strength.*

to transform the fatigue data into equivalent static strength data and examining the scatter in that data. The solid line in Fig. 2 shows the result of using Eq. 3 with $S = S_o$ (where S_o is the value of the slope parameter giving the tightest equivalent static strength distribution) and $C = 1$. As can be seen from the figure, the line corresponding to S_o seems to fit the fatigue data and gives a tight equivalent static strength distribution as shown by the solid curve on the ordinate axis. The dashed line, corresponding to $S < S_o$, seems to result in an equivalent static strength distribution with more scatter (shown by the dashed curve on the ordinate axis) and a poor fit of the fatigue data. A similar situation exists for $S > S_o$ and it is shown by the dotted line and dotted curve on the ordinate axis. Hence, these considerations suggest the required data-fitting procedure.

For the heuristic example, the procedure consists of (1) assuming a value of S, transforming the fatigue data into equivalent static strength data, and fitting a two-parameter Weibull distribution to the equivalent static strength data; and (2) repeating the procedure of Step (1) until the largest value of the maximum-likelihood estimate of the shape parameter (or minimum scatter in the σ_e data) is obtained. In other words, the procedure consists of maximizing the maximum-likelihood estimate of the shape parameter for the equivalent static strength distribution as a function of S. This procedure is readily generalized to cover more complex fatigue models.

Data-Fitting Procedure

The general procedure for fitting a fatigue model to experimental data consists of the following steps:

1. Set up a data traceability system to be used to identify which equivalent static strength value comes from a particular fatigue data point (maximum applied cyclic stress or load, number of cycles, and residual strength). It will be used in the analysis of the data to identify which σ_e-values are to be censored during the maximum-likelihood estimation of the Weibull parameters for the σ_e distribution. It will also be used in performing the goodness-of-fit test for the fatigue model.

2. Select the fatigue model to be used in fitting the fatigue data. The model will depend on a number of parameters. For the wearout model, the parameters are C and S.

3. Either assume or estimate from the fatigue data initial values of the fatigue model parameters. For the wearout model, initial estimates of the model parameters can be obtained as follows. If static strength data are available, let σ_1 be the lowest static strength value. Otherwise, let σ_1 be the highest maximum applied cyclic stress value. Let N_1 be the least number of cycles corresponding to σ_1. Let σ_3 be the lowest maximum applied cyclic stress value. Let N_3 be the least number of cycles corresponding to σ_3. Let σ_2

be an intermediate maximum applied cyclic stress level. The value of σ_2 should be larger than $(\sigma_1 + \sigma_3)/2$. Let N_2 be the least number of cycles corresponding to σ_2. Let

$$S_1 = (\log \sigma_2 - \log \sigma_3)/(\log N_3 - \log N_2) \tag{11}$$

$$S_0 = (\log \sigma_1 - \log \sigma_3)/(\log N_3 - \log N_1) \tag{12}$$

$$C_0 = \exp[(S_1/S_0) \log N_3 - \log N_3] \tag{13}$$

Equations 12 and 13 are the required initial estimates for the wearout model parameters.

4. Transform the fatigue data into equivalent static strength data by using the fatigue model with the assumed values of the model parameters. Make sure that the σ_e-values are traceable to the fatigue data.

5. Fit a two-parameter Weibull distribution to the σ_e data by using the maximum-likelihood method with progressive censoring. This requires the solution of nonlinear algebraic equation. The recommended procedure for doing this is described in the Appendix. Store the values of the shape and scale parameters.

6. Select a new set of model parameters and repeat Steps 4 through 6 until the maximum value of the shape parameter estimate is obtained. The maximum value of the shape parameter and the corresponding scale and fatigue model parameters constitutes the best fit of the fatigue data. Note that standard implicit function differentiation techniques can be used for efficiently selecting the new values of the model parameters in this iterative model fitting procedure.

7. Using the σ_e data corresponding to the best model parameters, prepare a probability-of-survival plot. Use different symbols for σ_e-values corresponding to the static strength and fatigue data at the different σ_a levels. If the different symbols on the probability-of-survival plot are well interspersed, then the fatigue model can be considered to fit the fatigue data well. If the different symbols are grouped in distinct regions of the probability-of-survival plot, then there is some question about the quality of the fit. A statistically rigorous test for goodness-of-fit can be performed by grouping the not-censored (nonrunout or tab failure) σ_e-values by fatigue test condition and comparing the resulting groups by using the Kruskal-Wallis H-test [9].

Application of Model-Fitting Procedure

Let us now consider the application of the fatigue model-fitting procedure to the wearout model. To make the parameter estimation easy, a Tektronix 4051 computer program was written to perform the iterative computations and prepare various plots of the data.

(0)$_8$ S2/5208 Glass-Epoxy Data

As the first example, consider the tension-tension fatigue data given in Table 1 for unidirectional S2/5208 glass-epoxy. The fatigue tests were conducted using constant-load-amplitude sinusoidal loading with a stress ratio of 0.1 and frequency of 3 Hz. The static tests were conducted at a loading rate comparable to the cyclic loading rate. The specimens came from a single panel with 68.3 percent fibers and 32.1 percent resin by volume as determined by the resin burnoff method. The best fit of the data using the present procedure is

$$S = 0.157, \quad C = 0.0485, \quad \hat{\alpha} = 19.54, \quad \hat{\beta} = 292.2 \text{ ksi} (2015 \text{ MPa}) \quad (14)$$

The experimental data and theoretical curves corresponding to $\sigma_e = \hat{\beta}$ are shown in Figs. 3 through 5. Figure 3 shows a conventional semilog plot of the S-N data, while Fig. 4 shows a log-log plot of the S-N data. Different symbols are used for the data points at the different fatigue test conditions. The data

TABLE 1—*Fatigue test results for (0)$_8$ S2/5208 glass-epoxy (stress ratio = 0.1, frequency = 3 Hz).*

Specimen No.	Maximum Cyclic Stress, ksi (MPa)	Cycles, N	Residual Strength, ksi (MPa)
22AA6	302 (2082)	1	...
22AA4	297 (2048)	1	...
22AB5	293 (2020)	1	...
22AB3	287 (1979)	1	...
22AA1	193 (1331)	153	...
22AA12	187 (1289)	267	...
22AB6	188 (1296)	319	...
22AB7	195 (1334)	436	...
22AA14	140 (965)	1 630	...
22AA9	140 (965)	1 330	...
22AB10	140 (965)	1 760	...
22AA10	140 (965)	1 220	...
22AA11	110 (758)	10 200	...
22AA2	110 (758)	9 000	...
22AB9	110 (758)	7 290	...
22AB12	110 (758)	6 750	...
22AA5	85 (586)	74 250	...
22AA2	85 (586)	67 490	...
22AB9	85 (586)	36 210	...
22AB12	85 (586)	49 800	...
22AA13	70 (483)	138 180	...
22AB11	70 (483)	93 880	...
22AB4	70 (483)	224 630	...
22AB14	70 (483)	55 780	...
22AA7	55 (379)	1 122 310	169 (1165)
22AA3	55 (379)	213 960	287 (1979)
22AB13	55 (379)	464 810	...
22AB2	55 (379)	211 800	254 (1751)

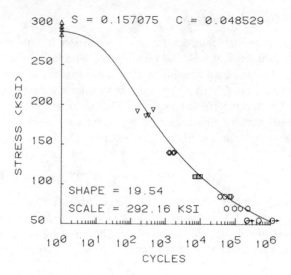

FIG. 3—S-N *curve for (0)₈ S2/5208 glass-epoxy (stress ratio = 0.1, frequency = 3 Hz).*

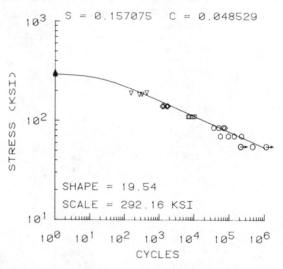

FIG. 4—S-N *curve for (0)₈ S2/5208 glass-epoxy (stress ratio = 0.1, frequency = 3 Hz).*

points with the arrows emanating from them show the number of cycles at which residual strength was measured. As can be seen from Figs. 3 and 4, the wearout model seems to give a good fit of the data. The reason that more data points lie below and to the left of the theoretical curve is that $\hat{\beta}$ is not the mean of the equivalent static strength data. Finally, Fig. 5 shows the probability of survival for the equivalent static strength data. In this figure, the

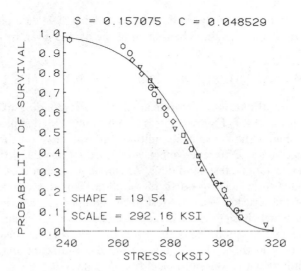

FIG. 5—*Probability of survival for the equivalent static strength data corresponding to the fatigue data in Figs. 3 and 4. Symbols are the same as in Fig. 3 and 4.*

symbols are assigned in the same manner as in Figs. 3 and 4. Hence, the distribution of the symbols in Fig. 5 gives a qualitative indication of the goodness-of-fit of the wearout model to the data. As can be seen from the figure, there is some tendency for the symbols to segregate. For example, all of the diamonds lie to the left of the triangles corresponding to the static strength data. This suggests that the wearout model may not, after all, give a good fit of the data. To obtain a quantitative measure of the goodness-of-fit, the equivalent static strength data were grouped by fatigue test condition, and the Kruskal-Wallis H-test [9] was applied to give $H = 7.5148$. Using the large sample approximation, this value of H is significant at approximately the 0.27 significance level. This indicates that the observed segregation in the data symbols can occur by chance about one time in four, which is not statistically significant. Hence, we may conclude that there is not statistical evidence that the wearout model does not fit the data.

[(0/45/90/—45)$_s$]$_s$ T300/934 Graphite-Epoxy Data [7]

As the second example, consider the tension-tension fatigue data given in Tables 8A, 1B, 7B, 8B, and 9B of Ref 7 for [(0/45/90/—45)$_s$]$_s$ T300/934 graphite-epoxy. The fatigue tests were conducted using constant-load-amplitude sinusoidal loading with a stress ratio of 0.0 and frequency of 10 Hz. The static tests were conducted at conventional loading rates. Since this laminate is not strain-rate sensitive, the static strength data (Ref 7, Table 8A) were combined with the fatigue data (Ref 7, Tables 1B, 7B, 8B, and 9B). The duplication in the fatigue data was eliminated in the analysis. The wearout model was fitted to the complete data set, consisting of 94 data points, by the procedure developed herein. The best fit of the data is given by the wearout model with

$$S = 0.0469, \quad C = 0.0181, \quad \hat{\alpha} = 23.02, \quad \hat{\beta} = 70.46 \text{ ksi (485.8 MPa)} \quad (15)$$

The experimental data and theoretical curves corresponding to $\sigma_e = \hat{\beta}$ are shown in Figs. 6 and 7. Figure 6 shows a conventional log-log plot of the S-N data, while Fig. 7 shows the probability-of-survival curve for the equivalent static strength data. Different symbols were used for σ_e-values corresponding to the different fatigue test conditions. As can be seen from these figures, the wearout model provides a good fit of the data. This was confirmed by application of the Kruskal-Wallis H-test to the grouped σ_e data. For this case, $H = 7.16$, which is significant at approximately the 0.3 significance level. This implies that the grouping of the symbols in Fig. 7 could occur by chance about one time in three. This is not statistically significant and, hence, we may conclude that the wearout model fits the data well.

Discussion

As pointed out in the previous section, the procedure for fitting fatigue models to fatigue data is applicable to a wide range of fatigue models based on the assumption that the equivalent static strength is two-parameter Weibull distributed. On occasion one may desire to use a fatigue model with

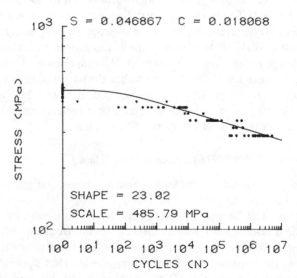

FIG. 6—S-N *curve for [(0/45/90/—45)ₛ]ₛ T300/934 graphite-epoxy (stress ratio = 0.0, fre-quency = 10 Hz) [7].*

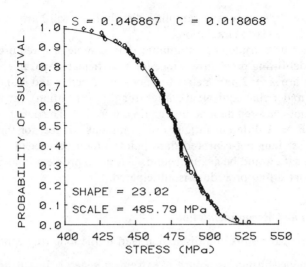

FIG. 7—*Probability of survival for the equivalent static strength data corresponding to the fatigue data in Fig. 6. Symbols are the same as in Fig. 6.*

a different underlying equivalent static strength distribution. For example, one may want to use a log-normal or a three-parameter Weibull distribution. In this situation, the procedure is exactly the same except that the maximum-likelihood estimate of the parameter that measures the data scatter would be optimized to give the least amount of scatter in the equivalent static strength distribution. Thus, for a log-normal underlying distribution, one would minimize the variance of the equivalent static strength distribution as a function of the model parameters.

Let us now consider the possible bias in the parameter estimates. As is well known, the maximum-likelihood estimator for the shape parameter of the two-parameter Weibull distribution is biased [10,11]. Hence, $\hat{\alpha}$ obtained by the present procedure will be biased. At present, the extent of the bias is not clear. It is conjectured that the bias in $\hat{\alpha}$ will depend only on the data sample size and extent of censoring, and not on the model parameters. If this conjecture turns out to be true, then $\hat{\alpha}$ can be unbiased by using tabulated unbiasing factors for the maximum-likelihood estimate of the shape parameter. Whether the model parameter estimates are biased can be determined only by Monte Carlo simulation.

In addition to the bias question, the sampling distribution for the model parameters must be determined for one to be able to compute confidence intervals on the parameter estimates and allowables. It is conjectured that the sampling distributions for all values of the model parameters will be the same as those for a standard set of fatigue model parameters. Moreover, the sampling distribution will depend only on the sample size and extent of censoring. Whether this conjecture is true can be verified by Monte Carlo simulation. Furthermore, the sampling distributions will have to be determined by Monte Carlo simulation.

Finally, let us consider the minimum data sample size required by the present model-fitting procedure. Since for a particular m-parameter fatigue model there are $m + 2$ parameters (m model parameters and α and β determining the underlying equivalent static strength distribution), the minimum number of not-censored data points required is $m + 2$. The data points must be at least $m + 1$ different fatigue test conditions. Hence, for the wearout model at least four not-censored data points would be needed. Normally, much more data would be available and, hence, no problem in applying the present model-fitting procedure is anticipated.

Conclusions and Recommendations

The theoretical results and discussion lead to the following conclusions:

1. The model-fitting procedure is extremely general. It will handle any fatigue model and it will work for a minimum set of fatigue data. This is in marked contrast to procedures proposed by other researchers (for example, see Ref 12).

2. For the wearout model, as few as four not-censored fatigue data points can be used to determine the model parameters. For small fatigue data sample sizes, the shape parameter estimate will be biased. The extent of the bias can be obtained by Monte Carlo simulation.

Based on experience with the model fitting procedure, the following recommendations are made:

1. The iterative computational procedure for maximizing $\hat{\alpha}$ for the σ_e distribution needs to be speeded up. This may be done by substituting the expression for σ_e in terms of σ_a, σ_r, n, and the model parameters into the equation defining $\hat{\alpha}$ and using implicit differentiation to obtain analytical expressions for the derivatives of $\hat{\alpha}$ with respect to the model parameters. This would eliminate the time-consuming numerical differentiation of $\hat{\alpha}$ with respect to the model parameters that currently is being used.

2. Monte Carlo simulations necessary for obtaining confidence bounds on the model parameters and allowables on the fatigue life should be performed.

APPENDIX

Maximum-Likelihood Estimators

Since tab failures and runouts can be expected to occur in the fatigue data set, the equivalent static strength data (σ_e-values) will normally include some data for which the statement "the actual σ_e-value is greater than the given value" holds. This type of datum is handled by progressive censoring when computing the maximum-likelihood estimates for the parameters of the two-parameter Weibull distribution. By using the traceability scheme introduced in the body of the paper, the σ_e data can be separated into valid (not censored) and censored groups. For the sake of clarity, assume that the σ_e data sample consists of m points of which k are censored. Arrange the data sample so that the last k data points are censored. To preserve computational accuracy and simplify the computations, normalize the σ_{ei}-values with respect to the geometric mean of the $(m$-$k)$ not-censored data points; that is, let

$$X_i = \sigma_{ei}/G \quad (i = 1,2,...,m) \tag{16}$$

where

$$G = \exp\left[1/(m - k)\right] \sum_{i=1}^{m-k} \log \sigma_{ei} \tag{17}$$

The maximum-likelihood estimate for the shape parameter is obtained by solving the following nonlinear algebraic equation

$$\sum_{i=1}^{m} [1 - \hat{\alpha} \log X_i] X_i^{\hat{\alpha}} = 0 \tag{18}$$

The maximum-likelihood estimate for the scale parameter is given by

$$\hat{\beta} = G\left\{[1/(m-k)] \sum_{i=1}^{m} X_i^{\hat{\alpha}}\right\}^{1/\hat{\alpha}} \tag{19}$$

The recommended procedure for solving Eq 18 is

1. Compute an initial estimate, $\hat{\alpha}_i$, for $\hat{\alpha}$ as

$$\hat{\alpha}_i = \log\{\log[m/(m+1)]/\log[1/(m+1)]\}/\log(X_{max}/X_{min}) \tag{20}$$

where X_{max} and X_{min} are the largest and smallest values of X_i in the normalized data sample.

2. Compute the sums

$$S_1 = \sum_{i=1}^{m} X_i^{\hat{\alpha}_i} \tag{21}$$

$$S_2 = \sum_{i=1}^{m} X_i^{\hat{\alpha}_i} \log X_i \tag{22}$$

$$S_3 = \sum_{i=1}^{m} X_i^{\hat{\alpha}_i} (\log X_i)^2 \tag{23}$$

3. Compute $\Delta\hat{\alpha}$ by using

$$\Delta\hat{\alpha} = (S_1 - \hat{\alpha}_i S_2)/(\hat{\alpha}_i S_3) \tag{24}$$

4. Compute $|\Delta\hat{\alpha}/\hat{\alpha}_i|$. If $|\Delta\hat{\alpha}/\hat{\alpha}_i| > 10^{-6}$, set

$$\hat{\alpha}_i = \hat{\alpha}_i + \Delta\hat{\alpha} \tag{25}$$

and repeat Steps 2 through 4. If $|\Delta\hat{\alpha}/\hat{\alpha}_i| < 10^{-6}$, set $\hat{\alpha} = \hat{\alpha}_i$ and the computations are completed.

This procedure converges very rapidly. Normally, fewer than 10 iterations are required. If the recommended procedure is used, the $\hat{\beta}$ is given by

$$\hat{\beta} = G\,[S_1/(m-k)]^{1/\alpha} \tag{26}$$

where S_1 is the most current value of the sum defined by Eq 21 in the iterative procedure.

References

[1] Halpin, J. C., Jerina, K. L., and Johnson, T. A. in *Analysis of Test Methods for High Modulus Fibers and Composites, ASTM STP 521,* American Society for Testing and Materials, 1973, pp. 5-64.
[2] Hahn, H. T. and Kim, R. Y., *Journal of Composite Materials,* Vol. 9, July 1975, pp. 297-311.
[3] Yang, J. N., *Journal of Composite Materials,* Vol. 12, Jan. 1978, pp. 19-39.
[4] Yang, J. N., Liu, M. D., *Journal of Composite Materials,* Vol. 11, April 1977, pp. 176-203.
[5] Chou, P. C. and Croman, R., *Journal of Composite Materials,* Vol. 12, April 1978, pp. 177-194.
[6] Yang, J. N., *Journal of Composite Materials,* Vol. 12, Oct. 1978, pp. 371-389.
[7] Ryder, J. T. and Walker, E. K., "Ascertainment of the Effect of Compression Loading on the Fatigue Lifetime of Graphite-Epoxy Laminates for Structural Applications," AFML-Tr-76-241, Air Force Material Laboratory, Wright-Patterson Air Force Base, Ohio, Dec. 1976.
[8] Little, R. E. and Jebe, E. H., *Statistical Design of Fatigue Experiments,* Halsted Press, New York, 1975.
[9] Hollander, M. and Wolfe, D. A., *Nonparametric Statistical Methods,* Wiley, New York, 1973.
[10] Thoman, D. R., Bain, L. J., and Antle, C. E., *Technometrics,* Vol. 11, No. 3, Aug. 1969, pp. 445-460.
[11] Billman, B. R., Antle, C. E., and Bain, L. J., *Technometrics,* Vol. 14, 1972, pp. 831-840.
[12] Whitney, J. M. in *Fatigue of Fibrous Composite Materials, ASTM STP 723,* American Society for Testing and Materials, 1981, pp. 133-151.

C. C. Chamis,[1] R. F. Lark,[1] and J. H. Sinclair[1]

Mechanical Property Characterization of Intraply Hybrid Composites

REFERENCE: Chamis, C. C., Lark, R. F., and Sinclair, J. H., **"Mechanical Property Characterization of Intraply Hybrid Composites,"** *Test Methods and Design Allowables for Fibrous Composites, ASTM STP 734,* C. C. Chamis, Ed., American Society for Testing and Materials, 1981, pp. 261–280.

ABSTRACT: An investigation was conducted to characterize the mechanical properties of intraply hybrids made from graphite fiber/epoxy matrix (primary composites) hybridized with varying amounts of secondary composites made from S-glass or Kevlar 49 fibers. The tests were conducted using thin laminates having the same thickness. The specimens for these tests were instrumented with strain gages to determine stress-strain behavior. The results show that the mechanical properties of intraply hybrid composites can be measured using available test methods such as the 10-deg off-axis method for intralaminar shear, and conventional test methods for tensile, flexure, and Izod impact properties. Intraply hybrids have linear stress-strain curves to fracture for longitudinal tension and nonlinear stress-strain curves for intralaminar shear.

The results also showed that combinations of high-modulus graphite/S-glass/epoxy matrix composites exist which yield intraply hybrid laminates with the "best" balanced properties: for example, 100 percent increase in impact resistance and 35 percent increase in tensile and flexural strengths, with no reduction in modulus compared with graphite fiber/epoxy matrix composites. In addition, the results showed that the translation efficiency of mechanical properties from the constituent composites to intraply hybrids may be assessed using a simple equation.

KEY WORDS: intraply hybrids, fiber composites, graphite fibers, S-glass fibers, Kevlar fibers, epoxy resins, elastic properties, tensile strength, flexural strength, intralaminar-shear strength impact resistance, property-translation efficiency, experimental data, stress-strain curves, composite materials

Intraply hybrid composites have two kinds of fibers embedded in the matrix, in general within the same ply. They have evolved as a logical sequence to conventional composites and to interply hybrids. Intraply hybrid composites have unique features that can be used to meet diverse and com-

---

[1]Aerospace and composite structures engineer, materials engineer, and aerospace materials engineer, respectively, NASA Lewis Research Center, Cleveland, Ohio 44135.

peting design requirements in a more cost-effective way than either advanced or conventional composites. Some of the specific advantages of intraply hybrids over other composites are balanced strength and stiffness, balanced bending and membrane mechanical properties, balanced thermal distortion stability, reduced weight or cost or both, improved fatigue resistance, reduced notch sensitivity, improved fracture toughness or crack-arresting properties or both, and improved impact resistance. By using intraply hybrids, it is possible to obtain a viable compromise between mechanical properties and cost to meet specified design requirements.

The available methodology for analysis and design of intraply hybrids, as well as areas that need further research, was covered in a recent review on hybrid composites in general [1].[2] Two of the areas identified in that reference are (1) the development of micromechanics equations for predicting the various mechanical and thermal properties of unidirectional intraply hybrids, and (2) the characterization of mechanical properties of intraply hybrid composites. Approximate equations based on the "rule of mixtures" were presented in Ref 2. Equations based on micromechanics concepts are described in Ref 3. Comparisons of properties using these micromechanics equations, linear laminate theory, and finite-element analysis are also given in Ref. 3. Verification of all these predictive methods requires measured properties obtained from the same laminate in order to minimize any effects that may be induced by processing and fabrication variables. The objective of this investigation was to determine whether available test methods for measuring mechanical properties such as longitudinal and transverse tensile, shear, flexural, and Izod impact strengths can be used for the mechanical property characterization of intraply hybrids using thin composite laminates. Another objective was to assess the load transfer efficiency from the constituent composites to the intraply hybrid using available equations.

Constituent Composites and Intraply Hybrids

The constituent composites used in this investigation were made from low- and high-modulus graphite fibers (AS and HMS), S-glass and Kevlar 49 fibers, and PR288 epoxy resin matrix. These constituent composites will be referred to, respectively, as AS/E, HMS/E, S-G/E, and KEV/E throughout the paper.

The unidirectional properties of the constituent composites that were used in this investigation are summarized in Table 1. The use of the properties in this table will be described later.

The intraply hybrids made from these constituent composites consisted of the following primary/secondary composite volume percentages: 90/10, 80/20, and 70/30 of AS/E with either S-G/E or KEV/E, and HMS/E with

---

[2]The italic numbers in brackets refer to the list of references appended to this paper.

TABLE 1—*Unidirectional properties of constituent composites, experimentally measured.*

Property	Composite			
	AS/E	HMS/E	S-G/E	KEV 49/E
Longitudinal strength, ksi	213.7	152.6	192.3	186
Transverse strength, ksi	10.4	2.88	11.2	4.1
Intralaminar shear strength, ksi	13.0	6.5	10.7	6.5
Longitudinal strain, %	1.12	0.535	2.84	1.73
Transverse strain, %	0.83	0.300	0.57	−0.76
Intralaminar shear strain, %	5.17	0.96	4.13	2.36[a]
Longitudinal modulus, 10^6 psi	18.2	26.5	6.95	11.2
Transverse modulus, 10^6 psi	1.28	0.95	2.17	0.80
Shear modulus, 10^6 psi	0.600	0.779	0.644	0.41
Major Poisson's ratio	0.32	0.25	0.30	0.44
Minor Poisson's ratio	0.05	0.022	0.075	0.029
Flexural strength (longitudinal), ksi	230.3	122.5	318	105
Flexural strength (transverse), ksi	17.8	7[a]	21.2	5.8
Izod impact (longitudinal), in.-lb/in.^{2}	241.3	84[a]	1260.0	790.8
Izod impact (transverse), in.-lb/in.^{2}	41.3	5[a]	69.6	25.2

[a]Estimated.
Most data based on average value of three specimens, two gages each, back-to-back.
Conversion factors: ksi = 6.89 MPa; 10^6 psi = 6.89 GPa; in.-lb/in.^{2} = 0.1458 cm N/cm^2.

either S-G/E or KEV/E. These intraply hybrids will be identified using the notation AS/E//S-G/E, As/E//KEV/E, HMS/E//S-G/E, and HMS/E//KEV/E.

Specimen Fabrication, Preparation, Instrumentation, and Testing

Constituents and intraply hybrid composite laminates were made by press-curing a total of eight unidirectional prepreg plies into laminates having a thickness of 0.10 cm (0.040 in.), a width of 15 cm (6 in.), and a length of 30 cm (12 in.). The constituent and intraply hybrid composite plies were made by combining continuous strands of fibers and a matrix resin, followed by staging to provide a prepreg material that could be cut and fitted into the laminate molds. The intraply hybrid composite plies were made by combining various percentages, by volume, of the primary composites with secondary, or hybridizing, composites in a "tow-by-tow" fashion (Fig. 1) that grouped the fibers in discrete bundles within the ply to give the volume percentages mentioned previously. A PR288 epoxy resin system (3M Co. designation) was used as the resin matrix for all of the laminates. The supplier's recommended curing procedure was used for fabrication of the laminates [2 h at 149°C (300°F)].

The laminates were cut into 1.27-cm-wide (0.5 in.) specimens by using a precision wafer-cutting machine equipped with a diamond wheel. A typical laminate specimen cutting plan is shown in Fig. 2.

The ends of the specimens subjected to tensile loading were reinforced with

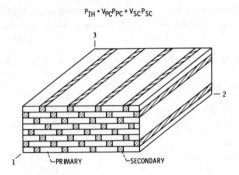

$$P_{IH} \approx V_{PC}P_{PC} + V_{SC}P_{SC}$$

FIG. 1—*Schematic of unidirectional intraply hybrid composite and approximate equation for property translation efficiency.*

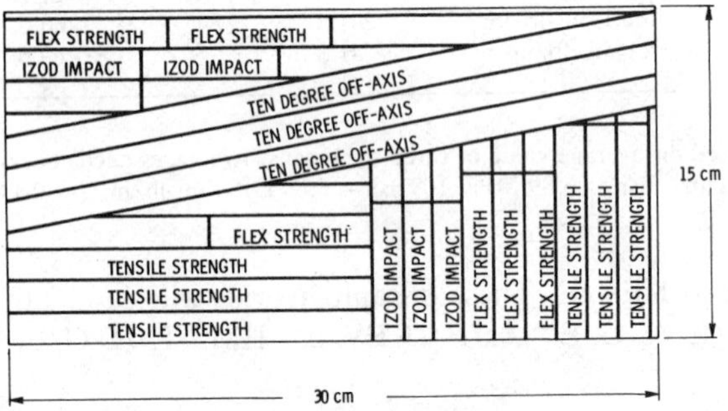

FIG. 2—*Laminate cutting plan (1cm = 0.4 in.).*

fiber glass/epoxy tabs adhesively bonded to the specimen surfaces. The longitudinal and transverse tension and 10-deg off-axis shear specimens were equipped with back-to-back strain gages. Details of the types and locations of the strain gages along with specimen dimensions are shown in Fig. 3.

Three replicates of tension specimens for longitudinal, transverse, and 10-deg off-axis properties were loaded to fracture using a mechanically actuated universal testing machine. The loading rate was 0.13 cm/min (0.05 in./min). Loading of all specimens was halted at periodic intervals so that strain-gage data could be obtained using a digital strain recorder. The digital data were processed using a strain-gage data reduction computer program [4] for stress-strain curves, moduli, and Poisson's ratios. This computer program also generates the intralaminar shear stress-strain curves and moduli from the 10-deg off-axis tensile data as described in Ref. 5.

The flexure specimens were tested for flexural strength in a mechanically actuated universal testing machine using a three-point loading system. The length of the specimens was 7.62 cm (3 in.). The span between supports was 5.08 cm (2 in.) or a span-to-depth ratio of about 51, which is considered more

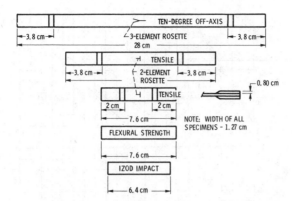

FIG. 3—*Specimen and instrumentation details (1 cm = 0.4 in.).*

than adequate for measuring flexural properties with negligible contribution from interlaminar shear. The flexural strength was calculated from the bending load at fracture using the simple beam equation.

The Izod impact specimens had a cantilever length of 3.2 cm (1.25 in.) and were tested using an Izod impact tester (TMI) equipped with a 0.9-kg (2 lb) hammer. The velocity of the hammer was approximately 3 m/s (10 ft/s). The data obtained were normalized with respect to the cross-sectional area of the specimens for convenience of comparison.

Results, Comparison, and Discussion

Typical stress-strain curves obtained from the reduction of the strain gage data are shown in Figs. 4–6. The curves in Figs. 4 and 5 show linear and approximately linear behavior to fracture for longitudinal and transverse tension. One conclusion from the curves in Figs. 4 and 5 is that the intraply hybrids exhibit "hybrid action." If this were not the case, the stress-strain curves would exhibit at least a bilinear behavior to fracture. The deviation from the first linear portion would occur after extensive fractures in the primary composite (AS/E or HMS/E). The intralaminar shear stress-strain curve in Fig. 6 is nonlinear, which should be expected since the corresponding curves of the constituents are also nonlinear. Photographs of typical fractured specimens are shown in Fig. 7. As can be seen, the specimens failed within the test gage section.

The measured results—averages of three replicates—for the mechanical properties of the various intraply hybrids are summarized in Tables 2–5. The mechanical properties for the AS/E//S-G/E hybrid are given in Table 2, for HMS/E//S-G/E in Table 3, for AS/E//KEV/E in Table 4, and for HMS/E//KEV/E in Table 5.

To facilitate comparisons and discussion, significant properties of the in-

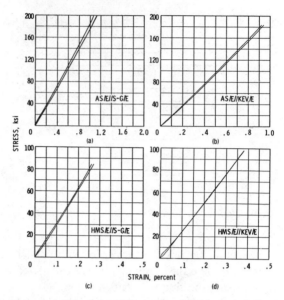

FIG. 4—*Longitudinal tensile stress-strain curves of intraply hybrid composites (80/20, volume percent of constituent composites; ksi = 6.89 MPa).*

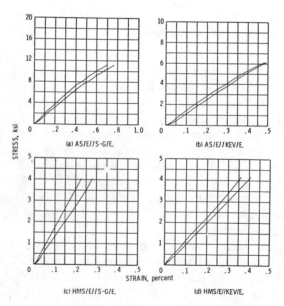

FIG. 5—*Transverse tensile stress-strain curves of intraply hybrid composites (80/20, volume percent of constituent composites; ksi = 6.89 MPa).*

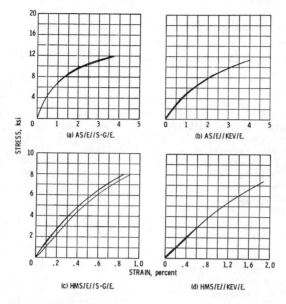

FIG. 6—*Intralaminar shear stress-strain curves (10-deg off-axis) of intraply hybrid composites (80/20, volume percent of constituent composites; ksi = 6.89 MPa).*

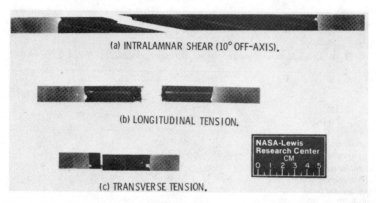

FIG. 7—*Fractured specimens of intraply hybrid composite (80/20 volume percent; AS/E//S-G/E).*

traply hybrids and the constituent properties are summarized in bar charts in Figs. 8–11. The bar chart summary for the tensile strength is shown in Fig. 8. It can be seen in this figure that the intraply hybrids from AS/E//S-G/E and AS/E//KEV/E utilize the tensile strength of the constituent composites effectively. That is, the tensile strength of these intraply hybrids is about equal to or greater than the lower property of the constituent composites (AS/E, S-G/E, or KEV/E). The tensile strength of the 90/10 AS/E//S-G/E is

TABLE 2—*Measured properties of intraply hybrids AS/E//S-G/E.*

Property	% Constituents (Primary/Secondary)	
	90/10	80/20
Longitudinal tensile strength, ksi	265	191
Transverse tensile strength, ksi	10.8	9.5
Intralaminar shear strength, ksi	14.4	12.3
Longitudinal tensile strain, %	1.3	1.06
Transverse tensile strain, %	0.74	0.63
Intralaminar shear strain, %	3.03	3.05
Longitudinal modulus, 10^6 psi	20	17.8
Transverse modulus, 10^6 psi	1.6	1.7
Shear modulus, 10^6 psi	1.12	0.925
Major Poisson's ratio	0.31	0.30
Minor Poisson's ratio	0.03	0.03
Flexural strength (longitudinal), ksi	263	275
Flexural strength (transverse), ksi	21.3	22.7
Izod impact (longitudinal), in.-lb/in.^{2}	388	522
Izod impact (transverse), in.-lb/in.^{2}	18.3	26.7

Conversion factors: ksi = 6.89 MPa; 10^6 psi = 6.89 GPa; in.-lb/in.^{2} = 0.1458 cm N/cm^2.

TABLE 3—*Measured properties on intraply hybrids HMS/E//S-G/E.*

Property	% Constituents (Primary/Secondary)		
	90/10	80/20	70/30
Longitudinal tensile strength, ksi	84.7	81.3	109
Transverse tensile strength, ksi	5.0	4.2	6.1
Intralaminar shear strength, ksi	8.15	8.09	9.5
Longitudinal tensile strain, %	0.38	0.31	0.45
Transverse tensile strain, %	0.36	0.34	0.35
Intralaminar shear strain, %	1.40	0.84	0.70
Longitudinal modulus, 10^6 psi	30.4	29.6	24.1
Transverse modulus, 10^6 psi	1.4	1.5	1.9
Shear modulus, 10^6 psi	0.87	1.38	1.3
Major Poisson's ratio	0.30	0.32	0.27
Minor Poisson's ratio	0.014	0.02	0.027
Flexural strength (longitudinal), ksi	109	148	153
Flexural strength (transverse), ksi	7.9	10.6	13.1
Izod impact (longitudinal), in.-lb/in.^{2}	324	453	618
Izod impact (transverse), in.-lb/in.^{2}	5.7	12.0	12.6

Conversion factors: ksi = 6.89 MPa; 10^6 psi = 6.89 GPa; in.-lb/in.^{2} = 0.1458 cm N/cm^2.

TABLE 4—*Measured properties of intraply hybrids AS/E//KEV/E.*

Property	% Constituents (Primary/Secondary)		
	90/10	80/20	70/30
Longitudinal tensile strength, ksi	196	204	205
Transverse tensile strength, ksi	8.4	6.7	5.4
Intralaminar shear strength, ksi	10.5	11.6	10.9
Longitudinal tensile strain, %	0.38	1.13	1.01
Transverse tensile strain, %	0.40	0.54	0.45
Intralaminar shear strain, %	2.72[a]	2.89	3.44[a]
Longitudinal modulus, 10^6 psi	18.5	17.8	16.8
Transverse modulus, 10^6 psi	1.4	1.4	1.2
Shear modulus, 10^6 psi	0.78	0.81	0.64
Major Poisson's ratio	0.32	0.33	0.30
Minor Poisson's ratio	0.015	0.045	0.03
Flexural strength (longitudinal), ksi	205	246	253
Flexural strength (transverse), ksi	7.4	12.9	10.1
Izod impact (longitudinal), in.-lb/in.^{2}	302	376	408
Izod impact (transverse), in.-lb/in.^{2}	11.1	9.4	9.6

[a]Estimated.
Conversion factors: ksi = 6.89 MPa; 10^6 psi = 6.89 GPa; in.-lb/in.^{2} = 0.1458 cm N/cm^2.

TABLE 5—*Measured properties of intraply hybrids HMS/E//KEV/E.*

Property	% Constituents (Primary/Secondary)		
	90/10	80/20	70/30
Longitudinal tensile strength, ksi	103	105	110
Transverse tensile strength, ksi	4.6	5.0	5.3
Intralaminar shear strength, ksi	7.99	7.97	7.52
Longitudinal tensile strain, %	0.37	0.38	0.43
Transverse tensile strain, %	0.40	0.43	0.52
Intralaminar shear strain, %	1.44	1.42	1.59
Longitudinal modulus, 10^6 psi	26.8	26.9	25.9
Transverse modulus, 10^6 psi	1.4	1.1	1.0
Shear modulus, 10^6 psi	0.745	0.549	0.659
Major Poisson's ratio	0.33	0.27	0.35
Minor Poisson's ratio	0.02	0.02	0.017
Flexural strength (longitudinal), ksi	205	130	130
Flexural strength (transverse), ksi	7.4	9.7	10.1
Izod impact (longitudinal), in.-lb/in.^{2}	190	196	177
Izod impact (transverse), in.-lb/in.^{2}	11.1	5.7	5.7

Conversion factors: ksi = 6.89 MPa; 10^6 psi = 6.89 GPa; in.-lb/in.^{2} = 0.1458 cm N/cm^2.

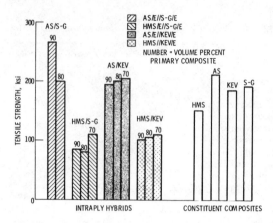

FIG. 8—*Tensile strength comparisons (ksi = 6.89 MPa).*

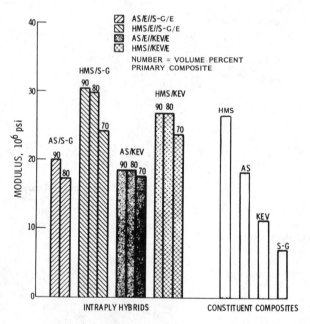

FIG. 9—*Tensile modulus comparisons (10^6 psi = 6.89 GPa).*

about 24 percent greater than the tensile strength of the AS/E constituent composite, indicating some synergistic effect.

The bar chart summary for tensile modulus is shown in Fig. 9. It can be seen in this figure that all intraply hybrids utilize the tensile modulus of the constituent composites effectively. The bar chart summary for flexural strength is shown in Fig. 10. Again, all the intraply hybrids utilize the flex-

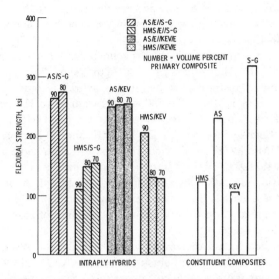

FIG. 10—*Flexural strength comparisons (ksi = 6.89 MPa).*

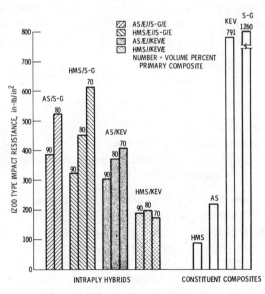

FIG. 11—*Izod-type impact comparisons (in.-lb/in.^{2} = 0.1458 cm N/cm^2).*

ural strength of the constituent composites effectively. The AS/E//KEV/E intraply hybrids show some 8 to 20 percent synergistic effect, while the 90/10 HMS/E//KEV/E intraply hybrid shows considerable (about 69 percent) synergistic effect.

The bar chart summary for thin specimen Izod longitudinal impact is shown in Fig. 11. The results in this figure show improvement in the longitudinal impact resistance of the intraply hybrids, relative to the primary composite (AS/E or HMS/E), as follows: (1) from 61 to 117 percent for the AS/E//S-G/E; (2) from 286 to 449 percent for the HMS/E//S-G/E; (3) from 25 to 169 percent for the AS/E//KEV/E; and (4) from 111 to 133 percent for the HMS/E//KEV/E. Note that the test data show a decrease for the 70/30 HMS/E//KEV/E intraply hybrid, which may indicate that an optimum hybridizing ratio exists for this class of intraply hybrids. The conclusion from these data is that hybridizing via the intraply hybrid is a very effective way for greatly improving the impact resistance of graphite fiber composites.

Taking the data for all the properties collectively, the AS/E//S-G/E intraply hybrids utilize the constituents most effectively. These intraply hybrids provide significant improvement in impact resistance, some improvement in tensile and flexural strengths, and negligible or no degradation in tensile modulus. Also, large improvements in impact resistance can be realized by hybridizing HMS/E with S-G/E.

The discussion thus far has been relative to comparisons of intraply hybrid properties with the properties of either one or both constituent composites. The anitcipated properties for intraply hybrids may be predicted from the constituent composite properties by using the following rule-of-mixtures equation

$$P_{HC} = P_{PC} + V_{SC} (P_{SC} - P_{PC}) \tag{1}$$

where

$$P = \text{property,}$$
$$V = \text{volume ratio, and}$$
$$\text{HC, PC and SC} = \text{hybrid composite, primary composite, and secondary composite, respectively.}$$

Detailed justifications for using Eq 1 are given in Ref 2 and 3. For the present discussion, it is sufficient to say that the derivation of Eq 1 requires complete hybrid response. This means (1) perfect bond between the constituent composites and (2) 100 percent property translation from the constituent composites to the intraply hybrid. Comparison of measured and predicted properties using Eq 1 provides an indication of the effectiveness of property translation and, indirectly, of the quality of the intraply hybrid.

Elastic and strength properties of the intraply hybrids predicted using Eq 1 are summarized in Tables 6-9. For convenience of comparison, the

TABLE 6—Comparison of measured and predicted properties for intraply hybrid AS/E//S-G/E.

Property	% Constituents (Primary/Secondary)								
	90/10			80/20			70/30		
	Measured	Predicted	%[a]	Measured	Predicted	%[a]	Measured	Predicted	%[a]
Modulus, 10^6 psi									
longitudinal tension	20.0	17.1	17.0	17.8	15.9	11.9	...	14.8	...
transverse tension	1.6	1.4	14.3	1.7	1.5	13.3	...	1.5	...
shear	1.12	0.60	86.7	0.925	0.61	51.6	...	0.61	...
Poisson's ratio	0.31	0.32	-3.1	0.30	0.32	-6.3	...	0.31	...
Strength, ksi									
longitudinal tension	265	212	25.0	191	209	-8.6	...	193	...
transverse tension	10.8	10.5	2.9	9.5	10.6	-10.4	...	10.6	...
intralaminar shear	14.4	12.8	12.5	12.3	12.5	-1.6	...	12.3	...
longitudinal flexure	263	239	10.0	275	248	-10.9	...	257	...
transverse flexure	21.3	18.1	17.7	22.7	18.5	22.7	...	18.8	...
Thin specimen									
Izod impact, in.-lb/in.^{2}									
longitudinal	328	343	-4.4	522	445	17.3	...	547	...
transverse	18.3	44.1	-58.5	26.7	47.0	-43.2	...	49.8	...

[a]With respect to predicted value.
Conversion factors: ksi = 6.89 MPa; 10^6 psi = 6.89 GPa; in.-lb/in.^{2} = 0.1458 cm N/cm^2.

TABLE 7—Comparison of measured and predicted properies for intraply hybrid HMS/E//S-G/E.

| | % Constituents (Primary/Secondary) | | | | | | | | |
| | 90/10 | | | 80/20 | | | 70/30 | | |
Property	Measured	Predicted	%[a]	Measured	Predicted	%[a]	Measured	Predicted	%[a]
Modulus, 10^6 psi									
longitudinal tension	30.4	24.5	24.1	29.6	22.6	31.0	24.1	20.6	17.0
transverse tension	1.4	1.1	27.3	1.5	1.2	25.0	1.9	1.3	46.1
shear	0.87	0.77	13.0	1.38	0.75	84.0	1.3	0.74	75.7
Poisson's ratio	0.30	0.32	−6.3	0.32	0.32	0.0	0.27	0.31	−12.9
Strength, ksi									
longitudinal tension	84.7	157	−46.1	81.3	161	−49.5	109	165	−33.9
transverse tension	5.0	3.7	35.1	4.2	4.5	−6.7	6.1	5.4	13.0
intralaminar shear	8.15	6.9	18.1	8.09	7.3	10.8	9.5	7.8	21.8
longitudinal flexure	109	142	−23.2	148	162	−8.6	153	181	−15.5
transverse flexure	7.9	8.4	−5.9	10.6	9.8	8.2	13.1	11.3	15.9
Thin specimen									
Izod impact, in.-lb/in.^{2}									
longitudinal	324	202	60.4	453	319	42.0	618	437	41.4
transverse	5.7	11.5	−50.4	12.0	17.9	−33.0	12.6	24.5	−48.6

[a]With respect to predicted value.
Conversion factors: ksi = 6.89 MPa; 10^6 psi = 6.89 GPa; in.-lb/in.^{2} = 0.1458 cm N/cm^2.

TABLE 8—Comparison of measured and predicted properties for intraply hybrid AS/E//KEV 49/E.

| | % Constituents (Primary/Secondary) | | | | | | | | |
| | 90/10 | | | 80/20 | | | 70/30 | | |
Property	Measured	Predicted	%[a]	Measured	Predicted	%[a]	Measured	Predicted	%[a]
Modulus, 10^6 psi									
longitudinal tension	18.5	17.5	5.7	17.8	16.8	6.0	16.8	16.1	4.3
transverse tension	1.4	1.2	16.7	1.4	1.2	16.7	1.2	1.1	9.1
shear	0.78	0.58	34.5	0.81	0.56	44.6	0.64	0.54	18.5
Poisson's ratio	0.32	0.33	-3.0	0.33	0.34	-2.9	0.30	0.36	-16.7
Strength, ksi									
longitudinal tension	196	211	-7.1	204	208	-1.9	205	205	0.0
transverse tension	8.4	9.8	-14.3	6.7	9.1	-26.4	5.4	8.5	-36.5
intralaminar shear	10.5	12.3	-14.6	11.6	11.7	-0.9	10.9	11.1	-1.8
longitudinal flexure	205	218	-6.0	246	205	20.0	253	193	31.1
transverse flexure	7.4	16.6	-55.4	12.9	15.4	-16.2	10.1	14.2	-289
Thin specimen									
Izod impact, in.-lb/in.^{2}									
longitudinal	190	296	-35.8	376	351	7.1	408	406	0.5
transverse	11.1	39.7	-72.0	9.4	38.1	-75.3	9.6	36.5	-73.7

[a]With respect to predicted value.

Conversion factors: ksi = 6.89 MPa; 10^6 psi = 6.89 GPa; in.-lb/in.^{2} = 0.1458 cm N/cm^2.

TABLE 9—Comparison of measured and predicted properties for intraply hybrid HMS/E//KEV 49/E.

| Property | % Constituents (Primary/Secondary) | | | | | | | | |
| | 90/10 | | | 80/20 | | | 70/30 | | |
	Measured	Predicted	%[a]	Measured	Predicted	%[a]	Measured	Predicted	%[a]
Modulus, 10^6 psi									
longitudinal tension	26.8	25.0	7.2	26.9	23.4	15.0	25.9	21.9	18.3
transverse tension	1.4	0.94	48.9	1.1	0.92	19.6	1.0	0.91	9.9
shear	0.745	0.742	0.4	0.549	0.705	−22.1	0.659	0.668	−1.3
Poisson's ratio	0.33	0.27	22.2	0.27	0.29	−6.9	0.35	0.31	12.9
Strength, ksi									
longitudinal tension	103	156	−34.0	105	159	−34.0	110	163	−32.5
transverse tension	4.6	3.0	53.3	5.0	3.1	61.3	3.2	3.2	0.0
intralaminar shear	7.99	6.5	22.9	7.97	6.5	22.6	7.52	6.5	15.7
longitudinal flexure	205	121	69.4	130	119	9.2	130	117	11.1
transverse flexure	7.4	6.9	7.2	9.7	6.8	42.6	10.1	6.6	53.0
Thin specimen									
Izod impact, in.-lb/in.^{2}									
longitudinal	190	155	22.6	196	225	−12.9	177	296	−40.2
transverse	11.1	7.0	58.6	5.7	9.0	−36.7	5.7	11.9	−52.1

[a]With respect to predicted value.

Conversion factors: ksi = 6.89 MPa; 10^6 psi = 6.89 GPa; in.-lb/in.^{2} = 0.1458 cm N/cm^2.

measured properties in these tables are normalized with respect to the corresponding predicted properties. The normalized results are summarized graphically in Fig. 12 for elastic properties and in Fig. 13 for strengths. The normalized results in these figures represent a measure of the efficiency of property translation from the constituent composites to the intraply hybrid as follows: (1) Unity values indicate 100 percent property translation (complete hybrid response); (2) greater-than-unity values indicate some "synergistic effect" for all the properties or a concentration of volume of the stronger constituent or both at the fracture surface for strengths; and (3) less-than-unity values indicate incomplete hybrid response (partial bond between constituents) for all the properties or a concentration of volume of the weaker constituent or both at the fracture surface for strengths.

It can be seen in Fig. 12 that the normalized results for the elastic properties lie either slightly below or above the unity value line in general. Therefore, the intraply hybrids exhibit complete hybrid response for elastic properties. The consistently higher-than-unity values for shear modulus (except for HMS/E//KEV/E) most probably indicate an S-glass-rich region at the strain-gage location.

The AS/E//S-G/E intraply hybrids show complete hybrid response (efficient property translation) for strengths except for transverse impact (TI) Fig. 13a). The low translation efficiency for TI may be, in part, due to the dynamic stress transfer at the interface of the constituent composites near the cantilever end of the Izod impact specimen. The HMS/E//S-G/E intraply hybrids show low efficiency in property translation for TI and longitudinal

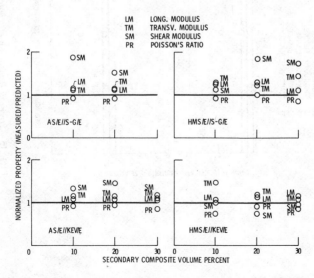

FIG. 12—*Elastic property translation efficiency summary of intraply hybrids (average of three replicates).*

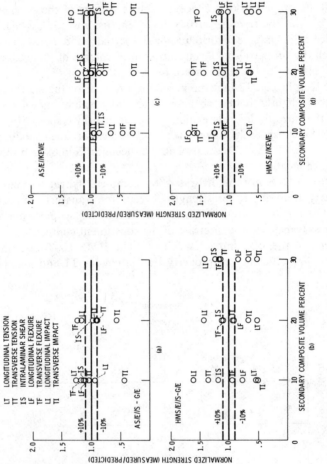

FIG. 13—*Strength translation efficiency summary of intraply hybrids (average of three replicates).*

tension (LT) strength (Fig. 13*b*). The reason, mentioned previously for the AS/E//S-G/E hybrid, is believed to be the cause for the low efficiency for TI. The low efficiency property translation for LT strength is due mainly to partial hybrid action caused perhaps by insufficient bond between the constituents at the interface. For example, the calculated longitudinal stress in the HMS/E composite at fracture is 503 MPa (73 ksi), which is about 48 percent of its unidirectional strength [1055 MPa (153 ksi), Table 1]. The AS/E//KEV/E intraply hybrid also shows low efficiency in property translation (Fig. 13*c*), while the HMS/E//KEV/E shows good efficiency except for TI (Fig. 13*d*). Taken collectively, the strength data in Fig. 13 show the following: (1) AS/E//S-G/E and HMS/E//KEV/E intraply hybrids have high efficiency in strength translation; (2) HMS/E//S-G/E and AS/E//KEV/E intraply hybrids are inefficient in strength translation; and (3) the intraply hybrids have poor transverse impact resistance.

Based on the correlation between measured data and Eq 1, it may be concluded that 8-ply-thick laminates can be used to characterize the tensile, flexural, and Izod impact properties of unidirectional intraply hybrids. Also, for the same reason, a specimen width of 1.27 cm (0.50 in.) appears to be sufficient. Specimens from the same thin laminate should be suitable for characterizing compressive properties of unidirectional intraply hybrids in compression test fixtures which provide lateral supports. Specimens from the same thin laminate should also be suitable for properties such as fatigue resistance, notch sensitivity, and environmental effects. Data from all these tests should provide a broad base to verify available predictive methods as well as to provide a basis for formulating new ones.

Summary of Results

An investigation was conducted to characterize the tensile, flexural, and Izod impact properties of intraply hybrid composites, and to assess the effective use of the constituent composites in the intraply hybrid as well as efficiency in property translation. The primary constituent composites were graphite fiber AS/epoxy PR288 and HMS/epoxy PR288 (AS/E and HMS/E). The secondary constituent composites were S-glass fiber/epoxy PR288 and Kevlar 49-fiber/epoxy PR288 (S-G/E and KEV/E). Intraply hybrids were made from the following volume percentages of primary/secondary composite: 90/10, 80/20, and 70/30 from combinations of (primary//secondary) AS/E//S-G/E, AS/E//KEV/E, HMS/E//S-G/E, and HMS/E//KEV/E. The major results from this investigation are as follows:

1. Thin laminates (8 plies thick) are suitable to characterize the tensile, flexural, and Izod impact properties of unidirectional intraply hybrids.

2. Stress-strain curves of these intraply hybrids exhibit linear or approx-

imately linear behavior to fracture for longitudinal and transverse tension and nonlinear behavior for intralaminar shear. Test specimens fractured within the test gage section.

3. Intraply hybrids utilize the constituents effectively; that is, the intraply hybrid property is greater than that of its weaker constituent.

4. Intraply hybrids exhibit complete hybrid response and show high translation efficiency (100 percent or greater) in elastic properties (moduli and Poisson's ratio).

5. Intraply hybrids AS/E//S-G/E and HMS/E//KEV/E show high translation efficiency in strength (except transverse Izod impact), while AS/E//KEV/E and HMS/E//S-G/E show low translation efficiency in some strengths based on predictions using approximate equations.

6. Intraply hybrids AS/E//S-G/E exhibit a synergistic effect in longitudinal tension (strength greater than either constituent); AS/E//KEV/E and HMS/E//KEV/E exhibit a synergistic effect in longitudinal flexure.

7. Intraply hybrids AS/E//S-G/E show appreciable improvements in longitudinal impact resistance (about 100 percent and greater compared with AS/E) accompanied by increases in longitudinal tensile and flexural strengths and no reduction in modulus or in intralaminar shear strength.

References

[1] Chamis, C. C. and Lark, R. F., "Hybrid Composites—State-of-the-Art Review: Analysis, Design, Application and Fabrication," NASA TM X 73545, National Aeronautics and Space Administration 1977; also in *Hybrid and Select Metal-Matrix Composites*, W. J. Renton, Ed., American Institute of Aeronautics and Astronautics, New York, 1979, pp. 13–51.

[2] Chamis, C. C. and Sinclair, J. H., "Prediction of Properties of Intraply Hybrid Composites," NASA TM 79087, National Aeronautics and Space Administration, 1979; also in *Proceedings,* 34th Annual Conference of the Society of the Plastics Industry Reinforced Plastics/Composites Institute, New York, 1979, Section 20-E, pp. 1–8.

[3] Chamis, C. C. and Sinclair, J. H., "Micromechanics for Predicting the Elastic and Thermal Properties of Intraply Hybrid Composites," NASA TM 79253, National Aeronautics and Space Administration, Washington, D.C., 1979; also "Modern Developments in Composites Materials and Structures" in *Proceedings,* American Society of Mechanical Engineers Winter Annual Meeting, New York, 2–7 Dec. 1979, pp. 253–267.

[4] Chamis, C. C., Kring, J. F., and Sullivan, T. L., "Automated Testing Data Reduction Computer Program," NASA TM X 68050, National Aeronautics and Space Administration, 1972.

[5] Chamis, C. C. and Sinclair, J. H., "10° Off-Axis Tensile Test for Intralaminar Shear Characterization of Fiber Composites," NASA TN D 8215, National Aeronautics and Space Administration, 1976; also, *Experimental Mechanics,* Vol. 17, No. 9, 1977, pp. 339–346.

G. C. Grimes[1]

Experimental Study of Compression-Compression Fatigue of Graphite/Epoxy Composites

REFERENCE: Grimes, G. C., **"Experimental Study of Compression-Compression Fatigue of Graphite/Epoxy Composites,"** *Test Methods and Design Allowables for Fibrous Composites, ASTM STP 734,* C. C. Chamis, Ed., American Society for Testing and Materials, 1981, pp. 281–337.

ABSTRACT: The objective of the study is to determine the static compression and $R = 10$ constant frequency compression-compression fatigue properties of Hercules AS/3501-6 tape graphite/epoxy composites. Environmental conditions selectively used are: (1) room temperature dry, (2) room temperature wet (1.1 percent by weight), (3) 103.3°C (218°F) dry, and (4) 103.3°C (218°F) wet (1.1 percent by weight). Ply orientations studied are $[0]_{nt}$, $[90]_{nt}$, $[\pm 45]_{ns}$, and $[(\pm 45)_5/0_{16}/90_4]_c$. All fatigue testing is at room temperature with subsequent residual strength testing on survivors at 103.3°C (218°F). New test methods are developed. Degradation is observed in the 1.25 million cycle endurance limit stresses and the compressive residual strength values compared with static properties.

KEY WORDS: graphite/epoxy, composite materials, compression, fatigue (materials), severe environment, residual strength, mechanical properties, stress and strain, stiffness, *f-N* curves, orthotropic, anisotropic, moisture effects, temperature effects

Nomenclature

CF   Compression fracture (failure mode)
SD   Standard deviation
CV   Coefficient of variation
DCS   Diagonal compression shear (failure mode)
LD   Longitudinal delamination (failure mode)
TB   Tab bondline (failure mode)
EOT   End of tab (location)
RTD   Room temperature dry: $22.8 \pm 1.1°C$ ($73 \pm 2°F$); dry meaning ex-

---

[1] Engineering specialist, Structural Mechanics Research Department, Northrop Corp., Hawthorne, Calif. 90250.

posed to laboratory ambient air-conditioned temperature/humidity environment for less than six months—moisture absorption during such periods ranges from 0 to 0.4 percent by weight

RTW   Room temperature wet: 22.8 ± 1.1°C (73 ± 2°F); wet meaning a laminate moisture content after moisture conditioning and during test of 1.1 ± 0.2 percent by weight

ETD   Elevated-temperature dry: 103.3 ± 2.8°C (218 ± 5°F); dry meaning exposed to laboratory ambient air-conditioned temperature/humidity environment for less than six months—moisture absorption during such period ranges from 0 to 0.4 percent

ETW   Elevated-temperature wet: 103.3 ± 2.8°C (218 ± 5°F); wet meaning a laminate moisture content after moisture conditioning and during test of 1.1 ± 0.2 percent by weight

GPa   Giga-pascals = 10^6 psi × 6.8947

MPa   Mega-pascals = ksi × 6.8947

ksi   1000 psi = MPa/6.8947

10^6 psi   GPa/6.8947

°C   (°F − 32) 5/9

ΔT, °C   (ΔT, °F) 5/9

Static compression testing of composites has evolved over the past 25 years from the ASTM Test for Compressive Properties of Rigid Plastics (D 695-77), used for woven fiber glass cloth-reinforced polymer composites, to honeycomb sandwich short columns (Mil-Std-401A) to honeycomb sandwich beams to a proliferation of platen and finger (in-plane) supported coupons to tubes and finally to the recent ASTM Test for Compressive Properties of Oriented Fiber Composites (D 3410-79) standard, which is an unsupported short column. These and other methods are still in use, resulting in a dearth of consistent, reliable, and duplicable data. Use of the ASTM Method D 3410-79 has improved the situation substantially for $[0]_{nt}$ specimens. Because of edge effects, laminate specimens which have some of the plies oriented in a nonzero direction cannot be used with this fixture to give reasonable results. No other universally accepted method(s) exists for such testing.

For compression-compression fatigue testing, there is no universally accepted test method, although a number of techniques have been used, such as short-column and platen-supported specimens and sandwich beams.

Because of these facts, two currently available static compression methods have been selected, one for $[0]_{nt}$ specimens and one for all other specimen orientations, and a new platen-supported compression-compression fatigue test method has been developed. With these methods the static compression and the $R = 10$ compression-compression fatigue properties are evaluated experimentally in an exploratory fashion for room and elevated temperatures at dry and 1.1 percent (by weight) wet conditions.

Experimental Methods

The three test methods used are the ASTM (Celanese) Method D 3410-79 (Part 36) for $[0]_{16T}$ static compression testing; the Northrop ETL method as shown in Fig. 1 for static compression testing of $[90]_{16T}$, $[\pm 45]_{4s}$, and $[(\pm 45)_5/0_{16}/90_4]_c$ laminates; and the newly developed Atmur test fixture detailed in Figs. 2-4 for compression-compression fatigue testing of $[0]_{24T}$, $[90]_{24T}$, $[\pm 45]_{6s}$, and $[(\pm 45)_5/0_{16}/90_4]_c$ laminates. In addition, residual static compression strength on surviving fatigue specimens is measured on a slightly modified version of the Atmur fixture as shown in Figs. 2, 4, and 5.

The ASTM D 3410-79 Celanese test, a short-column method used for $[0]_{16T}$ tape composites, provides load introduction by wedge grips through bonded-on tabs. The ETL method is a clamped-end, platen-supported method in which the specimen does not have tabs, so the load introduction is through bearing on the clamped ends. The Atmur specimen and fixture are much larger than either of the others used, being 5.08 cm (2 in.) wide by 20.32 cm (8 in.) long with fiber glass tabs bonded onto the specimen. The load introduction from fixture to specimen occurs partially through bearing on the ends and partially through shear in the tabs, both resulting from the way in which the fixture clamps the ends. Lateral support of the specimen by the Atmur fixture occurs in the gage length by discontinuous platens and by floating supports in the beveled area of the tab. Back-to-back longitudinal strain-gaging of all compression specimens is necessary to monitor bending and make fixture or setup adjustments to keep strains in balance within ±5 percent of the mean value.

Static Compression Data

Lamina static data in the 0-deg direction is obtained from $[0]_{16T}$ Laminate A of 62.5 percent fiber volume using ASTM Method D 3410-79. Steel tabs are bonded onto the specimen for gripping. Thickness tolerance over the tabs is extremely important because the test fixture will not load the specimen properly if these dimensions are not held. The ultimate strength data generated are summarized in Fig. 6 and presented in detail in the Appendix, Tables 6 and 7. Several specimens tested in the hot, wet, and hot-wet conditions exhibited tab-to-specimen bond failure but are included in the uncensored data. Removing these specimens with bond failures from the data results in the censored data with less scatter.[2] Mean wet (1.1 percent by weight) strength at room temperature is 89 percent of the dry value of −1400 MPa (−203 ksi), and the mean elevated temperature[3] strength in the dry condition is 84 percent of this room temperature value. Elevated tempera-

---

[2] Scatter is indicated by the coefficient of variation (CV) value shown in Fig. 6.
[3] Short-time (10 min) exposure at temperature; see Nomenclature for definitions of RTD, RTW, ETD, and ETW.

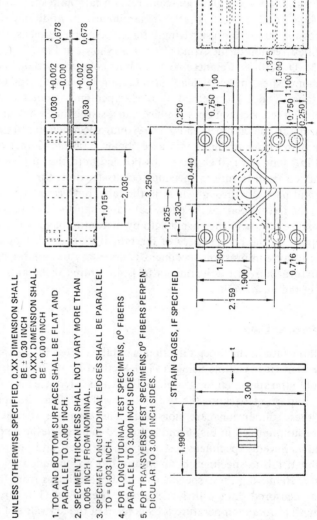

UNLESS OTHERWISE SPECIFIED, 0.XX DIMENSION SHALL
BE ± 0.30 INCH
0.XXX DIMENSION SHALL
BE ± 0.010 INCH

1. TOP AND BOTTOM SURFACES SHALL BE FLAT AND
PARALLEL TO 0.005 INCH.

2. SPECIMEN THICKNESS SHALL NOT VARY MORE THAN
0.005 INCH FROM NOMINAL.

3. SPECIMEN LONGITUDINAL EDGES SHALL BE PARALLEL
TO = 0.003 INCH.

4. FOR LONGITUDINAL TEST SPECIMENS, 0° FIBERS
PARALLEL TO 3.000 INCH SIDES.

5. FOR TRANSVERSE TEST SPECIMENS, 0° FIBERS PERPEN-
DICULAR TO 3.000 INCH SIDES.

STRAIN GAGES, IF SPECIFIED

FIG. 1—*ETL static compression specimen and test fixture.*

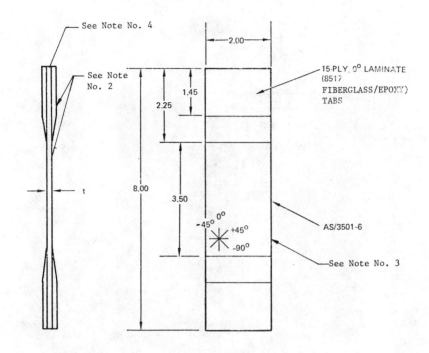

1. BOND 8517 TABS WITH AF-143 ADHESIVE.

2. SPECIMEN THICKNESS SHALL NOT VARY MORE THAN ±0.005 INCH
   FROM NOMINAL.

3. SPECIMEN LONGITUDINAL EDGES SHALL BE PARALLEL TO 0.005 INCH.

4. TOP END AND BOTTOM END SURFACES SHALL BE FLAT AND PARALLEL
   TO 0.001 INCH.

FIG. 2—*Compression-compression fatigue specimen for Atmur fixture.*

ture[3] mean strength in the wet (1.1 percent by weight) condition is 91 percent of the room temperature value. The elevated temperature wet (1.1 percent by weight) strength is 81 percent of that in the room temperature dry condition. The cause of this drop in strength appears to be a reduction in the fiber microbuckling or lamina shear buckling strength induced by a substantial change in the resin matrix properties in the hot-wet condition. Ultimate strain values are shown in Fig. 6 with reductions corresponding to the strength reductions. Modulus of elasticity and Poisson's ratio remain essentially unchanged for these environmental conditions at average values of $E_{xc} = 112.7$ GPa (16.35×10^6 psi) and $\nu_{xyc} = 0.34$; all the stress-strain curves are linear to near (>90 percent ultimate) failure.

Static ultimate strength data on the 90-deg-direction lamina are obtained from $[90]_{16T}$ Laminate A of 62.5 percent fiber volume using the platen-

FIG. 3—*Atmur test fixture setup for compression-compression fatigue.*

supported ETL fixture and specimen shown in Fig. 1. A summary of the data is shown in the left half of Fig. 7 with detailed data given in the Appendix, Table 8. At room temperature the mean wet (1.1 percent by weight) strength is 77 percent of the dry value of −258.6 MPa (−37.5 ksi) even though the corresponding change in the mean ultimate strain level is insignificant because of the reduction in modulus of elasticity. Under dry conditions the elevated temperature mean strength is 72 percent of the room temperature

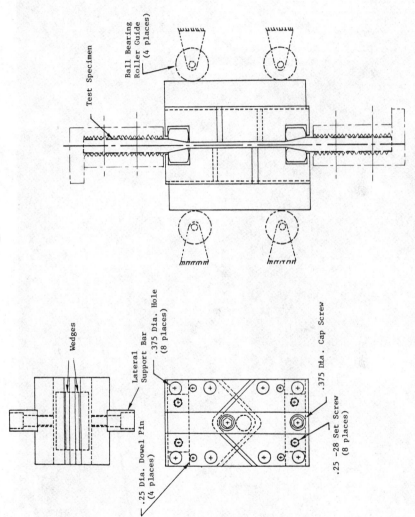

FIG. 4—*Schematic of Atmur compression fatigue and residual strength test fixture.*

FIG. 5—*Atmur test fixture setup for residual static strength.*

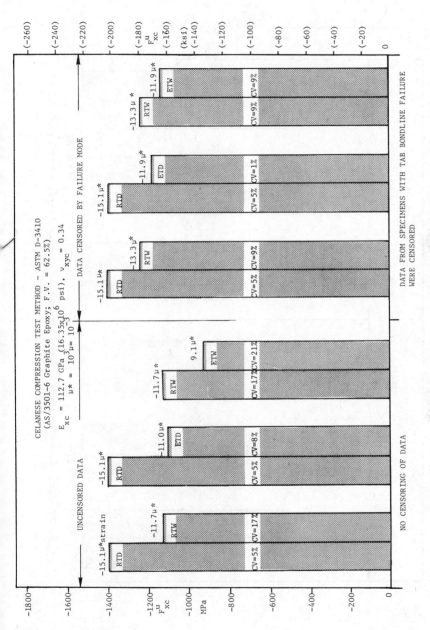

FIG. 6—*0-deg static compression strength of Laminate A,* $[0]_{16T}$.

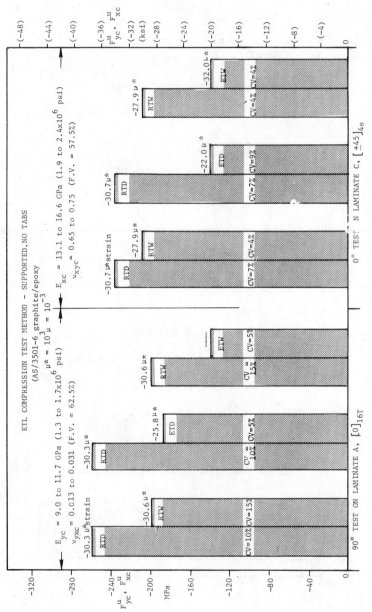

FIG. 7—Static compression strength, 90-deg Laminate A, [0]₁₆T and 0-deg Laminate B, [±45]₄s.

value with a corresponding drop in the mean ultimate strain value, indicating little change in modulus of elasticity. Under wet (1.1 percent by weight) conditions the elevated temperature mean strength is 71 percent of the room temperature value of -199.3 MPa (-28.9 ksi). Strain levels are not comparable because the large scatter in the strain data makes it questionable (see Appendix, Table 8). Elevated temperature mean wet (1.1 percent by weight) strength is 54 percent of the room temperature dry value. Both moisture and elevated temperature have significant derogatory effects on strength values that are directly cumulative; that is, the sum of the individual strength reductions caused by moisture and temperature separately can be added together to approximate the combined effect. If the wet reduction of 23 percent is added to the elevated temperature reduction of 28 percent, the total is 51 percent reduction; the hot-wet testing gave a reduction of 46 percent. In contrast, stiffness is degraded by moisture but not by elevated temperature.

Measuring 0-deg-direction laminate properties from $[\pm45]_{4s}$ Laminate C was accomplished to obtain the lamina shear characteristics indirectly per ASTM Recommended Practice for In-plane Shear Stress-Strain Response of Unidirectional Reinforced Plastics (D 3518-76) as well as for the laminate properties themselves. Strength data are shown in the right half of Fig. 7 with detailed data presented in the Appendix, Table 9. At room temperature the wet strength is 89 percent of the dry value of -235.8 MPa (-34.2 ksi). Under dry conditions the elevated temperature strength is 61 percent of the dry value. Under wet conditions the elevated temperature strength is 68 percent of the room temperature value of -210.9 MPa (-20.9 ksi). This hot-wet strength is 61 percent of the room temperature dry value. At room temperature, wet strength or strain levels show small reductions compared with dry values. Under dry conditions, elevated temperature strength and strain values show large reductions. Elevated temperature appears to be the principal cause of the large reductions in strength, not moisture. Elevated-temperature dry modulus values are 92 percent of the room temperature values of 16.6 GPa (2.4×10^6 psi), whereas RTW values are the same as RTD values. Under hot-wet conditions the modulus of elasticity drops to 79 percent of the RTD and RTW values. The combined effects of temperature and moisture are greater than the sum total of the individual effects. Ultimate strain levels are significantly affected by elevated temperature, whereas stiffness is significantly affected by the combined effects of elevated temperature and moisture.

Calculated lamina shear properties per ASTM Method D 3518 based on these $[\pm45]_{4s}$ laminate properties are given in Table 1.

Properties of a typical structural laminate with $53\frac{1}{3}$ percent 0-deg plies, $33\frac{1}{3}$ percent 45-deg plies, and $13\frac{1}{3}$ percent 90-deg plies are presented in Fig. 8 for $[(\pm45)_5/0_{16}/90_4]_c$ Laminate E[4] with detailed data presented in

---

[4]Actual lamination sequence is $[-45/0/90/0_3/\pm45/0/\pm45]_s$.

TABLE 1—*Calculated lamina shear properties—ASTM Method D 3518-76*
(using [±45]$_{4s}$ compression data).

| Condition | Ultimate Shear Stress, MPa (ksi) | Proportional Limit | | |
		Shear Stress, MPa (ksi)	Shear Strain, 10^{-3} units	Shear Modulus, GPa (10^6 psi)
RTD	±117.9 (±17.1)	±38.6 (±5.6)	±8.6	9.5 (1.4)
RTW	±105.4 (±15.3)	±39.3 (±5.7)	±8.3	9.6 (1.4)
ETD	±72.0 (±10.4)	±31.0 (±4.5)	±8.2	9.2 (1.3)
ETW	±71.7 (±10.4)	±19.6 (±2.8)	±4.7	7.8 (1.1)

the Appendix, Table 10. Such a laminate might be used as wing or empennage skin material with or without stiffeners or honeycomb stabilizing material. At room temperature the wet strength is 91 percent of the dry strength of −723.9 MPa (−105.0 ksi), whereas the elevated temperature dry strength is 85 percent of the room temperature dry strength. Under wet conditions the elevated temperature strength is 79 percent of the room temperature wet value of −661.2 MPa (−95.9 ksi), but only 72 percent of the room temperature dry value. Moisture and elevated-temperature effects are cumulative. Corresponding strain levels show approximately the same reduction as the strength values. Modulus values do not change significantly with these environmental conditions and average 71.2 GPa (10.33 × 10^6 psi). Poisson's ratio changes from a room temperature value of 0.32 to an elevated temperature value of 0.24. Proportional limit stresses are observed at 44 to 46 percent of the ultimate strength with some scatter, but the nonlinearity in the related stress-strain curves above this point is small. At room temperature the wet proportional limit stress is 89 percent of the dry value of −335.1 MPa (−48.6 ksi). For dry conditions the elevated-temperature proportional limit stress level is 83 percent of the dry value. Under wet conditions the elevated-temperature proportional limit stress is 78 percent of the room temperature value of −297.2 MPa (−43.1 ksi). The elevated-temperature wet proportional limit stress is 69 percent of the room temperature dry value. Related proportional limit strain values exhibit approximately the same reduction as the stresses.

Laminate E with its 53⅓ percent 0-deg plies is a fiber-dominated laminate. Or is it? A comparison of the $[(±45)_5/0_{16}/90_4]_c$ Laminate E 0-deg-direction ultimate strain levels with those of 0-deg direction $[0]_{16T}$ Laminate A is made in Fig. 9. The Laminate E strain levels at failure are 71 to 81 percent of the related Laminate A values. Differences in fiber volume, within the range shown, would not cause a reduction. The strain levels at failure for Laminate E remain at 78 to 81 percent of those occurring in Laminate A for the RTD, RTW, and ETD conditions, but drops to 71 percent for the ETW

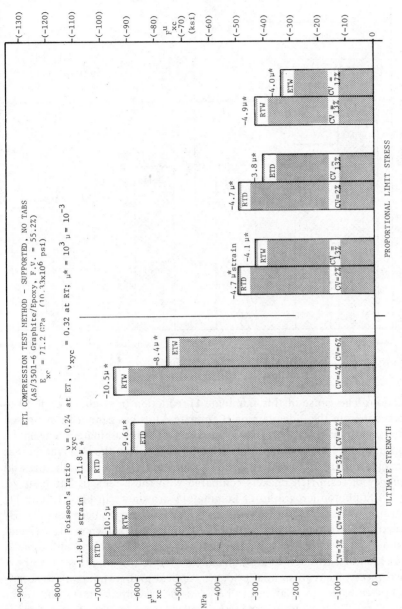

FIG. 8—0-deg static compression strength of Laminate E, $[(\pm45)_5/0_{16}/90_4]_c$.

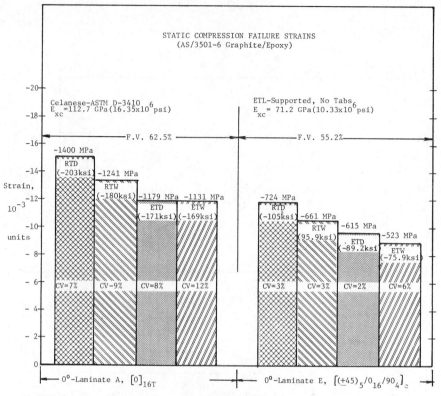

FIG. 9—*Strain level comparison: Laminates A, [0]$_{16T}$ and E, [(±45)$_5$/0$_{16}$/90$_4$]$_c$.*

condition. The cause of this Laminate E reduction in strain level of 19 to 29 percent from the 0-deg lamina (Laminate A) values cannot be explained except that the ETL support fixture and specimen for Laminate E may not be an optimum test method. The only other difference is in the observation that the microstructure of Laminate E shows a substantial increase in the size and frequency of resin-rich pockets compared with those seen in Laminate A. The ×100 photomicrographs of Laminate A are shown in Figs. 10 and 11, illustrating the small resin-rich pockets seen in some areas between plies. By contrast, the ×100 photomicrographs of Laminate E, shown in Figs. 12 and 13, show an increase in size and frequency of resin-rich pockets with some of them extending through one ply thickness. While this might seem to be a logical piece of circumstantial evidence, optical and scanning electron microscopy revealed no relationship between the fracture surfaces and resin-rich pockets. Resin-rich pockets in angle-plied laminates are characteristic of the Hercules AS/3501-6 graphite/epoxy material system; however, based on the limited observations of this study, they do not appear to be the cause of the reduced Laminate E strain capability compared with that of Laminate A.

FIG. 10—*Laminate A,* ×*100 transverse cross section photomicrograph.*

Nor do they appear to affect any other mechanical property measured in this study. One hypothesis to explain this phenomenon is that the multidirectional Laminate E requires more matrix stiffness and strength to stabilize the fibers against buckling than does the all 0-deg Laminate A.

With the kind of data generated in compression, that is, from both lamina and laminate specimens, a comparison of proportional limit strains is interesting from a cause-and-effect standpoint. These data, taken from the Appendix, Tables 8, 9, and 10, are summarized in Table 2 and by bar graph in Fig. 14. It is immediately apparent that the RTD and RTW proportional limit strains of 0-deg-direction Laminate E, $[(\pm45)_5/0_{16}/90_4]_c$, are similar in magnitude to those of 0-deg-direction Laminate C, $[\pm45]_{4s}$. For the ETD and ETW conditions, both 90-deg-direction Laminate A, $[0]_{16T}$, and 0-deg-direction Laminate C, $[\pm45]_{4s}$, proportional limit strains are somewhat similar in magnitude to those of 0-deg-direction Laminate E, $[(\pm45)_5/0_{16}/90_4]_c$. Study of the related detailed data, Tables 8, 9, and 10 of the Appendix, reveals that the (*) data in Table 2 have large scatter. Elimination of these "out-lier" data was accomplished in the Appendix Table 11 by censoring the affected test series as shown with the resulting censored data summarized in Table 3. Now it can be observed that (1) RTD and RTW proportional limit strains of 0-deg-direction Laminate E are similar in magnitude to those of

FIG. 11—*Laminate A, ×100 longitudinal cross section photomicrograph.*

0-deg-direction Laminate C; (2) ETD proportional limit strains of 0-deg-direction Laminate E are similar in magnitude to those of 90-deg-direction Laminate A; and (3) ETW proportional limit strains of 0-deg-direction Laminate E are similar in magnitude to 90-deg-direction Laminate A and 0-deg-direction Laminate C.

Compression-Compression Fatigue Data

At RTD conditions, thirteen 0-direction $[0]_{24T}$ specimens were fatigue tested at a stress ratio of $R = 10$, as were ten 0-direction $[0]_{24T}$ specimens at RTW conditions. These data are tabulated in the Appendix, Table 12, and plotted as an *f-N* curve in Fig. 15. The wet and dry data are combined in Fig. 15, which visually illustrates that within the scatterband recorded there is little difference in the RTD and RTW data. Combined wet and dry static data used in the fatigue plot are calculated in Table 13 of the Appendix. The 1.25×10^6 cycle runout endurance limit occurs at 814 MPa (118 ksi) or 61 percent of the 1338 MPa (194 ksi) average static value. The cause of such wearout in a $[0]_{24T}$ orientation is not fully understood at present, but it appears to be related to degradation in the resin matrix or fiber/matrix interface leading to fiber instability.

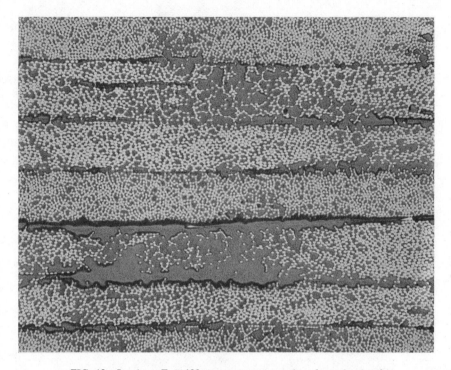

FIG. 12—*Laminate E, ×100 transverse cross section photomicrograph.*

Ten RTD and 6 RTW $R = 10$ 90-deg-direction $[0]_{24T}$ fatigue specimens were tested at 10 Hz to the 1.25×10^6-cycles runout endurance limit or failure; detailed data are tabulated in Table 14 of the Appendix. The RTD data are plotted with the average shown by the solid line in Fig. 16 as an *f-N* diagram. Static data are presented in the Appendix, Table 8, and in Fig. 7. At RTD conditions the *f-N* curve plots as a straight line with runout at -126.9 MPa (-18.4 ksi) or 49 percent of the static strength value of -258.6 MPa (37.5 ksi). For RTW conditions a dotted straight line is drawn in Fig. 16 between the static value and the highest 1.25×10^6 cycle runout stress. No fatigue failures were generated. The accuracy of such a dotted-line estimate is unknown, but, if not accurate, the error is probably on the conservative side because of the gradually decreasing slope, shape of *f-N* curves. The RTW runout stress level of -90.0 MPa (-13.0 ksi) is 45 percent of the static value of -199.3 MPa (-27.0 ksi).

Twelve $R = 10$ fatigue specimens were tested at RTD conditions and 10 at RTW conditions to failure or runout at the 1.25×10^6-cycles endurance limit for the $[\pm 45]_{6s}$ Laminate D. The cyclic rate was 1 to 10 Hz. Detailed fatigue data are tabulated in Table 15 of the Appendix and plotted as an *f-N* diagram in Fig. 17. Note that the RTD and RTW data plot together with the average shown by a solid straight line. The RTD and RTW static data were

FIG. 13—*Laminate E, ×100 longitudinal cross section photomicrograph.*

TABLE 2—*Static compression proportional limit strains (10^{-3} units) (ETL fixture).*

Test Environmental Conditions	0-Deg-Direction Laminate E, $[(\pm45)_5/0_{16}/90_4]_c$	90-Deg-Direction Laminate A, $[0]_{16T}$	0-Deg-Direction Laminate C, $[\pm45]_{4s}$
RTD	−4.7	−9.4	−4.9
RTW	−4.9	−12.5	−4.8
ETD	−3.8	−4.8 [a]	−5.0
ETW	−4.0 [a]	−3.8 [a]	−2.8

[a] Scatter is high.

NOTE—Boxed numbers denote line values in the last two columns which are similar to those in the first column.

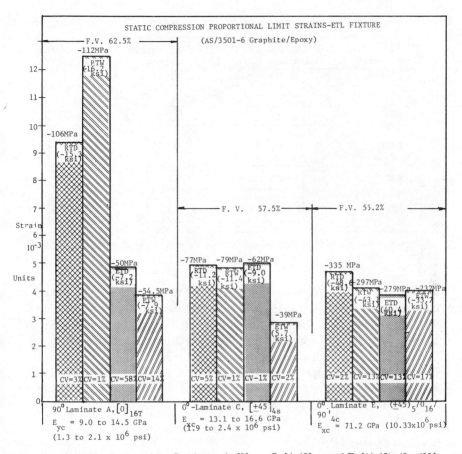

FIG. 14—*Strain level comparison: Laminates A, [0]$_{16T}$; C, [±45]$_{4S}$; and E, [(±45)$_5$/0$_{16}$/90]$_c$.*

taken from Appendix Table 9 and combined in the Appendix Table 16. The −117.2 MPa (−17.0 ksi) runout at 1.25×10^6 cycles is 52 percent of the average static strength value of −223.4 MPa (−32.4 ksi). In Fig. 17 the straight line average runs from approximately 10^2 cycles to 1.25×10^6 cycles. From 1 cycle (static) to 10^2 cycles the apparent slope of the straight line is different. The cause of this is not known and it may be within reasonable experimental scatter. However, the straight-line extension (dashed) of the fatigue average line of Fig. 17 seems to indicate a slightly higher static average value. This deviation could be the result of the extrapolation method of obtaining the static ultimate strength of these laminates (see Table 9 of the Appendix).

Fatigue data run at $R = 10$ and at 1 to 7 Hz on [(±45)$_5$/0$_{16}$/90$_4$]$_c$ Laminate E are tabulated in Table 17 of the Appendix, which shows 12 speci-

TABLE 3—*Censored static compression proportional limit strains (10^{-3} units) (ETL fixture).*

Test Environmental Conditions	0-Deg-Direction Laminate E, $[(\pm45)_5/0_{16}/90_4]_c$	90-Deg-Direction Laminate A, $[0]_{16T}$	0-Deg-Direction Laminate C, $[\pm45]_{4s}$
RTD	−4.7	−9.4	−4.9
RTW	−4.9	−12.5	−4.8
ETD	−3.8	−3.0	−5.0
ETW	−3.1	−2.9	−2.8

NOTE—Boxed numbers denote line values in the last two columns which are similar to those in the first column.

mens at the RTD condition and 10 at the RTW condition. These data are plotted as an *f-N* diagram in Fig. 18. Note that there is little difference in the RTD and RTW data. Degradation under fatigue load appears to start at between 10^4 and 10^5 cycles at near the static strength loading levels. The 1.25×10^6-cycles runout endurance limit is −547.4 MPa (−79.4 ksi) or 79 percent of the average static strength value of −692.9 MPa (−100.5 ksi). The combined RTD and RTW static strength data taken from Appendix Table 10 are shown in Appendix Table 18.

Residual Strength Data

All static compression residual strength tests were performed at ET conditions on specimens that survived fatigue testing to 1.25×10^6 cycles (runout) at room temperature. Analysis of the residual strength data from each laminate orientation follows.

Table 19 of the Appendix presents in detail residual strength data in the 0-deg and 90-deg test direction of Laminate B, $[0]_{24T}$. A summary graph of the 0-deg-direction data is presented in Fig. 19, plotting the maximum applied compressive fatigue loading at room temperature against the ultimate residual compressive strength at ET. In the minigraph in the upper-left corner of Fig. 19 the related maximum applied compressive fatigue strain is plotted against the ultimate residual compressive strain. Because of the limited data, firm conclusions cannot be drawn, but trends are evident and an hypothesis can be formed from these trends. At a given applied fatigue stress level there appears to be a 42 percent wide scatterband on the mean residual strength of survivors, but when the applied fatigue stress levels are increased, the residual strength levels of the survivors increase directly

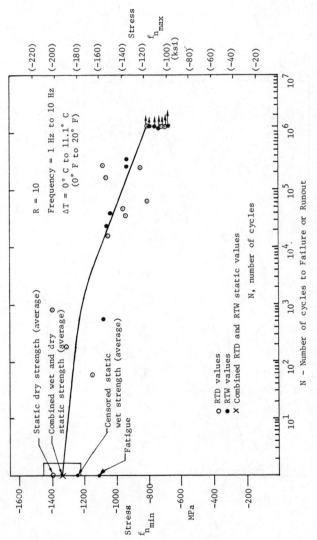

FIG. 15—*0-deg compression fatigue of Laminate B, [0]₂₄ₜ—RTD and RTW.*

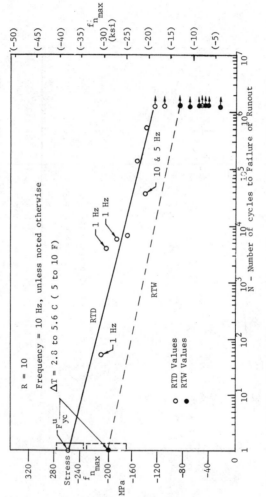

FIG. 16—*90-deg compression fatigue of Laminate B, [0]_{24T}—RTD and RTW.*

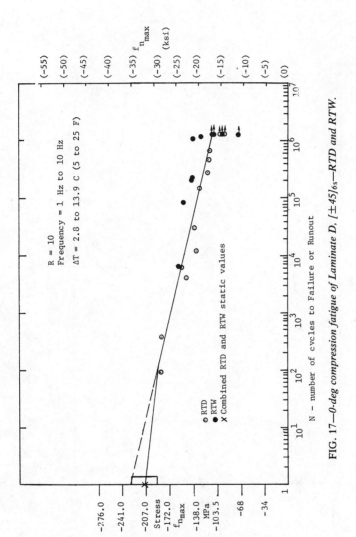

FIG. 17—0-deg compression fatigue of Laminate D, [±45]₆ₛ—RTD and RTW.

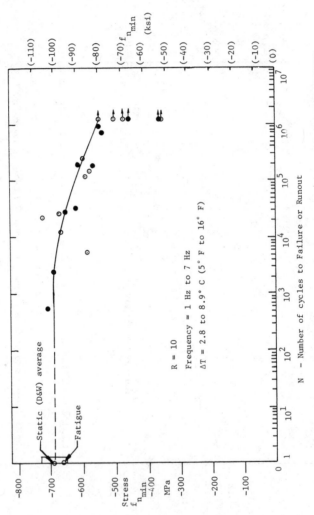

FIG. 18—0-deg compression fatigue of Laminate E, [(±45)₅/0₁₆/90₄]ₖ—RTD and RTW.

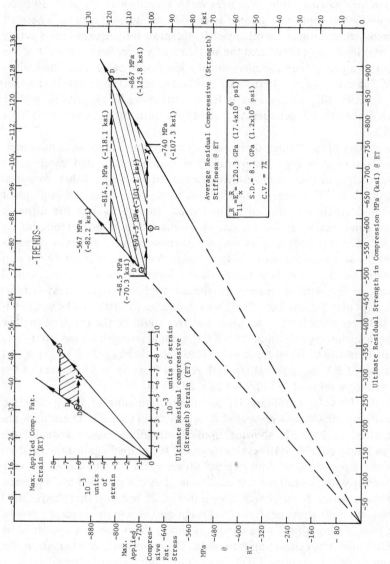

FIG. 19—*0-deg-direction RT maximum compression fatigue stress @ 1.25 × 10⁶-cycles runout versus residual compression strength @ ET. Laminate B. [0]₂₄T.*

although the number of survivors decreases. Since the modulus of elasticity does not change, the applied fatigue versus residual static strain levels show the same trend. One hypothesis, based on these experimentally observed trends, is that residual static compressive strength (strain) of surviving 0-deg-direction lamina composite structures varies directly with the $R = 10$ maximum applied compressive fatigue stress (strain) up to the point where all specimens fail in fatigue. In other words, the prior fatigue exposure is acting like a multicycle proof test, and the survivors of a more severe proof test are stronger as a group than the survivors of a less severe proof test. The limited data of Table 19 and Fig. 19 indicate residual strength (and strain) values for 0-deg-direction $[0]_{24T}$ laminates at ET conditions to be between 42 and 75 percent of the related combined dry/wet virgin static strength of -1155 MPa (-168 ksi).

A summary of the 90-deg-direction residual strength data on Laminate B, $[0]_{24T}$, is shown in Fig. 20, plotting the maximum applied compressive fatigue loading at room temperature against the ultimate residual compressive strength at ET. Figure 20 shows, in the upper-left corner, a minigraph of the related maximum applied fatigue strain plotted against the ultimate residual compressive strain. Because of the limited data, only trends can be observed. Scatterbands for dry and wet specimens have different domains both in the stress level and strain level plots. Within these separate dry and wet scatterbands, there is a trend toward increasing residual strength of survivors at ET with increasing maximum applied fatigue stress (strain) levels, although the number of survivors decreases. Scatter for the wet specimens is larger than for the dry ones, but this could be the result of too few data points. For dry conditions at ET the residual strength ranges from 77 to 94 percent of the static strength value of -186.2 MPa (-27.0 ksi). For wet conditions at ET the residual strength ranges from 58 to 97 percent of the static strength value of -140.7 MPa (-20.4 ksi).

For the $[\pm 45]_{6s}$ Laminate D, the limited amount of 0-deg-direction residual strength data is presented in Appendix Table 20. These data are summarized in Fig. 21, showing applied maximum compressive fatigue stress or strain at 1.25×10^6-cycles runout at RT plotted against the residual compressive proportional limit stress or strain at ET. Stiffness does not vary significantly with the fatigue test conditions, but has dropped to 78 percent of the static value. Scatterbands shown in Fig. 21 indicate that the residual proportional limit stress or strain level is not sensitive to the applied stress or strain level. Although the data sample is small, there does not appear to be any difference between wet and dry data. Ultimate static residual strengths were not recorded because of ETL test fixture limitations (discussed earlier). The residual proportional limit stress at ET ranged from -37.9 to -49.6 MPa (-7.2 to -5.5 ksi), which is 75 to 98 percent of the ETD and ETW averaged static proportional limit stress value (see Appendix Table 9) of -50.7 MPa (-7.4 ksi). However, these extreme values represent the wet

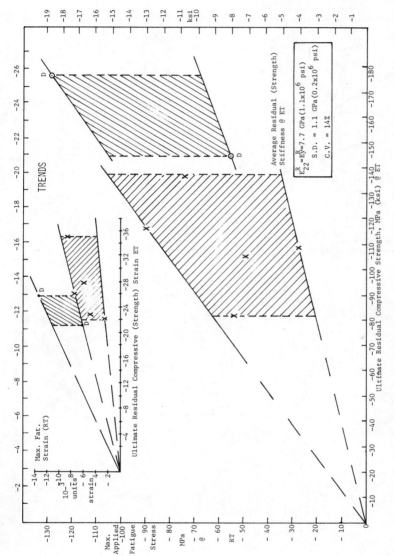

FIG. 20—*90-deg-direction RT maximum compression fatigue stress @ 1.25 × 10⁶-cycles runout versus residual compression strength @ ET. Laminate B, [0]₂₄T.*

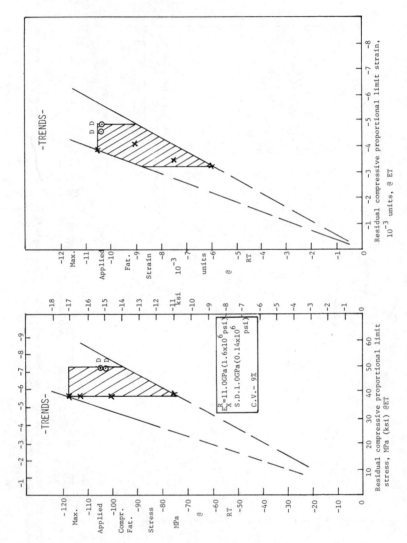

FIG. 21—*0-deg-direction RT maximum compression fatigue stress at 1.25 × 10⁶-cycles runout versus residual compression proportional limit stress, @ ET, Laminate D, [±45]6s.*

residual proportional limit stress level at the lower end and the related dry value at the upper end. On this basis, that is, using the ETW and ETD static proportional limit stress values, the wet residual proportional limit stress level at ET is 96 percent of the related wet static value of -39.3 MPa $(-5.7$ ksi), while the dry residual proportional limit stress level at ET is 80 percent of the related dry static value of -62.1 MPa $(-9.0$ ksi).

Residual strength data on $[(\pm45)_5/0_{16}/90_4]_c$ Laminate E are detailed in Appendix Table 21. A summary of these data is presented in Fig. 22, which plots the maximum applied compressive fatigue stress at RT against the residual compressive strength at ET. While stiffness values do not change due to the environment of fatigue loading conditions relative to the static values, the scatter does increase. Wet and dry data appear to be in two different domains. Wet residual strength values at ET average -417.2 MPa $(-60.5$ ksi) or 80 percent of the related static value of -523.3 MPa $(-75.9$ ksi), whereas dry residual strength values at ET average -595.9 MPa $(-86.4$ ksi) or 97 percent of the related static value of -615.0 MPa $(-89.2$

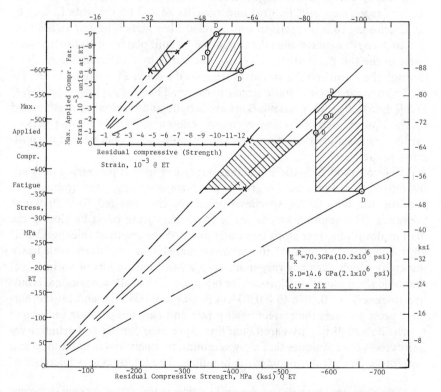

FIG. 22—*0-deg-direction RT maximum compression fatigue stress at 1.25×10^6 cycles runout versus residual compression strength @ ET, Laminate E, $[(\pm45)_5/0_{16}/90_4]_c$.*

ksi). Strain values show the same trends as the stress values, as can be seen in the miniplot in the upper-left corner of Fig. 22.

Summary

Experimental Methods

Two existing test methods are used to measure static compression properties, and a new method is developed to perform the $R = 10$ compression-fatigue testing. ASTM Method D 3410-79 (Celanese) is used to measure the $[0]_{16T}$ static properties of the AS/3501-6 graphite/epoxy tape material being evaluated. Load introduction is by shear through the wedge-gripped tabs. For the $[90]_{16T}$, $[\pm45]_{4s}$, and the $[(\pm45)_5/0_{16}/90_4]_c$ orientations the Northrop ETL method, which consists of a 5.08-cm-wide (2 in.) by 7.32-cm-long (3 in.) untabbed specimen with clamped split platen supports, is used. Load introduction is by bearing on the clamped ends of the composite. The new compression-compression fatigue method developed is designated the Atmur method (specimen and fixture) and consists of a 5.08-cm-wide (2 in.) by 20.32-cm-long (8 in.) specimen with beveled fiber glass tabs and an 8.89-cm (3.5 in.) length between the tabs supported by split platens with floating supports in the tab bevel area. Load introduction is by a combination of shear through the clamped tabs and by bearing on the ends. This Atmur method is used also in the residual static strength testing. The $[0]_{24T}$, $[90]_{24T}$, $[\pm45]_{6S}$, and $[(\pm45)_5/0_{16}/90_4]_c$ orientations are fatigue and residual strength tested by the Atmur method. These methods, chosen selectively for the various loading modes and orientations, give reasonable though not necessarily optimum properties.

ASTM Method D 3410-79 (Celanese) requires special processing in bonding on the tabs and a high degree of precision in testing. One problem was bonding the tabs to the specimen panels to the required final thickness tolerance. The solution was to use a hot-plate curing press by closing the platens down to spacer stops especially made to the required thickness. This approach assumes that (1) the laminate panel that one starts with has a thickness which does not vary more than a few thousandths of an inch, (2) available sheet metal[5] thicknesses for tab material will allow a nominal bondline thickness[6] of 0.0076 to 0.0203 cm (0.003 to 0.008 in.), and (3) the hot-plate press provides the correct heatup rate and cure temperature uniformity to provide a high-quality-cured bondline. Specimen cutting and grinding requires tools or techniques that allow a minimum amount of heat buildup and peel forces on the tab-to-composite bondline. The tab thickness tolerance of

---

[5] Where fiber glass tabs are desired, they can be molded to any required thickness or co-cured to the graphite/epoxy with the adhesive.

[6] It is assumed that a film adhesive is used that has acceptable performance within this range of bondline thicknesses.

0.3988 ± 0.0051 cm (0.157 ± 0.002 in.) is critical and necessary for the proper seating of the fixture wedge grips for uniform load introduction.

The Northrop ETL method and the Atmur method require that the load bearing ends be ground flat and parallel within 0.0025 cm (0.001 in.). The ETL method is limited in testing the $[\pm 45]_{4s}$ laminates because the large amount of lateral specimen expansion results in interference with the side of the fixture before ultimate load is reached; that is, ultimate strengths obtained for this orientation are obtained by extrapolation.[7] However, a better method is not presently known. Neither of these two platen supported methods is capable of developing ultimate material strength of $[0]_{16T}$, $[0]_{24T}$, or similar type laminates.

The minimum instrumentation for all three of these compression test methods is back-to-back longitudinal gages. Setup and specimen tolerance are such that the back-to-back longitudinal strain readings are within ±5 percent of the mean value up to 90 percent of ultimate loading. To achieve this, strain readings are monitored up to 40 percent of ultimate loading on both static compression test specimens and on the static compression setup of the fatigue specimens. The strain gages used are 350-Ω-resistance Type 03 for maximum accuracy (Micro-Measurements Co. terminology).

Static Compression Data

For $[0]_{16T}$ specimens the RTD static ultimate strength of -1400 MPa (203 ksi) is reduced to 89 percent of that value by RTW conditions and to 84 percent of that value by ETD conditions. ETW conditions reduce the strength to 81 percent of the RTD value. The cause of this drop in strength appears to be a reduction in fiber or lamina buckling stability that is induced by a substantial change in the resin matrix or fiber/matrix interface properties. Ultimate strain values show similar trends since moduli of elasticity and Poisson's ratio values, which average 112.7 GPa (16.35 × 10⁶ psi) and 0.34, respectively, are unaffected by these environmental conditions and the stress-strain curves are linear elastic to approximately 90 percent of ultimate. The RTW ultimate strain level is 88 percent of the RTD value of -15.1 × 10^{-3} units, whereas both the ETD and ETW strain values are 79 percent of the RTD value. It is obvious that ET conditions have a greater derogatory effect on the strength then does moisture at the 1.1 percent level, although the combined effect of moisture and temperature is the most degrading.

Specimens of the $[90]_{16T}$ orientations have an RTW average strength of 77 percent of the RTD value of -258.6 MPa (-37.5 ksi), whereas the ETD average strength is 72 percent of this value. At ETW conditions the average strength is 54 percent of the RTD value. Moduli and ultimate strain levels

---

[7] Straight-line extrapolation of the end of the decreasing slope stress-strain curve to the ultimate measured strain value gives an "effective" ultimate stress.

vary with the environmental condition; that is, their trends are not the same as the strength value. At RTW conditions the ultimate strain level is the same as the RTD value of -30.3×10^{-3} units. The RTW modulus value is 76 percent of the RTD value of 11.76 Pa (1.7×10^6 psi) while the RTW Poisson's ratio is 59 percent of the RTD value of 0.024. At ETD the strain level has dropped to 85 percent of the RTD/RTW value but the modulus values are about the same as those at RTD conditions, with Poisson's ratio being 17 percent above the RTD value. All strain, moduli, and Poisson's ratio values at ETW conditions have a large amount of scatter and, therefore, cannot be compared unless some censoring is done (see Appendix Table 11). From these data it is apparent that the effects of moisture and temperature together are much more pronounced than either one by itself, or both of them added together mathematically.

For $[\pm 45]_{4s}$ specimens the RTW ultimate strength is 89 percent of the RTD value of -235.8 MPa (-34.2 ksi), whereas at ETD conditions the strength drops to 61 percent of the RTD value. At ETW conditions the strength remains at 61 percent of the RTD condition value. The RTW ultimate strain value is 91 percent of the RTD value of -30.7×10^{-3} units, whereas the ETD value is 72 percent of the RTD value. At ETW conditions the ultimate strain value is 4 percent above the RTD value. The modulus value at RTD is 16.6 GPa (2.4×10^6 psi) and remains essentially unchanged through RTW and ETD conditions, but drops to 79 percent of the RTD value at ETW conditions. At RTW conditions the Poisson's ratio is 96 percent of the RTD value of 0.75, but at ETD conditions it drops to 87 percent of the RTD value. At ETW conditions it goes up slightly to 91 percent of the RTD value. From these data it is apparent that reductions in the RTD strength and ultimate strain values are greater for ETD conditions than for RTW conditions. ETW conditions do not cause further strength reductions over the ETD conditions, while the ultimate strain level goes back up to approximately the RTD value. The ETW modulus of elasticity is degraded compared with the RTD value.

The static strength of $[(\pm 45)_5/0_{16}/90_4]_c$ specimens at RTW conditions is 91 percent of the RTD condition value of -723.9 MPa (-105.0 ksi), with the ETD strength being 85 percent of the RTD value. At ETW conditions the laminate strength is 72 percent of the RTD value, an indication that the effects of wet conditions and elevated temperature exposure are cumulative. Corresponding ultimate strain levels show similar trends, with the RTW value being 89 percent of the RTD value of -11.8×10^{-3} units of strain. The ETD value of ultimate strain is 81 percent of the RTD value, but at ETW conditions the ultimate strain level is 71 percent of the RTD value, showing, again, the cumulative effects of moisture and temperature. Moduli of elasticity values remain nearly the same throughout these environmental conditions with an average value of 71.2 GPa (10.33×10^6 psi). While moisture does not affect Poisson's ratio, temperature does. At RTD/RTW conditions Poisson's ratio is equal to 0.32, whereas at ETD/ETW conditions it drops to 0.24. Proportional limit stress levels range from 44 to 46 percent of the ultimate strength values, and related proportional limit strain levels range from 37 to 40 percent of the ultimate strain values. Nonlinearities in the stress-strain curves above the proportional limit are small but significant. This laminate, with 53⅓ percent 0-deg plies, 33⅓ percent 45-deg plies, and 13⅓ percent 90-deg plies, is normally considered to be a fiber-dominated laminate; however, its ultimate strain levels are 71 to 81 percent of those of the $[0]_{16T}$ laminate for the various environmental conditions studied. The cause of this has not been determined, but some possibilities are: (1) the ETL test method used on the multidirectional laminate specimen, (2) variations in fiber properties, (3) some microfailure phenomenon peculiar to this particular orientation, and (4) multidirectional laminates such as Laminate E may be more sensitive to matrix strength and stiffness properties than $[0]_{16T}$, orientations such as Laminate A. At the proportional limits in the multidirectional laminate, the strain levels correspond to the proportional limit strain values of the $[90]_{16T}$ laminate at the ETD and ETW conditions and to the proportional limit strain values of the $[\pm 45]_{4s}$ laminate at the RTD, RTW, and ETW conditions.

A summary of the mean values of all static test results is presented in Table 4.

Fatigue Data

Differences in RTD and RTW static and fatigue data are found to be small for each of the $[0]_{24T}$, $[\pm 45]_{6s}$, and $[(\pm 45)_5/0_{16}/90_4]_c$ laminates so that the f-N data are plotted together. In contrast, the static and fatigue data of the $[90]_{24T}$ laminate are different in the RTW condition compared with the RTD condition.

A summary of the fatigue endurance limits is given in Table 5 along with the related static strengths. Specimens are tested in $R = 10$ compression-compression fatigue at 1 to 10 Hz constant frequency with a sawtooth waveform. Testing is carried out to 1.25×10^6 cycles. The specimen within each group/environmental condition for each orientation which survived this number of cycles with the largest applied fatigue stress (strain) level are used to define the endurance limit. The endurance limit stress level for the $[0]_{24T}$ laminate at RTD/RTW conditions is 61 percent of its static strength of -1338 MPa (-194 ksi), whereas the endurance limit strain level is 52 percent of its related static ultimate value of -14.4×10^{-3} units. For $[90]_{24T}$ laminates the RTD endurance limit stress is 49 percent of the static strength of -258.6 MPa (-37.5 ksi), with the endurance limit strain level being 45 percent of the static ultimate value of -30.3×10^{-3} units. At RTW conditions the $[90]_{24T}$ endurance limit stress level is 45 percent of the static ultimate value of -199.3 MPa (-28.9 ksi), whereas the endurance limit

TABLE 4—*Mean static compression properties of AS/3501-6[a] graphite/epoxy laminate.*

Environmental Condition	Ultimate		Proportional Limit		Modulus of Elasticity, GPa (10^6 psi)	Poisson's Ratio	Source
	Stress, MPa (ksi)	Strain, 10^{-3} Units	Stress, MPa (ksi)	Strain, 10^{-3} Units			
$[0]_{16T}$ Laminate A (Fatigue Value = 62.5%)							
RTD	−1400 (−203)	−15.1	... (...)	...	109.6 (15.9)	0.34	Appendix Table 7
RTW	−1241 (−180)	−13.3	... (...)	...	111.0 (16.0)	0.34	Appendix Table 7
ETD	−1179 (−171)	−11.9	... (...)	...	115.8 (16.8)	0.35	Appendix Table 7
ETW	−1131 (−164)	−11.9	... (...)	...	114.5 (16.6)	0.33	Appendix Table 7
$[90]_{16T}$ Laminate A (Fatigue Value = 62.5%)							
RTD	−258.6 (−37.5)	−30.3	−105.5 (−15.3)	−9.4	11.7 (1.7)	0.024	Appendix Table 8
RTW	−199.3 (−28.9)	−30.6	−111.7 (−16.2)	−12.5	8.8 (1.3)	0.013	Appendix Table 8
ETD	−186.2 (−27.0)	−25.8	−33.1 (−4.8)	−3.0	11.7 (1.7)	0.028	Appendix Tables 8, 11, and Text Table 3
ETW	−140.7 (−20.4)	−8.6	−50.6 (−7.4)	−2.9	12.4 (1.8)	0.088	Appendix Tables 8, 11, and Text Table 3
$[\pm45]_{4S}$ Laminate C (Fatigue Value = 57.5)							
RTD	−235.8 (−34.2)	−30.7	−77.2 (−11.2)	−4.9	16.6 (2.4)	0.75	Appendix Table 9
RTW	−210.9 (−30.6)	−27.9	−78.6 (−11.4)	−4.8	16.6 (2.4)	0.72	Appendix Table 9
ETD	−144.1 (−20.9)	−22.0	−62.1 (−9.0)	−5.0	14.7 (2.1)	0.65	Appendix Table 9
ETW	−143.4 (−20.8)	−32.0	−39.3 (−5.7)	−2.8	13.1 (1.9)	0.68	Appendix Table 9
$[(\pm45)/0_{16}/90_4]_c$ Laminate E (Fatigue Value = 55.2%)							
RTD	−723.9 (−105.0)	−11.8	−335.1 (−48.6)	−4.7	71.7 (10.4)	0.32	Appendix Table 10
RTW	−661.2 (−95.9)	−10.5	−297.2 (−43.1)	−4.1	69.6 (10.0)	0.32	Appendix Table 10
ETD	−615.0 (−89.2)	−9.6	−278.6 (−40.4)	−3.8	72.4 (10.5)	0.23	Appendix Table 10
ETW	−523.3 (−75.9)	−8.4	−232.4 (−33.7)	−3.1	71.7 (10.4)	0.24	Appendix Tables 10, 11

[a]Tape prepreg made by Hercules, Inc.

TABLE 5—Summary of R = 10 compression-compression fatigue endurance limits for AS/3501-6[a] graphite/epoxy laminates (1 to 10 Hz constant-frequency, sawtooth waveform).

Static Ultimate		Maximum[a] Applied Fatigue @ 1.25 × 10⁶ Cycles Runout				
Strength, MPa (ksi)	Strain, 10^{-3} Units	Stress, MPa (ksi)	As a % of Static Ultimate Strength	Strain, 10^{-3} Units	As a % of Static Ultimate Strain	Environmental Condition
[0]₁₆T Static and [0]₂₄T Fatigue Laminates A and B (Fatigue Value = 62.6%)						
−1338 (−194)	−14.4	−814.3 (−118.1)	61	−7.5	52	RTD/RTW
[90]₁₆T Static and [90]₂₄T Fatigue Laminates A and B (Fatigue Value = 62.6%)						
−258.6 (−37.5)	−30.3	−126.9 (−18.4)	49	−13.5	45	RTD
−199.3 (−28.9)	−30.6	−90.0 (−13.0)	45	−9.0	29	RTW
[±45]₄S Static and [±45]₆S Fatigue Laminates C and D (Fatigue Value = 57.5%)						
−223.4 (−32.4)	−29.7	−117.2 (−17.0)	52	−10.5	35	RTD/RTW
[(±45)₅/0₁₆/90₄]C Static and Fatigue Laminate E (Fatigue Value = 55.2%)						
−692.9 (−100.5)	−11.2	−547.4 (−79.4)	79	−9.0	81	RTD/RTW

[a]Tape prepreg made by Hercules, Inc.
[b]Maximum in this case is minimum algebraic value.

strain level is 29 percent of the static ultimate value of -30.6×10^{-3} units. For the $[\pm45]_{6s}$ laminate the RTD/RTW endurance limit stress level is 52 percent of the static ultimate strength value of -223.4 MPa (-32.4 ksi), but the endurance limit strain level is 35 percent of the static ultimate value of -29.7×10^{-3} units. The $53\frac{1}{3}$ percent 0-deg fiber laminate with the $[(\pm45)_5/0_{16}/90_4]_c$ orientation has an endurance limit stress at RTD/RTW conditions of 79 percent of the static ultimate value of -692.9 MPa (-100.5 ksi), with the related endurance limit strain value being 81 percent of the static ultimate strain level of -11.2×10^{-3} units.

It is evident that wearout occurs in all the laminates studied, whether fiber-dominated or matrix-dominated. Wearout is greater in the highly matrix-dominated $[\pm45]_{6s}$ and $[90]_{24T}$ laminates than in the highly fiber-dominated ones, as would be expected. Wearout in the $[(\pm45)_5/0_{16}/90_4]_c$ laminate at the endurance limit (21 percent reduction in stress level, 19 percent reduction in strain level) compared with the static ultimate value is less than that of the other highly fiber-dominated laminate orientation studied, that is, $[0]_{24T}$ laminate. This probably attests to the stabilizing ability of the basic $0/90/\pm45$ orientation on fiber microbuckling. Another difference is that this laminate (E) is 30-ply while the balance of the fatigue laminates are 24-plies, although all were platen-stabilized in the same manner. One might question why the endurance limit of the $[0]_{24T}$ laminate drops so much from the static value (reduction of 39 percent in strength and 48 percent in strain level). Again, the stability of fibers or laminas is suspected with the probable cause related to a degradation in the matrix or fiber/matrix interface. This degradation is apparently greater for the $[0]_{24T}$ laminate than for the $[(\pm45)_5/0_{16}/90_4]_c$ laminate at the endurance limit because there is less difference in their strain levels at this point than under static test conditions. The $[90]_{24T}$ laminate endurance limit strength degradation observed (a reduction of 51 to 55 percent in stress level and 55 to 71 percent in strain level) compared with the static ultimate values (RTD and RTW) is obviously the result of matrix or fiber/matrix interface degradation. The $[\pm45]_{6s}$ laminate endurance limit strength reduction (48 percent in stress level and 65 percent in strain level) compared with the static strength is caused by a combination of fibers or laminas microbuckling and matrix or fiber/matrix interface weakening, with the latter matrix degradation contributing to the former buckling resistance reduction. Such matrix or fiber/matrix degradation also reduces interlaminar shear strength. What happens in the wearout of the $[(\pm45)_5/0_{16}/90_4]_c$ laminate is probably a combination of these modes with the orientation and thickness contributing to stabilization and to a delay in the start of the microfailure mechanisms discussed in the preceding.

Residual Strength

Residual strength values appear to be scattered beyond reason, and the limited amount of data generated compounds the problem. However, some trends can be discussed. First, recall that the all residual strength static testing was performed at ET conditions, whereas the fatigue testing was done at RT conditions. For $[0]_{24T}$ laminates there is a trend toward higher residual strengths of surviving specimens tested at higher applied maximum fatigue stress (strain) levels, with wet or dry conditions not being significant, although the number of survivors decreases as the applied fatigue stresses increase. Essentially a proof testing phenomenon is observed with the fatigue exposure being the proof loading. For the $[90]_{24T}$ laminates the same general trend of increasing residual strength with increasing applied maximum fatigue stress (strain) level is evident, but the dry specimens have higher residual strength and lower residual strain levels than the wet ones. Again, the same general trend of increasing residual strength and strain levels with increasing maximum applied fatigue stresses (strains) for the $[\pm45]_{6s}$ laminates is evident, with no difference seen between dry and wet conditions. For the $[(\pm45)_5/0_{16}/90_4]_c$ laminates the trend observable from the residual strength (strain) data is that there is a significant difference between the wet and dry values.

Stiffness values measured during residual strength testing show no degradation for the $[0]_{24T}$ and the $[(\pm45)_5/0_{16}/90_4]_c$ laminates; however, the $[90]_{24T}$ and the $[\pm45]_{6s}$ laminates show degradations of 37 and 22 percent, respectively. The $[90]_{24T}$ ET residual strength modulus value is 63 percent of the ETD/ETW static value of 12.07 GPa (1.75×10^6 psi), whereas the $[\pm45]_{6s}$ ET residual strength modulus value is 78 percent of the ETD/ETW static value of 14.13 GPa (2.05×10^6 psi).

Conclusions

Experimental Methods

The following conclusions can be drawn from work on the experimental methods.

1. Three different methods are required to measure the static compression and compression-compression fatigue properties of the $[0]_{nT}$, $[90]_{nT}$, $[\pm45]_{nS}$, and $[(\pm45)_5/0_{16}/90_4]_c$ orientation graphite/epoxy laminates made with AS/3501-6 (Hercules, Inc. prepreg.) tape material.

2. Two existing methods are found to be satisfactory for use in static testing: ASTM Method D 3410-79 (Celanese) for $[0]_{16T}$ laminates and the Northrop ETL clamped-end platen-supported method for $[90]_{16T}$, $[\pm45]_{4s}$, and $[(\pm45)_5/0_{16}/90_4]_c$ laminates.

3. A new test method (specimen and fixture) is developed to measure the constant frequency $R = 10$ compression-compression fatigue properties of these laminates. It is designated the Atmur fixture.

4. All experimental methods used are satisfactory, though not optimum.

Static Properties

Conclusions about static properties are delineated in the following:

1. Elevated-temperature [102°C (218°F)] exposure results in more degradation of the $[0]_{16T}$ laminate than does the 1.1 percent wet moisture condition, but the combination of elevated temperature and wet conditions is the worst. ETW strength is reduced to 81 percent of the RTD strength value of -1400 MPa (-203 ksi). The modulus and Poisson's ratio are unaffected by the environmental conditions used.

2. Both elevated-temperature and 1.1 percent wet conditions cause substantial reductions from RTD strength of the $[90]_{16T}$ laminate, and the combination of elevated temperature and moisture is cumulative in its effect. ETW strength is 54 percent of the RTD value of -258.6 MPa (-37.5 ksi). Moisture appears to degrade the modulus of elasticity somewhat, but its value is not affected by the elevated temperature. The combination of elevated temperature and moisture does not appear to have any effect on the modulus of elasticity. Poisson's ratio values vary significantly with the different environmental conditions studied herein.

3. For the $[\pm 45]_{4s}$ laminate, moisture has little derogatory effect, but elevated temperature has substantial degrading effects. ETD and ETW condition strengths are 61 percent of the RTD value of -235.8 MPa (-34.2 ksi); however, the proportional limit, modulus, and Poisson's ratio values are reduced somewhat more at the ETW conditions.

4. The $[(\pm 45)_5/0_{16}/90_4]_c$ typical aircraft laminate shows worse degradation due to elevated temperature than due to the 1.1 percent wet condition, but the combined effects are the worst and are cumulative. The ETW strength is 72 percent of the RTD value of -723.9 MPa (-105.0 ksi). The modulus of elasticity is unaffected by these environmental conditions, but the Poisson's ratio at ETD and ETW conditions is reduced to 75 percent of its value at RTD and RTW conditions.

5. Resin-rich pockets observed in the laminates are more frequent and larger as the orientation becomes more complex; however, they do not appear to have any deleterious effects on the static properties.

Fatigue Properties

Conclusions for $R = 10$ compression-compression fatigue properties under constant-frequency loading are itemized in the following for a 1.25×10^6 cycles runout.

1. Laminates with $[0]_{24T}$ orientations have RTD/RTW runout endurance limit stresses that are 61 percent of their static strength value of -1338 MPa (-194 ksi).

2. The $[90]_{24T}$ laminates have an RTD endurance limit of 49 percent of

their static ultimate strength value of -258.6 MPa (-37.5 ksi), with the RTW value being 45 percent of the static ultimate strength value of -199.3 MPa (-28.9 ksi).

3. The RTD/RTW endurance limit stress for the $[\pm 45]_{6s}$ laminates is 52 percent of the static ultimate strength value of -223.4 MPa (-32.4 ksi).

4. At RTD/RTW conditions the $[(\pm 45)_5/0_{16}/90_4]_c$ laminates have an endurance limit stress of 79 percent of the static ultimate strength value of -692.9 MPa (-100.5 ksi).

5. The resin-rich pockets observed in the laminates do not appear to have any effect on the wearout seen in these fatigue tests.

6. Wearout appears to be the result of matrix or fiber/matrix interface degradation during cycling.

Residual Strength

The following trends are observed for laminate residual strength after surviving $R = 10$ constant-frequency fatigue cycling to 1.25×10^6 cycles. All residual strength testing is performed at elevated temperature [103°C (218°F)].

1. The $[0]_{24T}$ and $[\pm 45]_{6s}$ laminates have a general trend of increasing residual strength with increasing applied maximum fatigue stress level, with wet or dry conditions making no difference to the results. Increasing applied maximum fatigue stress level results in fewer survivors until a point is reached at which all specimens are consumed in fatigue; that is, the proof test being performed becomes increasingly difficult to pass.

2. For $[90]_{24T}$ laminates, the same general trend of increasing residual strength with increasing applied maximum fatigue stress level is observed, but the wet and dry data are significantly different.

3. Residual strength data from the $[(\pm 45)_5/0_{16}/90_4]_c$ laminates do not show any strength variation trends except that the data in the wet condition are significantly different from those in the dry condition.

4. No moduli of elasticity changes were observed in the residual strength data on $[0]_{24T}$ and $[(\pm 45)_5/0_{16}/90_4]_c$ laminates.

5. For the $[\pm 45]_{6s}$ laminates, the modulus of elasticity of the residual strength specimens is reduced by 22 percent compared with the static values.

6. For the $[90]_{24T}$ laminates, the modulus of elasticity reduction in the residual strength tests is 37 percent below the static value.

APPENDIX

Experimental Data Tables

TABLE 6—0-deg static compression test data, Laminate A. [0]$_{16T}$.

Specimen No.	Stress, MPa (ksi)	Strain, 10^{-3} units	Modulus of Elasticity, GPa (10^6 psi)	Poisson's Ratio	Failure Mode/Location[a]
			Test Series I—RTD		
RTD-1	-1469 (-213)	-16.6	105.5 (15.3)	0.34	CF, LD, DCS: EOT
RTD-2	-1276 (-185)	-13.8	108.3 (15.7)	0.34	DCS, LD: EOT
RTD-3	-1400 (-203)	-16.0	105.5 (15.3)	0.33	CF, LD, DCS: EOT
RTD-4	-1393 (-202)	-14.8	10.83 (15.7)	0.36	DCS, LD: EOT
RTD-5	-1448 (-210)	-14.6	122.0 (17.7)	0.36	CF, LD, DCS: EOT
Mean	-1400 (-203)	-15.1	109.6 (15.9)	0.34	CF and DCS with LD @ EOT
SD	-76 (-11)	-1.1	6.9 (1.0)	0.01	
CV	5%	7%	6%	4%	
			Test Series II—RTW		
RTW-1	-965 (-140)	-9.8	104.8 (15.2)	0.34	DCS, TB: EOT
RTW-2	-910 (-132)	-9.1	113.1 (16.4)	0.34	DCS, TB: EOT
RTW-3	-1379 (-200)	-14.6	115.1 (16.7)	0.33	CF, LD, DCS: EOT

				r
RTW-4	−1172 (−170)	−12.4	110.3 (16.0)	0.34	CF, LD, DCS: EOT
RTW-5	−1186 (−172)	−12.8	111.7 (16.2)	0.34	CF, LD, DCS: EOT
Mean	−1124 (−163)	−11.7	111.0 (16.0)	0.34	DCS/TB and CF/LD/DCS @ EOT
SD	−186 (−27)	−2.3	4.1 (0.6)	0.01	
CV	17%	19%	4%	2%	

Test Series III—ETD

ETD-1	−1062 (−154)	−10.2	118.6 (17.2)	0.38	TB: EOT
ETD-2	−1179 (−171)	−11.8	115.8 (16.8)	0.36	TB: EOT
ETD-3	−1165 (−169)	−11.2	114.5 (16.6)	0.32	DCS, LD: EOT
ETD-4	−1186 (−172)	−12.6	110.3 (16.0)	0.38	DCS, LD: EOT
ETD-5	−965 (−140)	−9.2	118.6 (17.2)	0.34	TB: EOT
Mean	−1103 (−160)	−11.0	115.8 (16.8)	0.35	TB, DCS, and LD @ EOT
SD	−90 (−13)	−1.3	3.5 (0.5)	0.03	
CV	8%	12%	3%	8%	

Test Series IV—ETW

ETW-1	−1200 (−174)	−12.9	107.6 (15.6)	0.30	CF, LD: EOT
ETW-2	−1062 (−154)	−10.9	111.0 (16.1)	0.32	CF, LD, DCS: EOT
ETW-3	−793 (−115)	−7.4	113.8 (16.5)	0.36	DCS, TB: EOT
ETW-4	−786 (−114)	−7.2	115.1 (16.7)	0.34	DCS, TB: EOT
ETW-5	−786 (−114)	−6.8	123.4 (17.9)	0.32	DCS, TB: EOT
Mean	−924 (−134)	−9.1	114.5 (16.6)	0.33	CF, LD and DCS, TB @ EOT
SD	−193 (−28)	−2.7	6.2 (0.9)	0.02	
CV	21%	30%	5%	7%	

[a] See Nomenclature.

TABLE 7—Censored 0-deg static compression test data summary, Laminate A, [0]$_{16T}$.

Statistical Values	Ultimate		Modulus of Elasticity, GPa (10⁶ psi)	Poisson's Ratio	Source
	Stress, MPa (ksi)	Strain, 10⁻³-units			
Test Series I—RTD					
Mean	−1400 (−203)	−15.1	109.6 (15.9)	0.34	Table 1 data, not censored
SD	−76 (−11)	−1.1	6.9 (1.0)	0.01	
CV	5%	7%	6%	4%	
No. of specimens	5	5	5	5	
Test Series II—RTW					
Mean	−1241 (−180)	−13.3	111.0 (16.0)	0.34	Table 1 data strength values, censored[a]
SD	−117 (−17)	−1.2	4.1 (0.6)	0.01	
CV	9%	9%	4%	2%	
No. of specimens	3	3	5	5	
Test Series III—ETD					
Mean	−1179 (−171)	−11.9	115.8 (16.8)	0.35	Table 1 data strength values, censored[a]
SD	−14 (−2)	−0.9	3.5 (0.5)	0.03	
CV	1%	8%	3%	8%	
No. of specimens	2	2	5	5	
Test Series IV—ETW					
Mean	−1131 (−164)	−11.9	114.5 (16.6)	0.33	Table 1 data strength values, censored[a]
SD	−97 (−14)	−1.4	6.2 (0.9)	0.02	
CV	9%	12%	5%	7%	
No. of specimens	2	2	5	5	

[a] All specimens with tab bond (TB) failures were censored.

TABLE 8—90-deg static compression test data:[a] Laminate A[0]₁₆T.

Specimen No.	Ultimate Stress, MPa (ksi)	Ultimate Strain, 10⁻³-units	Proportional Limit Stress, MPa (ksi)	Proportional Limit Strain, 10⁻³-unit	Modulus of Elasticity, GPa (10⁶ psi)	Poisson's Ratio
			Test Series VII—RTD			
VII-A-22	−266.8 (−38.7)	...	... (...)	...	... (...)	...
VII-A-23	−274.4 (−39.8)	−32.8	−103.4 (−15.0)	−9.2	12.4 (1.8)	0.024
VII-A-24	−284.1 (−41.2)	−33.1	−103.4 (−15.0)	−9.2	11.7 (1.7)	0.030
VII-A-25	−241.3 (−35.0)	−25.0	−109.6 (−15.9)	−9.8	11.0 (1.6)	0.019
VII-A-72X	−224.8 (−32.6)	...	... (...)	...	... (...)	...
Mean	−258.6 (−37.5)	−30.3	−105.5 (−15.3)	−9.4	11.7 (1.7)	0.024
SD	−24.8 (−3.6)	−4.6	3.5 (−0.5)	−0.3	0.7 (0.1)	0.006
CV	10%	15%	3%	4%	5%	23%
			Test Series VIII and X—RTW			
VIII-A-26	−165.5 (−24.0)	...	... (...)	...	... (...)	...
VII-A-27	−176.5 (−25.6)	...	... (...)	...	... (...)	...
X-A-38	−198.6 (−28.8)	−26.3	−112.4 (−16.3)	−12.2	9.2 (1.3)	0.016
X-A-39	−222.0 (−32.2)	−31.1	−111.0 (−16.1)	−12.8	8.5 (1.2)	0.012
X-A-40	−235.1 (−34.1)	−34.3	−111.0 (−16.1)	−12.5	8.6 (1.2)	0.011
Mean	−199.3 (−28.9)	−30.6	−111.7 (−16.2)	−12.5	8.8 (1.3)	0.013
SD	−29.7 (−4.3)	−4.0	−0.7 (−0.1)	−0.3	0.4 (0.1)	0.003
CV	15%	13%	1%	2%	5%	20%
			Test Series IX—ETD			
IX-A-31	−180.0 (−26.1)	...	... (...)	...	... (...)	...
IX-A-32	−175.8 (−25.5)	...	... (...)	...	... (...)	...
IX-A-33	−177.2 (−25.7)	−22.3	−35.2 (−5.1)	−3.1	11.0 (1.6)	0.026
IX-A-34	−189.6 (−27.5)	−24.6	−31.0 (−4.5)	−2.8	12.4 (1.8)	0.030

TABLE 8—Continued.

Specimen No.	Ultimate		Proportional Limit		Modulus of Elasticity, GPa (10⁶ psi)	Poisson's Ratio
	Stress, MPa (ksi)	Strain, 10^{-3}-units	Stress, MPa (ksi)	Strain, 10^{-3}-unit		
IX-A-35	−200.0 (−29.0)	−30.5	−83.4 (−12.1)	−8.4	9.7 (1.4)	0.037
IX-A-73X	−193.7 (−28.1)	...	... (...)	...	... (...)	...
Mean	−186.2 (−27.0)	−25.8	−49.6 (−7.2)	−4.8[b]	11.0 (1.6)	0.031
SD	−9.7 (−1.4)	−4.2	−29.0 (−4.2)	−3.1	1.4 (0.2)	0.006
CV	5%	16%	58%	66%	10%	18%
Test Series X and VIII—ETW						
X-A-36	−144.8 (−21.0)	...	... (...)	...	... (...)	...
X-A-37	−140.0 (−20.3)	...	... (...)	...	... (...)	...
VIII-A-28	−148.9 (−21.6)	−8.9	−49.6 (−7.2)	−3.2	13.1 (1.9)	0.102
VIII-A-29	−128.9 (−18.7)	−8.4	−51.7 (−7.5)	−2.6	19.3 (2.8)	0.075
VIII-A-30	−140.0 (−20.3)	−25.1	−63.4 (−9.2)	−5.6	11.7 (1.7)	0.052
Mean	−140.7 (−20.4)	−14.1[b]	−54.5 (7.9)	−3.8[b]	14.5 (2.1)[b]	0.076[b]
SD	−7.5 (−1.1)	−9.5	−7.6 (−1.1)	−1.6	4.1 (0.6)	0.025
CV	5%	67%	14%	42%	27%	33%

[a] ETL fixture used.
[b] Strain data for ETD and ETW conditions are questionable because of scatter.

TABLE 9—0-deg static compression test data;[a] Laminate C [±45]₄ₛ.

| Specimen No. | Effective Ultimate | | Proportional Limit | | Modulus of Elasticity, GPa (10⁶ psi) | Poisson's Ratio |
	Stress, MPa (ksi)[b]	Strain, 10⁻³-units	Stress, MPa (ksi)	Strain, 10⁻³-units		
			Test Series XIII—RTD			
XIII-C-1	... (...)	...	... (...)	...	... (...)	...
XIII-C-2	... (...)	...	... (...)	...	... (...)	...
XIII-C-3	−220.6 (−32.0)	−32.7	−73.1 (−10.6)	−4.6	15.9 (2.3)	0.71
XIII-C-4	−255.1 (−37.0)	−26.6	−79.3 (−11.5)	−4.7	17.9 (2.6)	0.78
XIII-C-5	−232.4 (−33.7)	−32.8	−80.0 (−11.6)	−5.3	15.2 (2.2)	0.76
Mean	−235.8 (−34.2)	−30.7	−77.2 (−11.2)	−4.9	16.6 (2.4)	0.75
SD	−17.2 (−2.5)	−3.6	−4.2 (−0.6)	−0.4	1.4 (0.2)	0.04
CV	7%	12%	5%	8%	8%	5%
			Test Series XIV and XVI—RTW			
XIV-C-6	−218.6 (−31.7)	−27.0	−79.3 (−11.5)	−4.7	15.9 (2.3)	0.63
XIV-C-7	−215.1 (−31.2)	−29.2	−78.6 (−11.4)	−5.0	15.9 (2.3)	0.68
XIV-C-8	−200.6 (−29.1)	−27.4	−77.9 (−11.3)	−4.5	19.3 (2.8)	0.84
XVI-C-20	... (...)	...	... (...)	...	... (...)	...
Mean	−210.9 (−30.6)	−27.9	−78.6 (−11.4)	−4.8	16.6 (2.4)	0.72
SD	−9.7 (−1.4)	1.1	0.7 (−0.1)	−0.2	2.1 (0.3)	0.11
CV	4%	4%	1%	5%	11%	15%
			Test Series XV—ETD			
XV-C-11	−160.0 (−23.2)	−24.9	−61.4 (−8.9)	−5.4	13.1 (1.9)	0.70
XV-C-12	−142.0 (−20.6)	−24.5	−62.1 (−9.0)	−4.6	15.9 (2.3)	0.57
XV-C-13	−132.4 (−19.2)	−16.5	−62.1 (−9.0)	−4.8	15.2 (2.2)	0.69

TABLE 9—Continued.

Specimen No.	Effective Ultimate		Proportional Limit		Modulus of Elasticity, GPa (10^6 psi)	Poisson's Ratio
	Stress, MPa (ksi)[b]	Strain, 10^{-3}-units	Stress, MPa (ksi)	Strain, 10^{-3}-units		
XV-C-14	... (...)	...	... (...)	...	... (...)	...
XV-C-15	... (...)	...	... (...)	...	... (...)	...
Mean	−144.1 (−20.9)	−22.0	−62.1 (−9.0)	−5.0	14.7 (2.1)	0.65
SD	−13.1 (−1.9)	−4.7	−0.7 (−0.1)	−0.4	1.5 (0.2)	0.08
CV	9%	22%	1%	9%	10%	12%
			Test Series XVI and XIV—ETW			
XVI-C-17	−147.6 (−21.4)	−33.4	−39.3 (−5.7)	−2.9	12.4 (1.8)	0.63
XVI-C-18	−137.2 (−19.9)	−28.6	−39.3 (−5.7)	−2.8	13.8 (2.0)	0.62
XVI-C-19	−145.5 (−21.1)	−34.0	−37.9 (−5.5)	−2.7	14.5 (2.1)	0.79
XIV-C-9	... (...)	...	... (...)	...	... (...)	...
XIV-C-10	... (...)	...	... (...)	...	... (...)	...
Mean	−143.4 (−20.8)	−32.0	−39.3 (−5.7)	−2.8	13.1 (1.9)	0.68
SD	−5.5 (−0.8)	−3.0	−0.9	−0.1	1.48 (0.2)	0.09
CV	4%	9%	2%	4%	8%	14%

[a] ETL fixture used.
[b] The effective ultimate stress is determined by extrapolating the secondary straight-line portion of the stress-strain curve to the ultimate strain level.

TABLE 10—0-deg static compression test data,[a] Laminate E, $[\pm 45)_5/0_{16}/90_4]_s$.

Specimen No.	Ultimate Stress, MPa (ksi)	Ultimate Strain, 10^{-3}-units	Proportional Limit Stress, MPa (ksi)	Proportional Limit Strain, 10^{-3}-units	Modulus of Elasticity, GPa (10^6 psi)	Poisson's Ratio
Test Series XIX—RTD						
XIX-E-1	−686.0 (−99.5)	...	(...)	...	.. (...)	...
XIX-E-2	−720.5 (−104.5)	...	(...)	...	.. (...)	...
XIX-E-3	−733.6 (−106.4)	−11.8	−329.6 (−47.8)	−4.6	72.4 (10.5)	0.34
XIX-E-4	−726.7 (−105.4)	−11.5	−334.4 (−48.5)	−4.5	72.4 (10.5)	0.31
XIX-E-5	−754.3 (−109.4)	−12.2	−340.6 (−49.4)	−4.8	70.3 (10.2)	0.32
Mean	−723.9 (−105.0)	−11.8	−335.1 (−48.6)	−4.7	71.7 (10.4)	0.32
SD	−24.8 (−3.6)	−0.4	−5.2 (−0.8)	−0.7	1.4 (0.2)	0.02
CV	3%	3%	2%	2%	2%	5%
Test Series XX—RTW						
XX-E-6	−652.2 (−94.6)	−10.8	−325.4 (−47.2)	−4.9	65.5 (9.5)	0.30
XX-E-7	−689.5 (−100.0)	−10.5	−255.1 (−37.0)	−3.2	69.0 (10.0)	0.29
XX-E-8	−675.7 (−98.0)	−10.2	−311.0 (−45.1)	−4.3	73.1 (10.6)	0.36
XX-E-9	−628.8 (−91.2)	...	(...)	...	.. (...)	...
XX-E-10	−661.2 (−95.9)	...	(...)	...	.. (...)	...
Mean	−661.2 (−95.9)	−10.5	−297.2 (−43.1)	−4.1	69.6 (10.0)	0.32
SD	−23.4 (−3.4)	−0.3	−37.2 (−5.4)	−0.9	3.5 (0.5)	0.04
CV	4%	3%	13%	22%	5%	12%
Test Series XXI—ETD						
XXI-E-11	−610.2 (−88.5)	...	(...)	...	.. (...)	...
XXI-E-12	−612.9 (−88.9)	−9.5	−255.1 (−37.0)	−3.5	73.1 (10.6)	0.24
XXI-E-14	−624.7 (−90.6)	−9.5	−260.0 (−37.7)	−3.5	72.4 (10.5)	0.25
XXI-E-15	−638.5 (−92.6)	−9.9	−320.6 (−46.5)	−4.4	72.4 (10.5)	0.22

TABLE 10—Continued.

Specimen No.	Ultimate		Proportional Limit		Modulus of Elasticity, GPa (10^6 psi)	Poisson's Ratio
	Stress, MPa (ksi)	Strain, 10^{-3}-units	Stress, MPa (ksi)	Strain, 10^{-3}-units		
XXI-E-21X	−659.8 (−95.7)	...	... (...)	...	... (...)	...
XXI-E-22X	−543.3 (−78.8)	...	... (...)	...	... (...)	...
Mean	−615.0 (−89.2)	−9.6	−278.6 (−40.4)	−3.8	72.4 (10.5)	0.23
SD	−39.3 (−5.7)	−0.2	−36.5 (−5.3)	−0.6	0 (0)	0.02
CV	6%	2%	13%	14%	0%	7%
Test Series XXII—ETW						
XXII-E-16	−515.7 (−74.8)	−8.0	−264.1 (−38.3)	−5.7	72.4 (10.5)	0.26
XXII-E-17	−555.0 (−80.6)	−9.0	−186.9 (−27.1)	−2.6	73.1 (10.6)	0.22
XXII-E-18	−517.1 (−75.0)	−8.1	−246.8 (−35.8)	−3.6	69.0 (10.0)	0.25
XXII-E-19	−475.7 (−69.0)	...	... (...)	...	... (...)	...
XXII-E-20	−551.6 (−80.0)	...	... (...)	...	... (...)	...
Mean	−523.3 (−75.9)	−8.4	−232.4 (−33.7)	−4.0	71.7 (10.4)	0.24
SD	−32.4 (−4.7)	−0.2	−40.7 (−5.9)	−1.6	2.1 (0.3)	0.02
CV	6%	6%	17%	40%	3%	10%

[a] ETL fixture used.

TABLE 11—*Censored static compression test data on 90-deg Laminate A and 0-deg Laminate E.*

Specimen No.	Ultimate		Proportional Limit		Modulus of Elasticity, GPa (10⁶ psi)	Poisson's Ratio	Reference Appendix Table No.	Laminate Description
	Stress, MPa (ksi)	Strain, 10^{-3} units	Stress, MPa (ksi)	Strain, 10^{-3} units	Modulus of Elasticity, GPa (10^6 psi)	Poisson's Ratio	Reference Appendix Table No.	Laminate Description
Test Series IX—ETD								
IX-A-31	−180.0 (−26.1)	...	... (...)	...	...	...	8	90-deg Laminate A
IX-A-32	−175.8 (−25.5)	...	... (...)	...	...	...	8	90-deg Laminate A
IX-A-33	−177.2 (−25.7)	−22.3	−35.2 (−5.1)	−3.1	11.0 (1.6)	0.026	8	90-deg Laminate A
IX-A-34	−189.6 (−27.5)	−24.6	−31.0 (−4.5)	−2.8	12.4 (1.8)	0.030	8	90-deg Laminate A
IX-A-35	−200.0 (−29.0)	−30.5 ←	omitted			→	8	90-deg Laminate A
IX-73X	−193.7 (−28.1)	...	... (...)	...	...	...	8	90-deg Laminate A
Mean	−186.2 (−27.0)	−25.8	−33.1 (−4.8)	−3.0	11.7 (1.7)	0.028		
SD[a]	−9.7 (−1.4)	−4.2	−3.0 (−0.4)	−0.2	1.0 (0.1)	0.003		
CV[b]	5%	16%	9%	7%	8%	10%		
Test Series X and VIII—ETW								
X-A-36	−144.8 (−21.0)	...	... (...)	...	...	...	8	90-deg Laminate A
X-A-37	−140.0 (−20.3)	...	... (...)	...	...	...	8	90-deg Laminate A
VIII-A-28	−148.9 (−21.6)	−8.9	−49.6 (−7.2)	−3.2	13.1 (1.9)	0.102	8	90-deg Laminate A
VIII-A-29	−128.9 (−18.7)	−8.4	−51.7 (−7.5)	−2.6 ←	omitted →	0.075	8	90-deg Laminate A
VIII-A-30	−140.0 (−20.3) ←	omitted		→	11.7 (1.7) ←	omitted →	8	90-deg Laminate A
Mean	−140.7 (−20.4)	−8.6	−50.6 (−7.4)	−2.9	12.4 (1.8)	0.088		
SD	−7.5 (−1.1)	−0.4	−1.5 (−0.2)	−0.4	1.0 (0.1)	0.019		
CV	5%	4%	3%	15%	8%	22%		
Test Series XXII—ETW								
XXII-E-16	−515.7 (−74.8)	−8.0	−264.1 (−38.3) ←	omitted →	72.4 (10.5)	0.26	10	0-deg Laminate E
XXII-E-17	−555.0 (−80.6)	−9.0	−186.9 (−27.1)	−2.6	73.1 (10.6)	0.22	10	0-deg Laminate E
XXII-E-18	−517.1 (−75.0)	−8.1	−246.8 (−35.8)	−3.6	69.0 (10.0)	0.25	10	0-deg Laminate E
XXII-E-19	−475.7 (−69.0)	...	... (...)	...	... (...)	...	10	0-deg Laminate E
XXII-E-20	−551.6 (−80.0)	...	... (...)	...	... (...)	...	10	0-deg Laminate E
Mean	−523.3 (−75.9)	−8.4	−232.4 (−33.7)	−3.1	71.7 (10.4)	0.24		
SD	−32.4 (−4.7)	−0.2	−40.7 (−5.9)	−0.7	2.1 (0.3)	0.02		
CV	6%	6%	17%	23%	3%	19%		

TABLE 12—*0-deg compression fatigue data, Laminate B, [0]$_{24T}$ (R = 10, constant frequency).*

Specimen No.	Stress-Algebraic Min Stress MPa (ksi)	Compression Max Strain 10^{-3} units	Cyclic Rate, Hz	No. of Cycles to Failure or Runout	Temperature Rise, ΔT, deg C (deg F)
		RTD			
V-B-1	−714.3 (−103.6)	−5.96	10, 5, 6, and 7	1 250 000[a]	6.7 (12)
V-B-2	−845.3 (−122.6)	−7.46	5, 6	249 760	6.1 (11)
V-B-3	−939.1 (−136.2)	−8.94	3–6	35 950	8.9 (16)
V-B-4	−1038.3 (−150.6)	−9.93	5	16 810	1.1 (2)
V-B-5	−1307.2 (−189.6)	−11.92	3	190	0 (0)
V-B-6	−1390.0 (−201.6)	−11.92	3	800	0.6 (1)
V-B-7	−1145.2 (−116.1)	−9.93	5	60	3.3 (6)
V-B-8	−950.1 (−137.8)	−8.94	5	49 290	3.9 (7)
V-B-9	−806.7 (−117.0)	−7.46	5	62 250	5.6 (10)
V-B-10	−686.7 (−99.6)	−6.00	5	1 250 000[a]	6.1 (11)
V-B-41X	−1075.6 (−156.0)	−10.00	5, 2, 1	270 280	11.1 (20)
V-B-42X	−1061.1 (−153.9)	−10.00	1, 2	166 680	2.8 (5)
V-B-43X	−814.3 (−118.1)	−7.50	5	1 250 000[a]	... (...)
		RTW			
VI-B-11	−734.3 (−106.5)	−6.00	5	1 234 110	b
VI-B-12	−939.1 (−136.2)	−9.00	5	327 840	b
VI-B-13	−1034.9 (−150.1)	−10.00	5	39 200	b
VI-B-14	−787.4 (−114.2)	−7.50	5	1 250 000[b]	b
VI-B-15	−1094.2 (−158.7)	−11.00	5	22 600	b
VI-B-16	−1053.5 (−152.8)	−10.00	5	22 600	b
VI-B-17	−691.5 (−100.3)	−6.00	5	1 250 000[a]	b
VI-B-18	−797.0 (−115.6)	−7.50	5	1 250 000[a]	b
VI-B-19	−937.7 (−136.0)	−9.00	5	263 390	b
VI-B-20	−1074.2 (−155.8)	−10.47	5	580	b

[a] No failure.
[b] Specimen had longitudinal splits about 6.35 mm (¼ in.) in from edge.
[c] Used air blowing on edges of specimen for cooling; ΔT = 2.8 deg C (5 deg F).

TABLE 13—*0-deg static compression at RTD and RTW, Laminate A, [0]$_{16T}$*
(reference Appendix Tables 6 and 7).

Test Series/Specimen Identification	Ultimate Strength, MPa (ksi)	Ultimate Strain, 10^{-3} units	Modulus of Elasticity, GPa (10^6 psi)	Failure Mode[a]
I-RTD-1	−1469 (−213)	−16.6	105.5 (15.3)	CF, LD, DCS
I-RTD-2	−1276 (−185)	−13.8	108.3 (15.7)	DCS, LD
I-RTD-3	−1400 (−203)	−16.0	105.5 (15.3)	CF, LD, DCS
I-RTD-4	−1393 (−202)	−14.8	108.3 (15.7)	DCS, LD
I-RTD-5	−1448 (−210)	−14.6	122.0 (17.7)	CF, LD, DCS
I-RTW-3	−1379 (−200)	−14.6	115.1 (16.7)	CF, LD, DCS
I-RTW-4	−1172 (−170)	−12.4	110.3 (16.0)	CF, LD, DCS
I-RTW-5	−1186 (−172)	−12.8	111.7 (16.2)	CF, LD, DCS
Mean	−1338 (−194)	−14.4	111.0 (16.1)	CF, LD, DCS
SD	−117 (−17)	−1.4	5.5 (0.8)	
CV	9%	10%	5%	

[a] See Nomenclature.

TABLE 14—*90-deg compression fatigue data, Laminate B, [0]$_{24T}$* *(R = 10, constant frequency).*

Specimen No.	Compression Max Stress-Algebraic Min, MPa (ksi)	Compression Max Strain, 10^{-3} units	Cyclic Rate, Hz	No. of Cycles to Failure or Runout	Temperature Rise, ΔT, deg C (deg F)
		RTD			
XI-B-21	−55.2 (−8.0)	−6.00	10	1 250 000[a]	[b]
XI-B-22	−111.7 (−16.2)	−12.00	10	1 250 000[a]	[b]
XI-B-23	−167.5 (−24.3)	−18.00	10	6 520	[b]
XI-B-24	−140.0 (−20.3)	−15.00	10	425 000	[b]
XI-B-25	−126.9 (−18.4)	−13.50	10	1 250 000[a]	[b]
XI-B-27	−155.8 (−22.6)	−16.50	10	137 800	[b]
XI-B-29	−186.2 (−27.0)	−19.50	1	5 790	[b]
XI-B-30	−200.0 (−29.0)	−21.00	1	3 850	[b]
XI-B-44X	−141.3 (−20.5)	−15.00	10. 5	34 660	[b]
XI-B-45X	−208.2 (−30.2)	−24.00	1	50	[b]
		RTW			
XII-B-35	−26.9 (−3.9)	−2.50	5	1 250 000	[c]
XII-B-36	−49.0 (−7.1)	−5.00	5	1 250 000	[c]
XII-B-37	−54.5 (−7.9)	−5.50	5	1 250 000	[c]
XII-B-38	−60.0 (−8.7)	−6.00	5	1 250 000	[c]
XII-B-39	−74.5 (−10.8)	−7.50	5	1 250 000	[c]
XII-B-40	−90.0 (−13.0)	−9.00	5	1 250 000	[c]

[a] No failure.
[b] ΔT = 5.6 deg C (10 deg F).
[c] ΔT = 2.8 deg C (5 deg F), used air blowing on edges of specimen for cooling.

TABLE 15—*0-deg compression fatigue data, Laminate D, [±45]$_{6s}$* *(R = 10, constant frequency).*

Specimen No.	Compression Max Stress-Algebraic Min, MPa (ksi)	Compression Max Strain, 10^{-3} units	Cyclic Rate, Hz	No. of Cycles to Failure or Runout	Temperature Rise, ΔT, deg C (deg F)
		RTD			
XVII-D-1	−136.5 (−19.8)	−12.54	10	154 220	[b]
XVII-D-2	−194.4 (−28.2)	−18.80	5	380	[b]
XVII-D-3	−165.5 (−24.0)	−15.67	5	6 230	13.9 (25)
XVII-D-4	−143.4 (−20.8)	−14.10	1, 2	30 500	[b]
XVII-D-5	−104.8 (−15.2)	−10.48	5	1 250 000[a]	[b]
XVII-D-6	−103.4 (−15.0)	−10.48	10	1 250 000[a]	[b]
XVII-D-7	−122.7 (−17.8)	−12.54	10. 5	282 530	[b]
XVII-D-8	−140.7 (−20.4)	−14.10	2	12 340	10.6 (19)
XVII-D-9	−156.5 (−22.7)	−15.67	1	4 200	[b]
XVII-D-10	−195.1 (−28.3)	−18.80	1	90	[b]
XVII-D-21X	−121.4 (−17.6)	−11.36	10	470 360	[b]
XVII-D-22X	−122.0 (−17.7)	−11.36	10	650 000	7.8 (14)
		RTW			
XVIII-D-11	−75.2 (−10.9)	−6.00	5	1 250 000[a]	[c]
XVIII-D-12	−99.3 (−14.4)	−7.50	5	1 250 000[a]	[c]
XVIII-D-13	−113.1 (−16.4)	−9.00	5	1 250 000[a]	[c]
XVIII-D-14	−117.2 (−17.0)	−10.50	5	1 250 000[a]	[c]
XVIII-D-15	−131.7 (−19.1)	−15.00	5	1 183 450	[c]
XVIII-D-16	−146.9 (−21.3)	−16.50	5	213 410	[c]
XVIII-D-17	−147.6 (−21.4)	−18.00	5	280 150	[c]
XVIII-D-18	−145.5 (−21.1)	−19.50	5	1 028 870	[c]
XVIII-D-19	−160.0 (−23.2)	−21.00	5	72 830	[c]
XVIII-D-20	−167.5 (−24.3)	−22.50	5	6 180	[c]

[a] Runout, no failure.
[b] ΔT ≤ 5.6 deg C (10 deg F).
[c] Used air blowing on edges of specimen for cooling. ΔT = ≤2.8 deg C (≤5 deg F).

TABLE 16—*0-deg static compression data at RTD and RTW, Laminate C, [±45]₄ₛ (reference Appendix Table 9).*

Test Series/Specimen Identification	Ultimate Strength, MPa (ksi)	Ultimate Strain, 10^{-3} units	Modulus of Elasticity, GPa (10^6 psi)	Poisson's Ratio
XIII-C-3D	−220.6 (−32.0)	−32.70	15.9 (2.3)	0.71
XIII-C-4D	−255.1 (−37.0)	−26.50	17.9 (2.6)	0.78
XIII-C-5D	−232.4 (−33.7)	−32.80	15.2 (2.2)	0.76
XIV-C-6W	−218.6 (−31.7)	−27.05	15.9 (2.3)	0.63
XIV-C-7W	−215.1 (−31.2)	−29.15	15.9 (2.3)	0.68
XIV-C-8W	−200.6 (−29.1)	−27.37	19.3 (2.8)	0.84
Mean	−223.4 (−32.4)	−29.27	16.6 (2.4)	0.74
SD	−18.6 (−2.7)	−2.85	1.4 (0.2)	0.07
CV	8%	10%	8%	10%

TABLE 17—*0-deg compression fatigue data, Laminate E, [(±45)₅/0₁₆/90₄]ᶜ (R = 10, constant frequency).*

Specimen No.	Compression Max Stress-Algebraic Min, MPa (ksi)	Compression Max Strain, 10^{-3} units	Cyclic Rate, Hz	No. of Cycles to Failure or Runout	Temperature Rise, ΔT, deg C (deg F)
		RTD			
XXIII-E-23	−359.2 (−52.1)	−6.00	5, 7	1 250 000ᵃ	b
XXIII-E-24	−475.0 (−68.9)	−7.50	5, 4	1 250 000ᵃ	7.2 (13)
XXIII-E-25	−577.8 (−83.8)	−9.00	3, 2	157 410	8.9 (16)
XXIII-E-26	−723.3 (−104.9)	−12.00	1	22 240	b
XXIII-E-27	−732.2 (−106.2)	−12.50	1	27 700	b
XXIII-E-28	−505.4 (−73.3)	−8.25	2, 5	1 250 000ᵃ	b
XXIII-E-29	−547.4 (−79.4)	−9.00	3, 5	1 250 000ᵃ	b
XXIII-E-30	−591.6 (−85.8)	−10.00	5	114 600	b
XXIII-E-31	−670.2 (−97.2)	−11.00	4	11 690	b
XXIII-E-32	−585.4 (−84.9)	−9.50	5	5 390	b
XXIII-43X	−597.8 (−86.7)	−10.00	5	256 870	b
XXIII-44X	−661.9 (−96.0)	−11.58	1	1	b
		RTW			
XXIV-E-33	−364.0 (−52.8)	−6.00	5	1 250 000ᵃ	b
XXIV-E-34	−549.5 (−79.7)	−9.00	5	953 060	b
XXIV-E-35	−693.6 (−100.6)	−12.00	5	2 430	b
XXIV-E-36	−622.6 (−90.3)	−10.50	5	34 940	b
XXIV-E-37	−537.8 (−78.0)	−9.00	5	737 990	b
XXIV-E-38	−655.0 (−95.0)	−12.00	5	28 960	b
XXIV-E-39	−612.3 (−88.8)	−10.50	5	190 160	b
XXIV-E-40	−712.9 (−103.4)	−12.50	5	540	b
XXIV-E-41	−566.1 (−82.1)	−9.50	5	193 840	b
XXIV-E-42	−460.6 (−66.8)	−7.50	5	1 250 000ᵃ	b

ᵃ Same as 1000 μ in./in.
ᵇ Runout, no failure.
ᶜ Used air blowing on edges for specimen cooling. $\Delta T < 2.8°C$ (≤5°F).

TABLE 18—*0-deg static compression data at RTD and RTW. Laminate E.* $[(\pm 45)_5/0_{16}/90_4]_c$
(reference Appendix Table 10).

Test Series/Specimen Identification	Ultimate Strength, MPa (ksi)	Ultimate Strain, 10^{-3} units	Modulus of Elasticity, GPa (10^6 psi)	Poisson's Ratio
XIX-E-1D	−686.0 (−99.5)	...	... (...)	...
XIX-E-2D	−720.5 (−104.5)	...	... (...)	...
XIX-E-3D	−733.6 (−106.4)	−11.79	72.4 (10.5)	0.34
XIX-E-4D	−726.7 (−105.4)	−11.47	72.4 (10.5)	0.31
XIX-E-5D	−754.3 (−109.4)	−12.19	70.3 (10.2)	0.32
XX-E-6W	−652.2 (−94.6)	−10.80	65.5 (9.5)	0.30
XX-E-7W	−689.5 (−100.0)	−10.47	68.9 (10.0)	0.29
XX-E-8W	−675.7 (−98.0)	−10.16	73.1 (10.6)	0.36
XX-E-9W	−628.8 (−91.2)	...	... (...)	...
XX-E-10W	−661.2 (−95.9)	...	... (...)	...
Mean	−692.9 (−100.5)	−11.15	70.3 (10.2)	0.32
SD	−40.4 (−5.8)	−0.79	2.8 (0.4)	0.03
CV	6%	7%	4%	8%

TABLE 19—Residual strength of fatigue specimens at ET,[a] Laminate B, [0]$_{24}$T.

Specimen No.	Load Direction, deg	Ultimate Strength, MPa (ksi)	Ultimate Strain, 10⁻³ units	Modulus of Elasticity, GPa (10⁶ psi)	1.25 Million Cycles Runout at Max Fatigue	
					Stress (RT), MPa (ksi)	Strain (RT), 10⁻³ units
			Dry			
V-B-1	0	−496.4 (−72.0)	−3.95	127.6 (18.5)	−714.3 (−103.6)	−5.96
V-B-10	0	−580.5 (−84.2)	−4.02	126.2 (18.3)	−686.7 (−99.6)	−6.00
V-B-43X	0	−866.7 (−125.7)	−8.46	110.3 (16.0)	−814.3 (−118.1)	−7.50
XI-B-21	90	−144.1 (−20.9)	−22.64	8.3 (1.2)	−55.2 (−8.0)	−6.00
XI-B-25	90	−175.8 (−25.5)	−27.69	8.3 (1.2)	−127.6 (−18.5)	−13.50
			Wet			
VI-B-17	0	−727.4 (−105.5)	−6.55	117.2 (17.0)	−691.5 (−100.3)	−6.00
XII-B-35	90	−107.6 (−15.6)	−23.85	8.3 (1.2)	−26.9 (−3.9)	−2.50
XII-B-36	90	−105.5 (−15.3)	−25.00	6.9 (1.0)	−49.0 (−7.1)	−5.00
XII-B-37	90	−81.4 (−11.8)	−29.89	5.5 (0.8)	−54.5 (−7.9)	−5.50
XII-B-39	90	−136.5 (−19.8)	−27.71	8.3 (1.2)	−74.5 (−10.8)	−7.50
XII-B-40	90	−116.5 (−16.9)	−34.37	8.3 (1.2)	−89.6 (−13.0)	−9.00

[a]Tested at 103.3 ± 2.8°C (218 ± 5°F) after 10 min of exposure.

TABLE 20—Residual strength of fatigue specimens at ET,[a] Laminate D, [±45]$_{6s}$ (0-deg load direction).

Specimen No.	Proportional Limit Stress, MPa (ksi)	Proportional Limit Strain, 10^{-3} units	Modulus of Elasticity, GPa (10^6 psi)	1.25 Million Cycles Runout at Max Fatigue	
				Stress (RT) MPa (ksi)	Strain (RT) 10^{-3} units
		Dry			
XVII-D-5	−49.6 (−7.2)	−4.56	11.0 (1.6)	−104.8 (−15.2)	−10.48
XVII-D-6	−49.6 (−7.2)	−4.78	10.3 (1.5)	−103.4 (−15.0)	−10.48
		Wet			
XVIII-D-11	−39.3 (−5.7)	−3.22	12.4 (1.8)	−75.2 (−10.9)	−6.00
XVIII-D-12	−37.9 (−5.5)	−3.44	11.7 (1.7)	−99.3 (−14.4)	−7.50
XVIII-D-13	−37.9 (−5.5)	−4.00	9.7 (1.4)	−113.1 (−16.4)	−9.00
XVIII-D-14	−37.9 (−5.5)	−3.84	11.0 (1.6)	−117.2 (−17.0)	−10.50

[a]Tested at 103.3 ± 2.8°C (218 ± 5°F) after 10 min at temperature.

TABLE 21—Residual strength of fatigue specimens at ET.[a] $[(\pm45)_5/0_{16}/90_4]_c$ (0-deg load direction).

Specimen No.	Ultimate Stress, MPa (ksi)	Ultimate Strain, 10^{-3} units	Modulus of Elasticity, GPa (10^6 psi)	1.25 Million Cycles Runout at Max Fatigue	
				Stress (RT) MPa (ksi)	Strain (RT) 10^{-3} units
Dry					
XXII-E-23	−656.4 (−95.2)	−12.12	59.3 (8.6)	−359.2 (−52.1)	−6.00
XXII-E-24	−561.2 (−81.4)	−9.31	71.0 (10.3)	−475.0 (−69.9)	−7.50
XXIII-E-28	−579.2 (−84.0)	...	... (...)	−505.4 (−73.3)	−8.25
XXIII-E-29	−586.7 (−85.1)	−10.10	62.1 (9.0)	−547.4 (−79.4)	−9.00
Wet					
XXIV-E-33	−413.7 (−60.0)	−4.83	95.2 (13.8)	−364.0 (−52.8)	−6.00
XXIV-E-42	−420.6 (−61.0)	−6.91	64.1 (9.3)	−460.6 (−66.8)	−7.50

[a] Tested at 103.3 ± 2.8°C (218 ± 5°F) after 10 min at temperature.

TABLE 22—Experimental values for static compression strain, AS/3501-6 (all strain in 10^{-3} units).

Test Environment	0-deg Laminate A, [0]$_{16T}$ Ultimate and Proportional Limit		90-deg Laminate A, [0]$_{16T}$ Ultimate and Proportional Limit		0-deg Laminate C, [±45]$_{4s}$ Ultimate and Proportional Limit		0-deg Laminate E [(±45)$_5$/0$_{16}$/90$_4$]$_c$ Ultimate and Proportional Limit		Strain-Limiting Plies at Proportional Limit
RTD	−15.1	...	−30.3	−9.4	−30.7	−4.9	−11.8 (78%)[a]	−4.7 (40%)[b]	±45
RTW	−13.3	...	−30.6	−12.5	−27.9	−4.8	−10.5 (79%)[a]	−4.1 (39%)[b]	±45
ETD	−11.9	...	−25.8	−4.8	−22.0	−5.0	−9.6 (81%)[a]	−3.8 (40%)[b]	90 and ±45
ETW	−11.9	...	−14.1	−3.8	−32.0	−2.8	−8.4 (71%)[a]	−4.0 (48%)[b]	±45 and 90
Fatigue Value	62.5%		62.5%		57.5%		55.2%		

[a] % of 0-deg Laminate A, [0]$_{16T}$ ultimate strain level.
[b] % of ultimate strain level.

Design Allowables for
Special Applications

C. T. Herakovich[1] and E. R. Johnson[1]

Buckling of Composite Cylinders Under Combined Compression and Torsion—Theoretical/Experimental Correlation

REFERENCE: Herakovich, C. T. and Johnson, E. R., **"Buckling of Composite Cylinders Under Combined Compression and Torsion—Theoretical/Experimental Correlation,"** *Test Methods and Design Allowables for Fibrous Composites, ASTM STP 734,* C. C. Chamis, Ed., American Society for Testing and Materials, 1981, pp. 341-360.

ABSTRACT: Elastic buckling loads for laminated composite circular cylindrical shells subject to combined compression and torsion are determined experimentally. Test specimens include both boron/epoxy and graphite/epoxy laminates with symmetric layups of $[\mp 45]_s$, $[-45_2/45_2]_s$, and $[0/\pm 45/90]_s$, and with the unsymmetric layup of $[-82.5/30/20/-82.5]$. Results are presented on dimensionless buckling interaction diagrams on which the compression load ratios versus torsional load ratios at buckling are plotted. Theoretically predicted buckling loads for the test specimens obtained from a computer program based on Flügge's cylindrical shell buckling equations are also plotted on the interaction diagrams. Good comparison between experiment and theory is achieved when the results are plotted on the interaction diagrams.

KEY WORDS: buckling, torsion, compression, combined loading, graphite/epoxy, boron/epoxy, laminated composites, theory, experiment, composite materials

Laminated composite shell structures are attractive in many applications where high strength and stiffness to weight are important. Because these structures usually are thin, elastic buckling is a major design consideration. The circular cylinder is perhaps the most practically important member of this class of composite shell structures. Consequently, the buckling of this shell has received the attention of many researchers, whose investigations through 1974 are summarized in review papers [1-3].[2] Much of this past research has been limited to buckling subject to a single simple load. Fewer

---

[1]Professor and assistant professor, respectively, Department of Engineering Science and Mechanics, Virginia Polytechnic Institute and State University, Blacksburg, Va. 24061.

[2]The italic numbers in brackets refer to the list of references appended to this paper.

investigators have considered buckling under combined loading, including a comparison of theory and experiment. Very recently Booton and Tennyson [4] have reported on experiments of combined torsion, external pressure, and axial compression buckling of $(-70/70/0)$ and $(45/0/-45)$ glass/epoxy laminated cylinders. They reported reasonably good agreement between experiment and a theory based on imperfect anisotropic cylinders. Germane to the present study are the experiments on combined torsion and axial compression buckling of $[\pm 45]_s$ and $[0/\pm 45]_s$ graphite/epoxy and boron/epoxy cylinders by Wilkins and Love [5]. Their experimental results do not agree very well with theory.

The present work consists of an experimental determination compared with theoretical predictions of the buckling of boron/epoxy and graphite/epoxy circular cylinders subject to combined axial compression and torsion (Fig. 1). Cylinders of each material were fabricated from $[\mp 45]_s$, $[-45_2/45_2]_s$, and $[0/\pm 45/90]_s$ symmetric laminates, and a $[-82.5/30/20/-82.5]$ unsymmetric laminate. The first entry in this laminate notation refers to the fiber angle in degrees, with respect to the shell generator, of the ply on the outside wall of the cylinder. The succeeding entries, separated by slashes, refer to the fiber angles of successive plies toward the inside wall. A positive fiber angle is shown in Fig. 1, and all plies are assumed to have the same thickness t. The radius to wall thickness ratio (R/h) of the specimens varied from 55 to 142, and the length to radius ratios (L/R) from 4.4 to 6.7 (Table 1). Theoretical buckling loads were obtained from a computer program developed by Wu [6] which is based on Flügge's cylindrical shell buckling equations. Both the experimental buckling loads and the theoretical buckling loads are plotted for comparison on a dimensionless load plane called a buckling interaction diagram. The ordinate in this plane is the axial load ratio and the abscissa the torsional load ratio.

Experimental Program

Specimens

The specimens used in this investigation were thin-walled cylinders made from boron/epoxy or graphite/epoxy prepreg tape. The boron/epoxy was AVCO 5505/4 and the graphite/epoxy Modmor I/Narmco 5208. All cylinders were approximately 15.24 cm (6 in.) in diameter and most were approximately 50.8 cm (20 in.) in length. Wall thickness varied with the number of plies in the laminate, ranging from a minimum of 0.053 cm (0.021 in.) to a maximum of 0.137 cm (0.054 in.). A detailed listing of specimen geometries is given in Table 1, and a typical specimen is shown in Fig. 2. Load was introduced through aluminum end fixtures which were bonded to the ends of the specimen. The fixture consisted of a stepped cylindrical base which extended 3.81 cm (1.5 in.) into the composite cylinder and an aluminum

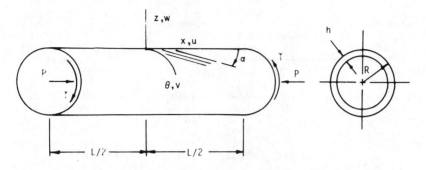

a)   GEOMETRY, COORDINATES, LOADS, AND FIBER ANGLE α

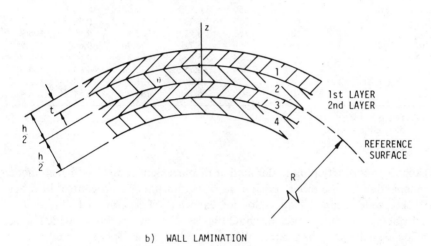

b)   WALL LAMINATION

FIG. 1—*Laminated circular cylindrical shell.*

collar around the outside of the cylinder. All surfaces were adhesively bonded
to provide fixed-end conditions. Twenty foil-type strain gages were bonded
to each specimen in the pattern shown in Fig. 3.

Loading

Most of the tests were conducted at Virginia Tech on an MTS axial-
torsional combined loader which had a rated capacity of 222.4 kN (50 kips)
axial and ±2.26 kN-m (±20 kip-in.) torsional. Cylinders with torsional
buckling loads in excess of 2.26 KN-m (20 kip-in.) were tested at National
Aeronautics and Space Administration (NASA) Langley Research Center on
a pure torsion machine. The combined loader at Virginia Tech had the

TABLE 1—*Specimen data.*

Laminate	Specimen No.	Radius, R (cm)^a	L/R	R/h	V_f	E_{11}, GPa^b
Boron/Epoxy						
$[0_8]$	2	7.536	4.719	79.97	0.57	217.6
$[\mp 45]_s$	1	7.587	6.654	135.8	0.473	181.1
$[-45_2/45_2]_s$	6	7.658	6.624	68.52	0.473	181.1
	19	7.587	6.716	78.37	0.547	209.0
$[0/\pm 45/90]_s$	13	7.559	6.720	74.40	0.52	198.8
	23	7.562	6.382	78.34	0.55	210.1
$[-82.5/30/20/-82.5]$	4	7.625	6.672	139.6	0.483	184.9
	12	7.579	6.702	142.1	0.495	189.4
Graphite/Epoxy						
$[\mp 45]_s$	11	7.587	4.394	114.9	0.46	106.9
$[-45_2/45_2]_s$	14	7.559	6.720	59.52	0.48	111.0
	18	7.559	6.720	55.11	0.44	103.0
	20	7.557	6.723	57.21	0.46	106.9
$[0/\pm 45/90]_s$	15	7.559	6.720	74.40	0.60	137.9
	22	7.554	5.674	58.31	0.60	137.9
	A&B	7.554	6.755	72.44	0.59	134.6
$[-82.5/30/20/-82.5]$	8	7.638	6.661	131.9	0.53	122.2
	10	7.633	6.156	134.2	0.54	123.4

^{a}2.54 cm = 1 in.
^{b}6.8944 GPa = 1 msi.

capability of functioning under load or displacement control with each mode independent of the other. Early tests were run under load control in order to set predetermined load ratios for each test. This proved to be an unsatisfactory loading mode for buckling studies on specimens which were to be tested more than once. When the machine is operated under load control, the load continues to increase at the specified loading rate throughout the test. As the buckling point is approached, the specimen cannot support the additional load and failure occurs without warning. Subsequent tests were run under displacement control or a combination of displacement and load control. The combined displacement and load control was used for the pure compression and pure torsion tests on cylinders which were made with unsymmetric laminates. This was done in order to insure that undesired loads were not introduced due to the coupling between in-plane forces and bending moments associated with unsymmetric laminates. For pure compression tests, the machine was operated in axial displacement control with torsional load control set to zero; pure torsion tests were conducted under torsional rotation control with the axial load control set to zero. All other tests were conducted under displacement control for both axial and torsional loading. These two loading methods (displacement control, and combined displacement and load control) proved to be more satisfactory for determining

FIG. 2—*Typical composite cylinder with end plugs.*

the buckling point without failing the specimen. A disadvantage to the combined displacement loading mode is that proportional loading cannot be insured.

Load and strain data were acquired using a variety of data-acquisition devices including strip charts, oscilloscopes, and microprocessor-controlled automatic data acquisition systems.

Test Procedures and Determination of the Buckling Point

It was desired to obtain a complete series of tests on each cylinder consisting of pure compression, pure (positive and negative) torsion, and a spectrum of combined compression-torsion loads. The specimens used in this investigation were very expensive and, therefore, it was necessary to conduct as many tests as possible with each cylinder in order to obtain a spectrum of loading combinations at a reasonable cost. This required that each specimen be loaded under a particular loading configuration up to the buckling point and then unloaded before any damage occurred. This proved to be a particularly troublesome task throughout the entire testing program. Buckling the cylinders under pure torsion was a fairly stable event and the deformation at the buckling torque could usually be controlled without damage to the specimen. In addition, the torsional buckling load was quite

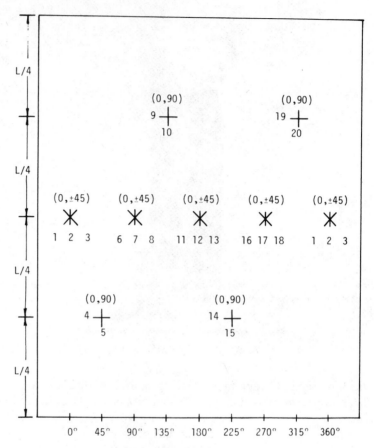

FIG. 3—*Strain-gage pattern.*

predictable and this was extremely helpful during the test program. Buckling under pure compression was very unpredictable and usually a much more dynamic event. Thus, it was difficult to obtain buckling loads for pure compression loading without damaging the specimens. As a result of the need to conduct many tests on a given specimen without inducing damage, a very conservative approach was adopted during the experimental program. Subsequent analysis of the experimental data indicated that in many cases the test was stopped prior to attainment of the buckling load. Such data points are not presented in this paper, but may be found in Ref 7. Even with a conservative approach, many of the specimens failed before the complete loading spectrum was attained.

The load and strain-gage output were monitored on an oscilloscope during the experiment. Prior to buckling, the load-strain relationship is linear, but begins to deviate from linearity at the onset of buckling. The load when this deviation from linearity occurs was taken to be the experimental buckling load. Of course, there is greater confidence in establishing (nondestructively) the experimental buckling load by this method when the postbuckling response is stable. Practically stable postbuckling states were achieved experimentally if the torsional load component significantly exceeded the axial load component.

Theoretical Predictions

Buckling Loads

The computer program developed by Wu [6] was used to predict the buckling loads of the cylinders in the experimental investigation. Mathematically the buckling of the anisotropic cylindrical shell is formulated as an eigenvalue problem. The Wu program uses a numerical iteration scheme to find the smallest eigenvalue such that the governing shell equations and boundary conditions are satisfied. The smallest eigenvalue is the proportional load magnitude at buckling.

The mathematical model for cylindrical shell buckling in the Wu program is based on Flügge's nonlinear equilibrium equations linearized to predict the onset of buckling, Flügge's linear strain-displacement relations, and linear elastic, homogeneous, anisotropic shell theory constitutive equations reflecting the arbitrary layering of perfectly bonded orthotropic lamina. The prebuckling equilibrium state is assumed to be momentless with spatially uniform axial and shear stress resultants equal to the applied compressive edge load P per unit edge length, and the applied edge shear load T per unit edge length, respectively (Fig. 1). Also, the cylinder is assumed to be geometrically perfect.

The method of solution is to first combine the governing shell equations into three homogeneous partial differential equations in the reference surface displacements u, v, and w. Nontrivial solutions to these equations are determined if the proportional load magnitude q, the axial buckling wave parameter λ, and the circumferential buckling wave number n satisfy a characteristic equation. This characteristic equation is expressed as an eighth-degree polynomial in λ whose coefficients are functions of q and n. A second condition among the parameters, q, λ, and n (actually eight values of λ for each q and n) is obtained from the buckling boundary conditions, which are restricted to be identical at the ends $X = \pm L/2$. Satisfying these homogeneous boundary conditions leads to a determinate which must vanish if q, the eight values of λ, and n form a solution. Since n is restricted to an integer value for compatible deformations, the iteration scheme in the Wu program begins by assuming n equal to two, guessing a value for q, determining the λ-values from the characteristic equation, and then testing to determine if the boundary determinate numerically vanishes. If this determinate does not vanish, then q is iterated until it does. Then n is increased by one and the process of iteration on q is repeated. The smallest of the q-values determined for each n considered is the eigenvalue sought.

The Wu program requires geometric, material, boundary condition, and loading input data. The geometric properties were determined from measurements made on each cylinder (Table 1). Material property data were selected as "typical" for boron/epoxy and graphite/epoxy, and these values are listed in Table 2. However, the values of the fiber direction moduli E_{11} used as input to the Wu program were adjusted from the values given in Table 2 to account for actual fiber volume fractions determined from each test specimen. Adjusting the fiber direction moduli was considered necessary to achieve a more accurate model of the laminate stiffness characteristics for each test specimen. The actual fiber volume fractions V_f were obtained from thickness measurements of each specimen, the known number of plies, and the typical ply thicknesses and fiber volume fractions listed in Table 2. The basic assumption in these calculations was that the actual volume of fibers in each lamina was the same as for the typical lamina in Table 2, and consequently volume differences in the actual lamina were due solely to matrix volume changes. The adjusted E_{11}-values are listed in Table 1 for each specimen.

At the onset of buckling, clamped-edge boundary conditions were used to compute the theoretical buckling loads. The Wu program, however, distinguishes four types of clamped boundary conditions based on various in-plane edge conditions. Here the C3 boundary conditions were specified, and the in-plane conditions for these are

$$u = 0, \qquad N_{x\theta} + \frac{1}{R} M_{x\theta} + P \frac{\partial v}{\partial x} = 0, \qquad x = \pm L/2$$

where u, $N_{x\theta}$, and $M_{x\theta}$ are the additional axial displacement, additional shear stress resultant, and additional twisting stress couple, respectively, due to buckling. The effect of a second set of in-plane edge conditions on the buckling loads was also examined to assess more rigid in-plane clamping. This set of boundary conditions is designated C2, and the in-plane conditions for it are

$$u = 0, \qquad v = 0, \qquad X = \pm L/2$$

where v is the additional tangential displacement due to buckling. For the test specimens in this investigation the theoretical buckling loads predicted for pure compression and pure torsion with the C2 boundary conditions differed by less than 3 percent from the corresponding buckling loads with the C3 boundary conditions. Hence more rigid in-plane edge conditions do not significantly affect the buckling loads for these relatively long cylinders.

The loading input data for the Wu program were selected to sufficiently

TABLE 2—*Material properties.*

Boron/Epoxy (AVCO 5505/4)

Nominal ply thickness[a] = 0.117 mm (0.0046 in.)
Assumed constituent moduli: E_f = 397 GPa (55 msi)
E_m = 3.38 GPa (0.49 msi)
For a fiber volume fraction[a] V_f = 0.57
E_{11} = 218 GPa (31.56 msi)[a] E_{22} = 18.6 GPa (2.7 msi)
G_{12} = 6.89 GPa (1.0 msi) ν_{12} = 0.21

Graphite/Epoxy (Modmor I fibers, Narmco 5208 resin)

Nominal ply thickness = 0.127 mm (0.0050 in.)
Assumed constituent moduli: E_f = 228 GPa (33 msi)
E_m = 3.38 GPa (0.49 msi)
For a fiber volume fraction V_f = 0.50
E_{11} = 139 GPa (20 msi) E_{22} = 11.7 GPa (1.7 msi)
G_{12} = 6.89 GPa (1.0 msi) ν_{12} = 0.21

[a]Obtained from measurements and tests on a $[0_8]$ tube specimen.

determine the theoretical buckling interaction diagram. This diagram is a dimensionless load plane on which the axial compression load ratio R_c is plotted on the ordinate and the torsional load ratio R_t on the abscissa. The axial compression load ratio is the axial buckling load in a combined loading mode divided by the axial buckling load for single compression loading. Similarly, the torsional load ratio is the torsional buckling load in a combined loading mode divided by the positive torsional buckling load (see Fig. 1) in single torsional loading. The theoretical buckling interaction diagrams for each test specimen are shown in Figs. 4 to 11. A striking feature of these diagrams is that positive and negative torsional buckling loads are unequal in magnitude, and thus the interaction diagram is unsymmetrical about the ordinate axis. This is contrary to the symmetrical interaction diagrams of isotropic cylinders. Asymmetry in the interaction diagrams occurs because each laminate in this investigation has plies with shear strain-normal stress coupling (that is, $\overline{Q}_{16}$, $\overline{Q}_{26} \neq 0$). For an element of the laminate subjected to pure shear (equivalent to pure torsion of the cylinder), this coupling results in a larger bending stiffness in the direction of equivalent compression (either ±45 deg with respect to the shear planes) for one sense of the shear as opposed to shear applied in the opposite sense. Hence the sense of the shear is equivalent to compression loading in directions of different bending stiffnesses, and this results in larger shear buckling loads for one sense of the applied torque than for the opposite sense. An examination of the interaction diagrams in Figs. 4–11 shows that the magnitude of the largest torsional buckling load occurs when the fibers of the outermost angle ply are in compression. Consequently, this is the direction of greater laminate bending stiffness when compared with a direction perpendicular to it.

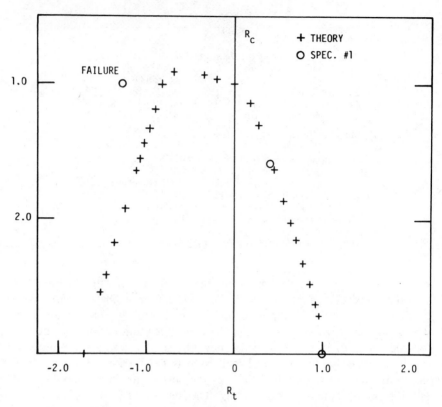

FIG. 4—*Buckling interaction diagram: [∓45]ₛ boron/epoxy.*

Laminate Moduli

To assess if the material properties used to calculate theoretical buckling loads are reasonable, axial-compression/axial-strain and torque/shear-strain data were obtained for each test specimen. These strains were measured by strain gages mounted on the outside of the specimen at its midlength so that the effect of the boundary restraints were small. Thus, as long as the applied loads are small enough to avoid buckling, the midsection of the cylinder is nearly in a membrane state of stress with spatially uniform axial and shear stress resultants equal to the applied edge loads. The experimental load-strain data, then, can be interpreted to give laminate axial stiffness E_x and shear stiffness $G_{x\theta}$. These stiffnesses can also be computed from lamination theory [8]. The details involve solving the anisotropic constitutive equations for the spatially uniform membrane axial and shear strains in terms of the appropriate applied stress resultant. In this analysis, the shell was assumed so thin ($R/h \gg 1$) that the shell constitutive equations

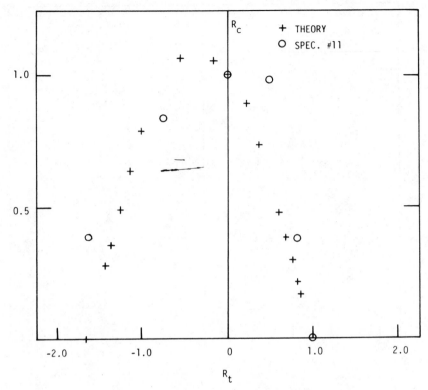

FIG. 5—*Buckling interaction diagram:* $[\mp45]_s$ *graphite/epoxy.*

could be replaced by flat-plate equations. Results of these calculations for each test specimen are given in Table 3.

Results and Discussion

Modulus Values

Correct material modulus values are important if reasonable theoretical estimates of the buckling loads are expected. A comparison of theoretical and experimental laminate moduli in axial compression (E_x) and shear ($G_{x\theta}$) is presented in Table 3. For the boron/epoxy test specimens the percentage difference between theory and experiment with respect to the theoretical values is less than 15 percent. Larger differences are exhibited by the graphite/epoxy specimens. For these graphite/epoxy specimens the largest difference in axial stiffness is 37 percent for Specimen 14, and the largest difference in the shear stiffness is 52 percent for Specimen 22.

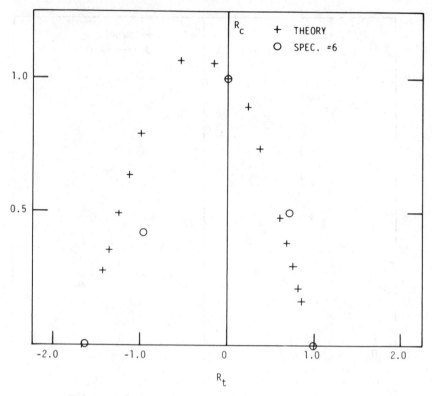

FIG. 6—*Buckling interaction diagram:* $[-45_2/45_2]_s$ *boron/epoxy.*

Pure Compression and Pure Torsion

Table 4 presents the comparison between experimental and theoretical buckling loads for pure compression and pure torsion. In compression, the experimental buckling loads range from 0.363 (Specimen 6) to 0.700 (Specimen 22) of the theoretical predictions. In torsion better agreement is achieved. Except for Specimens 2, 10, and 12, experimental torsional buckling loads range from 0.819 (Specimen 18) to 1.164 (Specimen 22) of the theoretical predictions.[3]

The large discrepancy between theory and experiment for single axial compression loading is not surprising since composite cylindrical shells are imperfection-sensitive in compression, whereas in single torsional loading these shells are considerably less imperfection-sensitive [1]. Consequently, better agreement is achieved for torsional loading. The fact that experimental torsional buckling loads exceed theoretical predictions—note especially the

---

[3]Specimen 2 was not tested in combined loading, and Specimen 10 exhibited cracking and failure at relatively low loads.

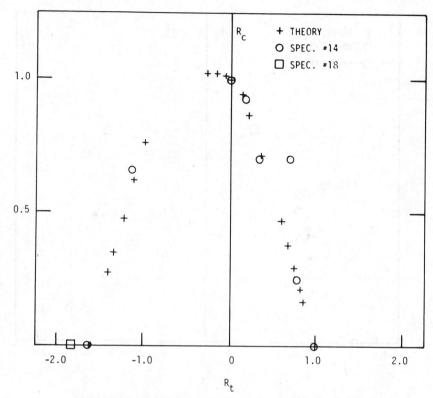

FIG. 7—*Buckling interaction diagram:* $[-45_2/45_2]_s$ *graphite/epoxy.*

$[0/\pm45/90]_s$ graphite/epoxy cylinder results—may be due to improper material specifications (Table 3) or neglect of prebuckling deformations. The nonlinear shear behavior of these materials (that is, the decrease in shear modulus at large shear stress) is not considered to be a large contributor to this discrepancy. For example, the shear stress at buckling for positive torsion of Specimen 15 is 63.2 MPa (9.17 ksi), and for graphite/epoxy this shear stress is small enough such that linear shear-stress shear-strain is reasonable.

Discrepancies between theory and experiment may be attributable to neglect of the prebuckling deformations, as indicated in the previous paragraph. The spatially uniform linear membrane prebuckling state assumed in the Wu program does not satisfy exact boundary conditions, nor does it account for the bending-extension coupling of unsymmetrically laminated shells. The effect of the exact prebuckling deformations on the buckling loads has been considered by Jones and Hennemann [9] and Booton and Tennyson [4]. Jones and Hennemann concluded that the effect of exact prebuckling deformations on the axial compression and lateral pressure

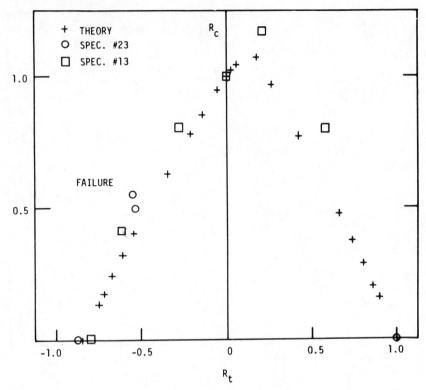

FIG. 8—*Buckling interaction diagram:* [0/±45/90]ₛ *boron/epoxy.*

buckling loads was small for simply supported cross-ply laminated cylindrical shells. For clamped ($-\theta$, 0, θ) glass/epoxy laminated circular cylindrical shells subject to axial compression, however, Booton and Tennyson cite as much as a 15 percent reduction in the buckling load depending on the fiber orientation θ if exact prebuckling effects are included. In extrapolating from these analyses, it appears that neglecting the exact prebuckling deformations in the present analysis would account for much of the discrepancy between theory and experiment for, torsion if reasonable material property characterization were assured. This, of course, cannot be said for axial compresssion loading, where the inclusion of initial shape imperfections is more significant than satisfying exact prebuckling deformations [4].

Combined Compression-Torsion Buckling

Comparisons of experimental and theoretical combined compression-torsion buckling loads are shown in Figs. 4–11. Since experimental and

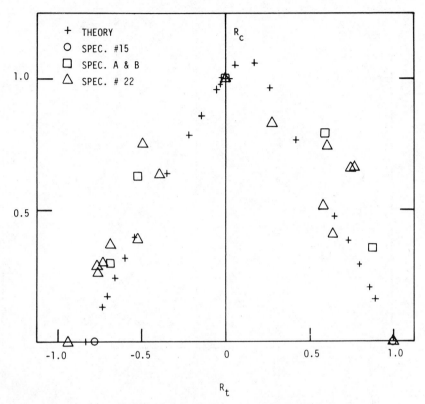

FIG. 9—*Buckling interaction diagram: [0/±45/90]$_s$ graphite/epoxy.*

theoretical buckling loads in combined loading are normalized by their respective values for single compression and single positive torsion, both experimental and theoretical interaction curves pass through (R_t, R_c) coordinates of (1,0) and (0,1). In cases where no experimental value for the pure axial compression buckling load P^* or pure torsional buckling load T^* were obtained, these values were estimated.

The interaction curves are grouped by laminate layup configuration, since this has a strong influence on the shape of these curves. Although the test specimens in this study are not geometrically identical, their curvature parameters (L^2/Rh) are in a moderately large range of values such that the shape of their interaction curves is assumed practically independent of geometry changes. For isotropic cylinders the buckling interaction curves are known not to change significantly for $L^2/Rh > 200$ [10]. Thus for a given laminate and material the experimental results for geometrically different specimens are grouped on one plot.

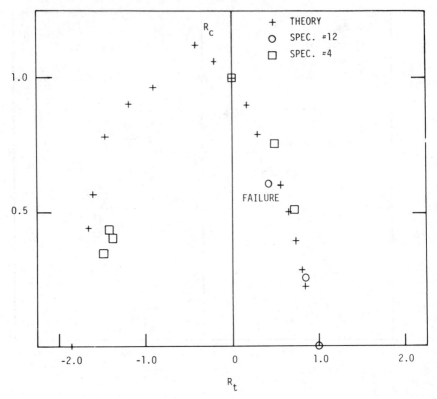

FIG. 10—*Buckling interaction diagram:* [−82.5/30/20/−82.5] *boron/epoxy.*

[∓45]ₛ Laminate: Figs. 4 and 5

The largest compressive buckling load occurs when the shell has a negatively applied torque, and the interaction curves are concave toward the origin. For the boron/epoxy specimen, only three experimental data points were obtained since the cylinder fractured in an early test with the loading machine in the load control mode. The experimental pure axial compression buckling load P* for the boron/epoxy specimen was estimated to be 17.8 kN (4.0 kips). More experimental data points were obtained for the graphite/epoxy specimen, and correlation between experiment and theory is good.

[−45₂/45₂]ₛ Laminate: Figs. 6 and 7

The shape of the interaction curves resembles the [∓45]ₛ laminate. More experimental data points were obtained for the graphite/epoxy Specimen 14 than for Specimens 18 (graphite/epoxy) and 6 (boron/epoxy). Again, agreement between theory and experiment is quite good.

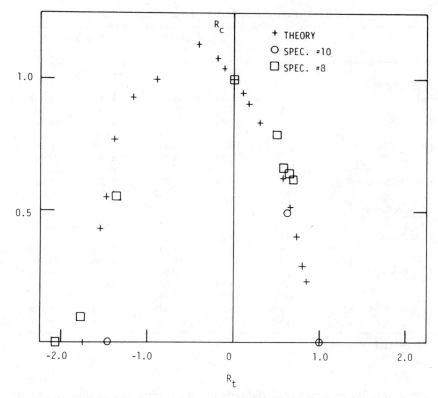

FIG. 11—*Buckling interaction diagram: [−82.5/30/20/−82.5] graphite/epoxy.*

[0/±45/90]ₛ Laminate: Figs. 8 and 9

The peak compression buckling loads on these interaction curves shift to the positive torsion side of the interaction diagram, since the +45-deg plies are farther from the shell middle surface than the −45-deg plies. For this "quasi-isotropic" laminate the positive and negative pure torsional buckling loads are nearly the same. The experimental data correlate well with theory. The positive pure torsional buckling load T^* for Specimen 13 was estimated to be 2.82 kN·m (25 kip-in.), and P^* for Specimens 15 and 23 was estimated to be 100 kN (22.5 kips) and 133 kN (30 kips), respectively.

[−82.5/30/20/−82.5] Laminate: Figs. 10 and 11

The shape of these interaction curves is quite interesting. The theoretical compressive buckling loads remain relatively high for a large range of negative torsional loads, then decrease rapidly near the negative pure torsional buckling load. Correlation between theory and experiment is good, but un-

TABLE 3—*Experimental and theoretical modulus values.*

Laminate	Speci-men No.	Experiment		Theory	
		E_x, GPa	$G_{x\theta}$, GPa	E_x, GPa	$G_{x\theta}$, GPa
Boron/Epoxy					
$[0_8]$	2	215.8	8.687	217.6	6.894
$[\mp 45]_s$	1	...	48.47	24.36	48.20
$[-45_2/45_2]_s$	6	22.96	48.40	24.36	48.20
	19	...	63.08	24.70	55.16
$[0/\pm 45/90]_s$	13	75.70	33.44	77.99	29.76
	23	75.70	33.44	77.99	
	23	...	...	81.70	31.16
$[-82.5/30/20/-82.5]$	4	34.61	10.96	37.86	10.78
	12	...	9.652	38.10	10.82
Graphite/Epoxy					
$[\mp 45]_s$	11	15.72	37.64	22.56	28.55
$[-45_2/45_2]_s$	14	14.20	30.54	22.70	29.57
	18	...	31.03	22.43	27.59
	20	...	40.40	22.56	28.55
$[0/\pm 45/90]_s$	15	...	28.20	55.49	21.60
	22	68.19	32.82	55.49	21.60
	A&B	62.26	27.92	54.39	21.19
$[-82.5/30/20/-82.5]$	8	28.34	8.825	30.43	9.556
	10	23.17	7.308	30.53	9.571

[a]6.8944 GPa = 1 msi.

fortunately no experimental data points were obtained in the region of negative torsion near the ordinate axis, where the theory predicts relatively large compressive buckling loads. $P*$ for Specimens 10 and 12 was estimated to be 20 kN (4.5 kips) and 26.7 kN (6.0 kips), respectively.

Concluding Remarks

Elastic buckling loads of laminated composite circular cylindrical shells subject to combined axial compression and torsional loading are plotted on a dimensionless load plane whose ordinate is the axial load ratio and whose abscissa is the torsional load ratio. Both theoretically predicted and experimentally determined load ratios at buckling are plotted on this load plane, and the locus of these points generates theoretical and experimental buckling interaction curves. For the boron/epoxy and graphite/epoxy test specimens in this investigation, generally good correlation is achieved between theoretical and experimental interaction curves. These interaction curves are asymmetric about the ordinate, appear concave toward the origin, and are strongly influenced by the laminate layup configuration. In single torsional loading the magnitudes of the positive and negative buckling loads may differ by as much as a factor or two for laminates having normal stress

TABLE 4—*Experimental and theoretical buckling loads.*

Laminate	Specimen No.	Experimental			Theory		
		P, kN[a]	T_+, kN·m[b]	T_-, kN·m	P, kN	T_+, kN·m	T_-, kN·m
Boron/Epoxy							
[0₈]	2	55.6	...	-1.01	121	1.55	-1.55
[∓45]ₛ	1	...	0.45	...	42.3	0.49	-0.82
[-45₂/45₂]ₛ	6	61.6	2.36	-3.93	170	2.56	-4.25
	19	...	1.90	-3.13	132	1.88	-3.13
[0/±45/90]ₛ	13	137	...	-2.24	207	2.85	-2.41
	23	...	2.69	-2.35	191	2.64	-2.24
[-82.5/30/20/-82.5]	4	25.8	0.71	-1.62	40.7	0.83	-1.54
	12	...	0.56	...	37.3	0.80	-1.49
Graphite/Epoxy							
[∓45]ₛ	11	18.2	0.59	-1.07	41.4	0.64	-1.08
[-45₂/45₂]ₛ	14	54.3	1.99	-3.62	156	2.30	-3.73
	18	...	2.19	-4.0	180	2.68	-4.27
	20	...	2.03	-3.76	165	2.47	-3.98
[0/±45/90]ₛ	15	...	2.30	-1.86	143	2.0	-1.66
	22	100	2.20	-1.92	143	1.99	-1.65
	A&B	85.6	2.28	-1.73	148	2.07	-1.69
[-82.5/30/20/-82.5]	8	20.1	0.63	-1.30	31.2	0.66	-1.16
	10	...	0.58	-0.81	29.7	0.65	-1.18

[a] 4.448 kN = 1 kip.
[b] 0.1130 kN·m = 1 kip·in.

to shear strain coupling. For the laminates in this investigation, the largest theoretical compressive buckling load occurs when the cylinder is simultaneously subjected to a small amount of torque. These features of the interaction curves suggest that a linear approximation connecting pure compressive and pure torsional buckling loads is "safe" as a design methodology. For some laminated composite cylinders, however, the linear approximation substantially underestimates their buckling resistance; for example, see Figs. 10 and 11.

The theoretical buckling loads obtained from a program developed by Wu [6] are based on Flügge's cylindrical shell equations, neglect exact prebuckling deformations, and do not include initial shape imperfections. For pure torsion there is good correlation between buckling loads from theory and experiment. For pure axial compression loading, however, there is poor correlation, since the initial shape imperfections of the test specimens are not modeled.

Acknowledgment

This work was originally suggested by Mr. H. Benson Dexter of NASA Langley Research Center. He is also responsible for the specimen fabrication and some test results. The authors gratefully acknowledge his contributions. Virginia Tech personnel who also made significant contributions are D. A. O'Brien, G. L. Farley, W. L. Unkenholz, P. R. Frosell, Jr., P. W. Hsu, D. A. Danello, G. G. Lough, and F. Carter.

The research herein was supported by NASA Contract NAS1-13175 and the NASA-Virginia Tech Composites Program, NASA Grant NGR 47-004-129.

References

[1] Tennyson, R. C., *Composites*, Jan. 1975, pp. 17–24.
[2] Bert, C. W. in *Composite Materials*, Vol. 7, Part I, C. C. Chamis, Ed., Academic Press, New York, 1975.
[3] Bert, C. W. and Francis, P. H., *Journal*, American Institute of Aeronautics and Astronautics, Vol. 12, No. 9, 1974, pp. 1173–1186.
[4] Booton, M. and Tennyson, R. C., *Journal*, Vol. 17, No. 3, March 1979, pp 278–287.
[5] Wilkins, D. J. and Love, T. S. in *Proceedings*, 15th AIAA/ASME/SAE, Structures, Structural Dynamics, and Materials Conference, Las Vegas, Nev., April 1974, Paper No. 74-379.
[6] Wu, C-H., "Buckling of Anisotropic Circular Cylindrical Shells," Ph.D. dissertation, Case Western Reserve University, Cleveland, Ohio, June 1971.
[7] Herakovich, C. T., "Theoretical Experimental Correlation for Buckling of Composite Cylinders Under Combined Compression and Torsion," VPI-E-78-14, Virginia Polytechnic Institute and State University (also NASA CR-157358), Blacksburg, Va., July 1978.
[8] Jones, R. M., *Mechanics of Composite Materials*, McGraw-Hill, New York, 1975.
[9] Jones, R. M. and Hennemann, J. C. F. in Proceedings, 19th AIAA/ASME Structures, Structural Dynamics, and Materials Conference, Bethesda, Md., 3–5 April 1978, Paper No. 78-516.
[10] Booton, M., "Buckling of Imperfect Anisotropic Cylinders Under Combined Loading," University of Toronto Institute for Aerospace Studies (UTIAS) Report No. 203, Toronto, Ont., Canada, Aug. 1976.

H. T. Hahn[1]

Proof-Load Determination for Pressure Vessels Wound with Aramid Fiber*

REFERENCE: Hahn, H. T., **"Proof-Load Determination for Pressure Vessels Wound with Aramid Fiber,"** *Test Methods and Design Allowables for Fibrous Composites, ASTM STP 734,* C. C. Chamis, Ed., American Society for Testing and Materials, 1981, pp. 361–375.

ABSTRACT: The paper presents a method of determining the proof pressure that can guarantee a design lifetime for Kevlar-49/epoxy pressure vessels. The method is based on the assumption that, when taken from the same population, a statically strong vessel is also strong in stress rupture. The lifetime data available at higher pressures are extrapolated to estimate the lifetime distributions at lower pressures. In so doing, we assume that the lifetime distribution is exponential and that the characteristic lifetime is related exponentially to the applied pressure divided by the average burst pressure. Two examples are given to illustrate the method.

KEY WORDS: proof-testing, pressure vessel, lifetime distribution, stress rupture, Kevlar 49/epoxy, composite materials

Nondestructive examination of composite structures as a means of quality control is still in the state of infancy as the failure of composites is not a result of a single, dominant crack growth. As a result, proof-testing has attracted serious attention as an alternative means of ensuring structural integrity.

In proof-testing of composites, we assess the structural integrity of a composite structure by analyzing its behavior during a quasi-static loading to a predetermined level. Acoustic emission measurements are usually taken to detect any subcritical damage that can occur during the test. In this way we can certainly screen out those structures whose strengths are weaker than the proof load. However, there is evidence, at least at the laboratory-specimen level, that, under simple load histories, static strength is related to lifetime in

---

*This work was performed under the auspices of the U.S. Department of Energy by Lawrence Livermore Laboratory under Contract No. W-7405-Eng-48.

[1] Mechanical engineer, Lawrence Livermore Laboratory, University of California, Livermore, Calif. 94550.

such a way that a statically strong specimen has a longer lifetime [1,2].[2] Therefore, by estimating a lower-bound strength through proof-testing, we should be able to guarantee a lower-bound lifetime.

The relation between static strength and lifetime can be established as follows. Two sets of specimens are selected from the same population. For convenience, the number of specimens in each set is the same. One set is tested for static-strength distribution and the other for lifetime distribution. The lowest strength is then matched with the shortest lifetime, the second lowest strength with the second shortest lifetime, and so forth. The relation obtained in this way forms the basis for determining the proof-load level required to guarantee a given design lifetime.

In the present paper, a strength-lifetime relationship is established for Kevlar-49[3]/epoxy pressure vessels. With the help of this relationship, a step-by-step procedure is described to determine the proof-load level for a given design lifetime.

Experimental Results

The data on which our analysis is based are for the spherical filament-wound pressure vessels reported in Ref 3. Since the publication of this report, additional failures were added to expand the lifetime data base. The pressure vessels were filament-wound with Kevlar-49 (380 denier) fiber coated with DER 332/T-403 epoxy resin over an aluminum mandrel. The 1100-0 (hydroformed) mandrel/liner had a 114-mm outside diameter and a 1.02-mm wall thickness (Fig. 1). The composite wall thickness was 1.1 mm, and the details of the winding pattern are reported in Ref 4. Macroscopically, the composite shell is quasi-isotropic.

A total of 29 vessels were tested to burst, and the burst pressures are reported in Ref 4. In the following, the data are analyzed by using a two-parameter Weibull distribution of the form.

$$R_s(X) = \exp\left[-\left(\frac{X}{X_0}\right)^{\alpha_s} \right] \qquad (1)$$

where

$R_s(X)$ = probability of the burst pressure being greater than or equal to X

---

[2] The italic numbers in brackets refer to the list of references appended to this paper.

[3] Reference to a company or product name does not imply approval or recommendation of the product by the University of California or the U.S. Department of Energy to the exclusion of others that may be suitable.

α_s and X_0 = shape parameter and characteristic strength, respectively, and

$R_s (\cdot)$ = static strength distribution.

The maximum likelihood estimates of the parameters (see, for example, Ref 5 for details of the method) have been determined to be

$$\alpha_s = 27.835, X_o = 35.111 \text{ MPa}$$

Note that the average burst pressure $\overline{X}$ is related to X_o by

$$\frac{\overline{X}}{X_o} = \Gamma \left(\frac{1}{\alpha_s} + 1 \right) \tag{2}$$

where $\Gamma (\cdot)$ is the gamma function.

The lifetime data at various applied stresses have also been analyzed by using a two-parameter Weibull distribution of the form

$$R_l (t) = \exp \left[- \left(\frac{t}{t_o} \right)^{\alpha_l} \right] \tag{3}$$

where $R_l (\cdot)$ and t_o are the lifetime distribution and characteristic lifetime, respectively. The maximum likelihood estimates of the parameters are listed in Table 1 and shown pictorially in Figs. 2 and 3. These figures also include

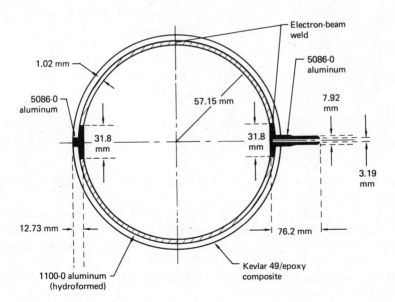

FIG. 1—*Sectional view of spherical vessel.*

TABLE 1—*Maximum likelihood estimates of Weibull parameters for lifetime distribution.*

Applied Pressure (% Average Burst Pressure)	Total No. of Vessels, n	No. of Failures, r	Last Failure Time, h	α_l	t_o, h
86	39	39	...	0.570	1.658×10^2
80	24	24	...	0.873	8.387×10^2
74	24	19	11.487×10^3	1.060	7.716×10^3
68	21	6	14.400×10^3	1.666	2.740×10^4

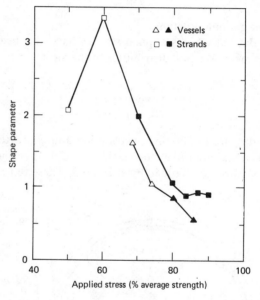

FIG. 2—*Shape parameters of lifetime distributions for vessels and strands. Shape parameter for vessels is smaller than for strands and increases with decreasing applied stress. (Closed symbols represent complete data; open symbols, censored data.)*

the corresponding parameters for the Kevlar-49/epoxy (ERL 2258/ZZL 0820) strands [3] for comparison purposes.

The shape parameter α_l of vessels increases with decreasing pressure level. As α_l changes from below unity to above unity, the corresponding failure process changes. Specifically, at higher pressures the failure process is characterized by a decreasing failure rate, whereas at lower pressures it is characterized by an increasing failure rate. Note that the same trend is observed in the lifetime data of Kevlar-49/epoxy strands.

Three typical failure modes of a strand that can result from a broken fiber are described in the following. When the stress is high, the stress concentra-

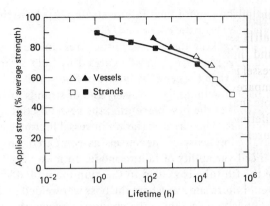

FIG. 3—*Characteristic lifetimes. Vessels enjoy longer characteristic lifetimes at every stress level than strands. (For symbols refer to Fig. 2.)*

tion on the neighboring fibers may be high enough to enable the initial crack to propagate across the specimen, leading to a catastrophic failure. Thus the failure rate is high initially. However, once the specimen survives the initial period of unstable crack-propagation failure, the original crack tends to grow along the broken fiber, relieving the stress concentration on the neighboring fibers. Thus, if we equate the normal crack-propagation mode with higher failure rate and the parallel crack-propagation mode with lower failure rate, we can easily understand the decreasing failure rate observed at higher stresses.

When the stress is lower, on the other hand, the normal crack propagation described earlier is less likely to occur; instead, the parallel crack growth prevails. Thus the final failure of the composite is in essence the stress rupture of fibers without stress concentration. Kevlar-49 is an organic fiber, and it seems to be susceptible to gradual strength degradation with time under load. Therefore, the failure rate increases with time.

The aging resulting from exposure to ultraviolet (UV) light apparently compounds such stress-induced aging. Kevlar fiber is known to absorb UV light and consequently suffer strength loss [6,7]. The strands in Figs. 2 and 3 were exposed to fluorescent lights for 24 h a day and, therefore, are believed to undergo strength degradation because of UV absorption [8,9]. Thus, the downturn of the stress-logarithmic lifetime curve near 10^4 h is attributed to such degradation.

The failure process of the composite shell in a vessel, of course, differs from that of composite strands. However, the ultimate failure of the vessel is controlled by the strand strength. As the pressure is increased in a vessel, the aluminum liner yields first because of its low yield strain (~ 0.05 percent). Upon further increase of pressure, matrix cracking occurs along the fibers because the transverse failure strain of unidirectional lamina is only ~ 7 percent of the longitudinal failure strain [10]. Note that the matrix cracking is detected by acoustic emission [11].

After the matrix cracking there is very little stress in the direction normal to the fibers and thus the load is mostly carried by fiber strands. Ultimate failure of the vessel is then triggered by failure of fiber strands.

Figure 4 compares the lifetimes of vessels and of strands at various failure percentiles. Note that at the first percentile, the vessel lifetimes are the same as the strand lifetimes. The larger variation in vessel lifetimes may be attributed to structural complexity of the vessels as compared with the strands; hence the enhanced variability of failure mode. In most vessels, failure was initiated where the fill tube is welded to the aluminum bladder (see Fig. 1). The next frequent failure site was where the boss was welded to the aluminum bladder at the opposite pole. Thus, the change in geometry and material at these welds is believed to contribute to the stress concentration in the composite and to limit the failure initiation to these areas. However, it is remarkable that the lifetimes of vessels are still comparable to those of strands at the first percentile.

Another point we should keep in mind when comparing the two sets of data in Fig. 4 is the difference in exposure to UV light. As mentioned before, the strands were exposed to fluorescent lights for 24 h a day. However, the vessels were kept in a steel enclosure and thus were shielded from fluorescent lights. Therefore, a downturn in the stress-logarithmic lifetime relation of the vessels would not be expected.

Strength-Lifetime Relationship

In homogeneous, especially brittle, materials both the strength and lifetime are determined by the growth of a single dominant flaw. Thus a relation-

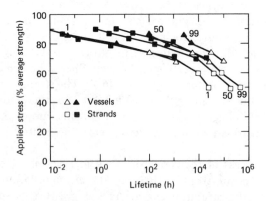

FIG. 4—*Comparison of lifetime distributions between vessels and strands. The numbers by the curves represent percentiles of failed specimens. The downturn in the strand data near 10^4 h is believed to be caused by the prolonged exposure to UV light. At the first percentile, the vessel data coincide with the strand data. (For symbols refer to Fig. 2.)*

ship between strength and lifetime can be obtained from knowledge of the flaw distribution and from a deterministic flaw growth equation [12].

The failure modes of composite laminates are quite different from those of homogeneous materials. Fortunately, however, recent investigations point up the possibility that lifetime is uniquely related to strength [1,2]. The reason is that both static and time-dependent strengths depend very much on the fiber strength, and, hence, a statically strong specimen is also likely to be strong in stress rupture.

Suppose we choose two sets of specimens from the same population. One set is tested for static strength and the other for lifetime under a sustained load. The strength and lifetime distributions thus obtained are described by Eqs 1 and 3, respectively.

The strength-lifetime relationship is then obtained from

$$R_s(X) = R_l(t) \tag{4}$$

Substitution of Eqs 1 and 3 into Eq 4 yields

$$\frac{t}{t_0} = \left(\frac{X}{X_0}\right)^{\alpha_s/\alpha_l} \tag{5}$$

Thus, if strength X is known, the corresponding lifetime t can be determined from Eq 5.

Now that we have a relationship between strength and lifetime, we can determine the minimum lifetime t_p corresponding to a proof stress σ_p from Eq 5

$$\frac{t_p}{t_0} = \left(\frac{\sigma_p}{X_0}\right)^{\alpha_s/\alpha_l} \tag{6}$$

Because the true strength is higher than σ_p, the true lifetime will be longer than t_p. Thus, t_p is the minimum predicted lifetime after proof-loading to σ_p.

Prediction of Lifetimes at Low Pressures

The procedure described in the preceding section is applicable if both distributions $R_s(X)$ and $R_l(t)$ are known *a priori*. However, in actual situations the service pressure is fairly low and the required design lifetime is much longer than the available lifetime data. Thus, we must predict the lifetime distribution at the service pressure from the available data at much higher pressures.

As discussed earlier, the shape parameter α_l increases with decreasing pressure. However, it is not likely to increase without bound. To gain further insight into the variation of α_l with applied pressure, we shall take advantage of the strength-lifetime relationship discussed in the preceding section.

Consider the ordered lifetimes $\{t_i \mid i = 1, 2 \ldots, n\}$ generated at an applied pressure S. If the static strength distribution, Eq 1, is known, then for each t_i the corresponding strength X_i is

$$X_i = X_o(-\ln R_i)^{\alpha_s} \tag{7}$$

where R_i is the median rank for t_i

$$R_i = 1 - \frac{i - 0.3}{n + 0.4} \tag{8}$$

Thus, for each t_i the ratio of the applied stress to the strength, called the homologous stress ratio, is S/X_i.

Figure 5 is a plot of log (S/X_i) versus log t_i. All the lifetime data at the four different pressure levels are shown in the figure. Interestingly, there is little difference in lifetime regardless of applied pressure if the applied pressure is the same fraction of strengths. That is, suppose $t_i^{(1)}$ is a lifetime at $S^{(1)}$ and $t_j^{(2)}$ at $S^{(2)}$. Then, if $S^{(1)}/X_i^{(1)} = S^{(2)}/X_j^{(2)}$, we have $t_i^{(1)} \approx t_j^{(2)}$.

In view of the foregoing observation, we can now pool all the data and fit them with a power law

$$t\left(\frac{S}{X}\right)^a = b \tag{9}$$

Values of a and b were determined initially for only the 86 percent data. Next, the 80 percent data were combined with the 86 percent data, and so on, until all the data were included. Table 2 lists values of a and b obtained in this manner.

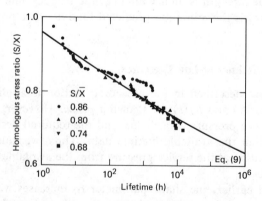

FIG. 5—*Homologous stress ratio versus lifetime for vessels. Lifetime depends on the applied pressure S through homologous stress ratio S/X.*

TABLE 2—*Parameters* a *and* b *in Eq 9 and the inferred shape parameter* α_l *in Eq 10.*

Applied Pressure[a] (% Average Burst Pressure)	Parameter		
	a	b	α_l
86	50.91	0.0305	0.5467
80	39.83	0.1140	0.6988
74	34.79	0.2398	0.8001
68	33.26	0.3055	0.8369

[a] The lowest pressure at and above which lifetime data are analyzed.

When Eq 9 is used, the resulting lifetime distribution is no longer arbitrary; substitution of Eq 9 into Eq 1 leads to a Weibull distribution of the same form as Eq 3. Furthermore, α_l and t_o are determined from

$$\alpha_l = \alpha_s/a \qquad (10)$$

$$t_o \left(\frac{S}{X_o}\right)^a = b \qquad (11)$$

Table 2 also lists the values of α_l according to Eq 10.

Because α_l increases toward unity as more data at lower pressures are included in the analysis, we now assume α_l to be unity at any pressure. That is, the lifetime distribution is of an exponential type

$$R_l(t) = \exp\left(-\frac{t}{t_o}\right) \qquad (12)$$

For an exponential distribution [5], the maximum-likelihood estimate of t_o simply becomes

$$\hat{t}_o = \frac{1}{r} \sum_{i=1}^{r} t_i + \frac{n-r}{r} t_{(r)} \qquad (13)$$

where r is the number of failures and $t_{(r)}$ the last failure time. Furthermore, the lower bound at the 95 percent confidence level is given by

$$\tilde{t}_o = \frac{2r\hat{t}_o}{X_\gamma^2(2r)} \qquad (14)$$

where $X_\gamma^2(2r)$ is the 100 γth percentile of the χ-square distribution with $2r$ degrees of freedom. Note that for the present case $\gamma = 0.95$. The values of $\hat{t}_o$ and $\tilde{t}_o$ are listed in Table 3.

TABLE 3—*Estimates of characteristic lifetime to form exponential distribution.*

Pressure (% Average Burst Pressure)	Maximum-Likelihood Estimate, $\hat{t}_o$ h	Lower Bound $\tilde{t}_o$ at 95% Confidence level, h
86	3.154×10^2	2.470×10^2
80	9.037×10^2	6.668×10^2
74	7.701×10^3	5.440×10^3
68	4.411×10^4	2.518×10^4

Figure 6 shows a comparison between t_o in Table 1 and $\tilde{t}_o$ in Table 3. Although it is difficult to tell, based on the available data alone, whether the stress-lifetime relationship follows a power law or an exponential law, we use an exponential law in order to be conservative at lower pressures. Thus, we choose

$$S/\overline{X} = c + d \log t_0 \tag{15}$$

$$S/\overline{X} = \tilde{c} + \tilde{d} \log \tilde{t}_o \tag{16}$$

The parameters $\tilde{c}$, d, $\tilde{c}$, and d were determined using the least-squares method and are listed in Table 4.

Equations 15 and 16 are shown pictorially in Fig. 6. Below ~80 percent of the average burst pressure, Eq 16 yields shorter lifetimes than Eq 15. Furthermore, since α_l is likely to be higher than unity at lower pressures, the exponential distribution with $\tilde{t}_o$ is definitely conservative for higher re-

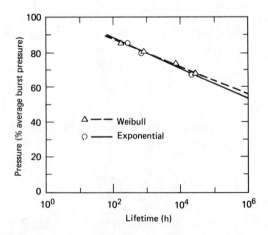

FIG. 6—*Characteristic lifetimes for Weibull and exponential distributions. The exponential distribution leads to conservative estimates of lifetimes at lower pressures.*

TABLE 4—*Parameters for stress-lifetime relations.*

Equation	c or $\tilde{c}$	d or $\tilde{d}$, $(\log h)^{-1}$	Coefficient of Correlation
Eq 15	1.0333	-7.8187×10^{-2}	0.9952
Eq 16	1.0557	-8.5469×10^{-2}	0.9919

liability, $R_l > e^{-1}$, compared with the Weibull distribution with t_0. Therefore, the lifetime distribution to be used is chosen as

$$R_l(t) = \exp\left(-\frac{t}{\tilde{t}_0}\right) \tag{17}$$

Determination of Proof Pressure

The equations necessary to determine proof pressure for any design lifetime and reliability are Eqs 1, 16, and 17. The following two examples show how these equations can be used to determine a proof pressure.

Example 1

A design lifetime t_d under a service pressure S is given. What is the expected reliability and what is the proof-pressure level required to guarantee such a lifetime?

The characteristic lifetime $\tilde{t}_0$ under the operating pressure S follows from Eq 16

$$\tilde{t}_0 = 10^{(S/\overline{X} - \tilde{c})/\tilde{d}}$$

The expected reliability R_l and the proof pressure σ_p are determined, respectively, from Eqs 17 and 14

$$R_l = \exp\left(-t_d/\tilde{t}_0\right)$$

$$R_l = R_s = \exp\left[-\left(\frac{\sigma_p}{X_0}\right)^{\alpha_s}\right]$$

that is

$$\frac{\sigma_p}{X_0} = (-\ln R_l)^{1/\alpha_s} = (t_d/\tilde{t}_0)^{1/\alpha_s}$$

Noting that the average burst pressure $\overline{X}$ is related to X_0 by Eq 2, we finally obtain

$$\frac{\sigma_p}{X} = \frac{(-\ln R_l)^{1/\alpha_s}}{\Gamma (1/\alpha_s + 1)} = \frac{(t_d/t_o)^{1/\alpha_s}}{\Gamma (1/\alpha_s + 1)}$$

Both $\sigma_p/\overline{X}$ and R_l are shown graphically as functions of $S/\overline{X}$ for two different design lifetimes, 5 years and 25 years, in Fig. 7. As the service pressure increases, the reliability decreases, and a higher proof pressure is required.

Example 2

Design reliability R_d under a service pressure S is given. What is the minimum expected lifetime and what is the proof pressure to guarantee such a lifetime?

The minimum lifetime t_p follows from Eq 17 as

$$t_p = \tilde{t}_o \, (-\ln R_d)$$

where $\tilde{t}_o$ is determined in the same way as in Example 1. The proof pressure in this case is independent of the service pressure and is given by

$$\frac{\sigma_p}{\overline{X}} = \frac{(-\ln R_d)^{1/\alpha_s}}{\Gamma (1/\alpha_s + 1)}$$

Figure 8 shows t_p and $\sigma_p/\overline{X}$ for the design reliability of $R_d = 0.9999$.

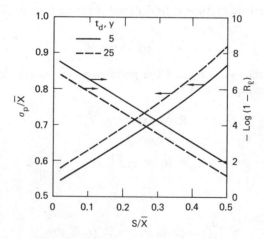

FIG. 7—*Design chart showing proof pressure* σ_p *and reliability* R_l *as functions of service pressures. The chart is for two different design lifetimes,* $t_d = 5$ *years and 25 years;* $\overline{X}$ *is the average burst pressure.*

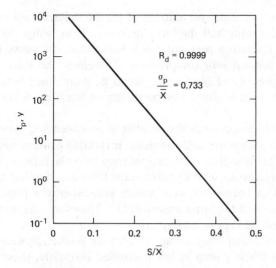

FIG. 8—*Minimum guaranteed lifetime* t_p *as a function of service pressures for a given design reliability* $R_d = 0.9999$. *The normalized proof pressure* $\sigma_p/\bar{X}$ *depends only on the design reliability.*

Discussion

A method has been presented for determination of the proof-pressure level that can guarantee a desired minimum lifetime for composite pressure vessels. The method is based on the following assumptions:

1. A statically strong vessel is also strong in stress rupture.
2. Proof-loading does not cause any damage that can lead to premature failure in stress rupture.
3. Lifetime distribution is exponential regardless of applied pressure level.
4. Logarithmic lifetime decreases linearly with increasing pressure level.

Of the foregoing assumptions, the second assumption in particular deserves further discussion. For the subject composite, as discussed earlier, subcritical matrix cracking will occur at pressures as low as ~7 percent of the burst pressure. However, such matrix damage does not seem to have any deleterious effect, because vessels enjoy longer median lifetimes than strands while lifetimes are comparable at the first-failure percentile. The question then arises, How high can the proof pressure be without damaging fibers? Unlike the matrix cracking, fiber damage can result in accelerated failure at operating pressures and, therefore, should be avoided. Unfortunately, however, such maximum pressure level is not known at present.

Another potential problem is with the aluminum liner. Because the longitudinal failure strain of Kevlar-49/epoxy is about 2 percent, the aluminum liner undergoes an extensive amount of plastic deformation even when the proof pressure is only half the burst pressure. Thus, when the vessel is unloaded to an operating pressure much lower than the proof pressure, the liner will be pressed into compression. Therefore, the liner should be of sufficient thickness and good bond should be maintained between the liner and composite skin to avoid any possibility of buckling as a result of such compression.

Another problem concerns the number of proof-loading cycles. The metal liners, such as aluminum and titanium, invariably contain welds and these welds are the favorite sites of crack initiation in cyclic fatigue. For example, graphite/epoxy vessels with an aluminum liner can develop a leak in less than ~ 60 cycles because of weld failure even when the maximum fatigue pressure is only half the burst pressure [13]. Therefore, the number of proof loadings should be kept to a minimum.

Assumptions 3 and 4 regarding the lifetime prediction were necessitated by the lack of lifetime data at low pressures. Naturally, these assumptions have to be verified by experimental data. Stress-rupture testing at low pressure levels is a really time-consuming task. Also, there is always the possibility of an endurance limit for the material system studied. Therefore, any design decision should take into account all of these aspects.

Finally, proof-testing of composite pressure vessels is an interdisciplinary approach. The determination of proof-pressure level as proposed in the present paper is only a first step toward ensuring satisfactory structural performance through proof-testing. In view of the earlier discussions, we recommend that the possible damages in structural details of vessels at the proposed proof pressure be investigated by using various nondestructive examination techniques.

References

[1] Hahn, H. T. and Kim, R. Y., *Journal of Composite Materials*, Vol. 9, 1975, pp. 297–311.
[2] Awerbuch, J. and Hahn, H. T. in *Fatigue of Filamentary Composite Materials, ASTM STP 636*, K. L. Reifsnider and K. N. Lauraitis, Eds., American Society for Testing and Materials, 1977, pp. 248–266.
[3] Toland, R. H. and Chiao, T. T., "Stress-Rupture Life of Kevlar/Epoxy Spherical Pressure Vessels," UCID-17755, Part 2, University of California, Lawrence Livermore Laboratory, Livermore, Calif., 1978.
[4] Toland, R. H., Sanchez, R. J., and Freeman, D., "Stress-Rupture Life of Kevlar/Epoxy Spherical Pressure Vessels," UCID-17755, Part 1, University of California, Lawrence Livermore Laboratory, Livermore, Calif., 1978.
[5] Mann, N. R., Schafer, R. E., and Singpurwalla, N. D., *Methods for Statistical Analysis of Reliability and Life Data*, Wiley, New York, 1974.
[6] Penn, L. and Larsen, F., *Journal of Applied Polymer Science*, Vol. 23, 1979, pp. 59–73.
[7] *Kevlar-49 Data Manual*, DuPont de Nemours & Co., Wilmington, Del., 1974.
[8] Chiao, T. T., Chiao, C. C., and Sherry, R. J., *Fracture Mechanics and Technology*, G. C. Sih and C. L. Chow, Eds., Sijthoff and Noordhoff, The Netherlands, Vol. 1, 1977, pp. 257–269.

[9] Hahn, H. T. and Gates, T., *Composites Technology Review*, Vol. 1, No. 4, 1979, p. 12.
[10] Clements, L. L. and Moore, R. L., *Society of Aerospace Material and Process Engineers Quarterly*, Vol. 9, 1977, pp. 6–12.
[11] Hamstad, M. A. and Chiao, T. T., "Structural Integrity of Fiber/Epoxy Vessels by Acoustic Emission: Some Experimental Considerations," UCRL-78198, University of California, Lawrence Livermore Laboratory, Livermore, Calif., 1976.
[12] Evans, A. G. and Wiederhorn, S. M., *International Journal of Fracture*, Vol. 10, 1974, p. 379.
[13] Chiao, T. T., Hamstad, M. A., Jessop, E. S., and Toland, R. H., "High-Performance Fiber/Epoxy Composite Pressure Vessels," UCRL-52533, NASA CR-159512, University of California, Lawrence Livermore Laboratory, Livermore, Calif., 1978.

R. L. Ramkumar[1]

Bolted Joint Design

REFERENCE: Ramkumar, R. L., **"Bolted Joint Design,"** *Test Methods and Design Allowables for Fibrous Composites, ASTM STP 734,* C. C. Chamis, Ed., American Society for Testing and Materials, 1981, pp. 376–395.

ABSTRACT: Bolts are commonly used in many structures to transfer loads from one component to another. When the structural components are fabricated using heterogeneous, fiber-reinforced composite materials and the magnitudes of the load transferred directly through the bolt vary, concerns regarding structural integrity arise. A change in the failure mode due to a change in the bolt load could be detrimental to the strength of the bolted joint. Results are available in the literature for the extreme cases where the bolt load is zero (unloaded hole) and where the total applied load is resisted by a bolt. This paper summarizes the results from an experimental-analytical program that investigated the effect of transferring an arbitrary fraction of the applied load through the bolt. In the analytical part of the program, a two-dimensional finite-element analysis was carried out, using NASTRAN, to predict the stress state around the bolted joint and the average stress failure criterion used to predict the joint strength and failure modes. The experimental program was conducted under room temperature dry and wet (1.2 percent moisture by weight) conditions. Experiments were conducted on AS/3501-6 graphite/epoxy specimens in test fixtures designed to transfer variable loads through the bolt. Identical specimens were subjected to static tensile and compressive loads, and to two lifetimes of tension and compression-dominated F-5 fatigue spectrum load conditions. Design curves were generated for the different test cases.

KEY WORDS: composite materials, AS/3501-6 graphite/epoxy, bolts, variable bolt loads, finite-element analysis, average stress failure criterion, tension, compression, static fatigue loads, spectrum fatigue loads

Composite aircraft structural components, connected through bolted joints to other parts of the airplane, transfer loads of large magnitudes. The introduction of local attachment loads into a structural panel can have a significant effect upon the overall allowable tensile or compressive stress of the panel. Previous work has shown that the allowable panel stress depends, among other things, upon how much of the panel load is taken out by the bolted joint (local bearing stress) and how much bypasses the joint area. The integrity of these joining areas is crucial to the overall performance of the airplane, measured by the static strength and fatigue life of

[1] Senior engineer, Structural Mechanics Research, Northrop Corp., Hawthorne, Calif. 90250.

the airplane. Empirical analytical procedures developed for metals cannot be extended directly to composite materials, which are brittle, anisotropic, and heterogeneous. The variable strengths and moduli of these materials cause them to exhibit failure modes (Fig. 1) that are different from those of metallic joints. Furthermore, as the bolt load is varied from zero (unloaded hole) to the total applied load (maximum local bearing stress), the failure modes and joint strengths exhibit marked differences. An experimental-analytical study conducted by Ramkumar [1][2] addresses this problem. The present paper summarizes the results obtained in Ref 1.

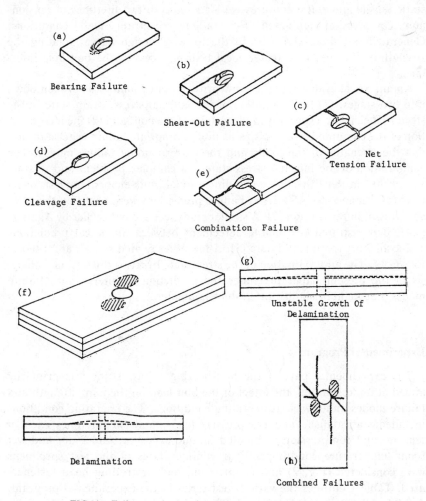

Bearing Failure

Shear-Out Failure

Net
Tension Failure

Cleavage Failure

Combination Failure

Unstable Growth Of
Delamination

Delamination

Combined Failures

FIG. 1—*Failure modes in advanced composite mechanical joints.*

[2] The italic numbers in brackets refer to the list of references appended to this paper.

In the experimental part of the program, quasi-isotropic laminates of AS/3501-6 graphite/epoxy material with 4.76-mm-diameter (3/16 in.) holes were subjected to static and fatigue tests for different proportions of the total applied load introduced through a bolt at the hole. A close fit between the bolt and the hole was achieved through careful machining. The bolts were 100-deg flush tension-head titanium bolts with the trade name HL13VAP6. Tension and compression tests were carried under room temperature conditions. Dry and wet (1.2 percent moisture by weight) situations were simulated. Test fixtures were designed and built to achieve variable load transfers at the joint. Identical specimens were subjected to static tensile and static compressive loads, and to two lifetimes of tension- and compression-dominated F-5 fatigue spectrum load conditions. Generated test data were cast in the form of design curves relating the strength of the specimen to the local bearing stress at the bolt-hole interface.

An analytical evaluation of the integrity of bolted joints was broken down into two stages: (1) the estimation of the internal stress/strain state in the structural element, especially in the vicinity of the bolt; and (2) the incorporation of the computed stresses/strains into appropriate failure criteria to predict the strengths of the joints and the corresponding failure modes. The computation of the internal stress state is a complex problem and was accomplished in Ref 1 through a two-dimensional finite-element analysis using the MSC version of NASTRAN. Failure predictions were based on the average stress failure criterion [2]. A similar study, carried out earlier by Agarwal [3,4], demonstrated reasonable agreement between analytical predictions and available experimental data [5]. A few other related studies are listed in Refs 6–14. The approximation of the stress state by a two-dimensional analysis in the present paper precludes the prediction of interlaminar failure modes that could be significant in some laminates.

Experimental Program

The experimental part of the bolt-bearing/bypass study was primarily aimed at understanding the effect of the bolt load on the joint strength and failure modes under various operating conditions. The operating conditions included static tensile and compressive loads, and two lifetimes each of tension- and compression-dominated fatigue exposure, under dry and wet room temperature conditions. Four configurations of the test specimens were considered (Fig. 2) and countersunk and protruding head fasteners used. Table 1 outlines the various test series in the experimental program. A total of 96 tests were conducted. The specimens were machined from 24-ply $[\pm45/0/90]_{3s}$ AS/3501-6 graphite/epoxy laminated panels.

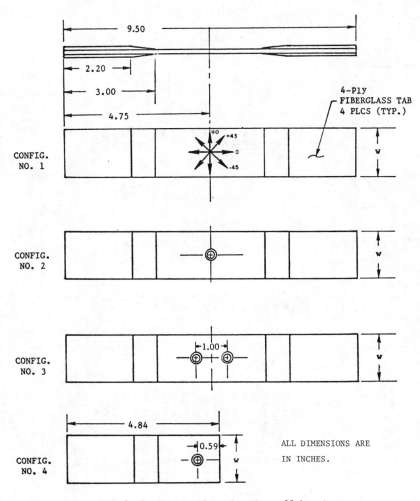

FIG. 2—*Specimen configurations (in. = 25.4 mm)*

The test fixture, designed to apply an arbitrary proportion of the total applied load at the bolt, was developed at Northrop [*15*]. The test machine used for static testing, with a specimen in the assembled test fixture, is shown in Fig. 3.

To simulate environmental conditions to which service aircraft are exposed, an average moisture absorption level of 1 percent by weight was induced in a few specimens (Table 1). This was achieved in the laboratory by exposing the specimens to 77°C (170°F) at 95 percent relative humidity for 40 days, followed by an exposure to 77°C (170°F) at 80 percent relative humidity for 24 days. This two-step exposure gives nearly uniform moisture

TABLE 1—Test series for bolt-bearing/bypass study.

(1) Test Series T, C	Specimen Description (2)				(3) Environmental Exposure	Load Description (4)			
	Configuration No.	Width, W in.	Hole Type	No. of Holes		Bearing to Total Load Ratio	Bypass to Bearing Ratio	Fatigue Exposure	No. of Replicates
1	1	1.50	...	0	RTD	...	...	none	2
2	1	1.50	...	0	RTD	...	...	none	2
3	1	1.50	...	0	RTW	...	...	none	2
4	1	1.50	...	0	RTW	...	...	none	2
5	2	1.50	3/16 TH	1	RTD	0.10	9	none	2
6	2	1.50	3/16 TH	1	RTD	0.20	4	none	2
7	4	1.50	3/16 TH	1	RTD	1.00	0	none	2
8	2	1.50	3/16 TH	1	RTW	0.10	9	none	2
9	2	1.50	3/16 TH	1	RTW	0.20	4	none	2
10	4	1.50	3/16 TH	1	RTW	1.00	0	none	2
11	2	1.50	3/16 PH	1	RTD	0.10	9	none	2
12	2	1.50	3/16 PH	1	RTD	0.20	4	none	2
13	4	1.50	3/16 PH	1	RTD	1.00	0	none	2
14	2	1.125	3/16 TH	1	RTD	0.133	6.5	none	2
15	2	1.125	3/16 TH	1	RTD	0.267	2.75	none	2
16	4	1.125	3/16 TH	1	RTD	1.00	0	none	2
17	3	1.50	3/16 TH	2	RTD	0.20	4	none	2
18	3	1.50	3/16 TH	2	RTD	0.40	1.5	none	2
19	2	1.50	3/16 TH	1	RTD	0.10	9	2 LT	2
20	2	1.50	3/16 TH	1	RTD	0.20	4	2 LT	2
21	4	1.50	3/16 TH	1	RTD	1.00	0	2 LT	2
22	2	1.50	3/16 TH	1	RTW	0.10	9	2 LT	2
23	2	1.50	3/16 TH	1	RTW	0.20	4	2 LT	2
24	4	1.50	3/16 TH	1	RTW	1.00	0	2 LT	2

Total specimens per test series = 48

NOTES:

(1) T = tension tests. C = compression tests.

(2) Specimen configurations are shown in Fig. 2. Laminate is 24-ply, $\pi/4$ orientation of AS/3501-6 graphite/epoxy. 3/16 TH = 0.194 to 0.190-in.-diameter hole. countersunk for HL13VAP6 fastener (100-deg flush head of 0.376 to 0.386-in. diameter). 3/16 PH = 0.194 to 0.190-in.-diameter hole. 1 in. = 25.4 mm.

(3) RTD = room temperature dry specimens. RTW = room temperature, moisture-conditioned specimens (wet).

(4) 2 LT = two lifetimes of tension- or compression-dominated fatigue exposure. Load in the fatigue spectrum to be determined.

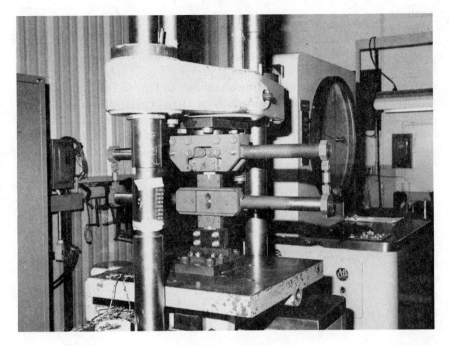

FIG. 3—*Static test setup.*

profiles for this laminate. Moisture control specimens and traveler coupons were used to ensure the desired moisture absorption in the test specimens.

Test Series 1 to 18 in Table 1 describe the static tests conducted under the present program. The variables in the various tests are the number of holes, type of bolt, specimen width, and N, the ratio of the bolt-bearing load to the total applied load. The bolts are torqued to the same level for all the tests. The basic unnotched laminate strength is measured using specimen configuration No. 1 in Fig. 2. Configuration 2 is used to estimate the stress concentration effect of the notch, and to measure the effect of N on the strength and failure mode in a single lap shear bolted joint. Configuration 3 is used to estimate the effect of N on two bolt holes in tandem in the load direction. Configuration 4 is used when the bolt carries the total applied load ($N = 1$) with no bypass load.

Test Series 19 to 24 in Table 1 describe the fatigue tests conducted under the present program. Three values of N were considered under dry and wet conditions. The fatigue exposure conditions were two lifetimes of tension-dominated (F-5) spectrum fatigue loads for one set of experiments and the corresponding compression-dominated fatigue loads for the second set of experiments. The maximum spectrum loads for the tension (compression)-dominated fatigue tests were chosen to be two thirds of the average static tension (compression) failure loads obtained from Test Series 5 to 10.

Analysis

The static failure analysis was carried out using a finite-element stress analysis and an average stress failure criterion. A detailed description of the procedure and computational details is presented in Ref *4*. A schematic of the single shear bolted joint specimen is shown in Fig. 4. The finite-element model of the specimen used in the present analysis is presented in Fig. 5. Isoparametric membrane elements were used, and the fastener load was modeled by applying a constant displacement at one end of the plate and imposing zero radial displacements on the load-carrying half-segment of the fastener hole surface. A typical NASTRAN pin-load distribution around the hole

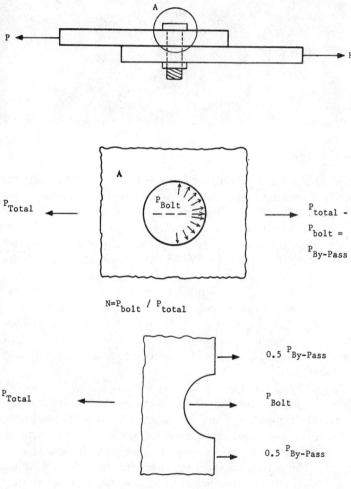

FIG. 4—*Single-shear bolted joint.*

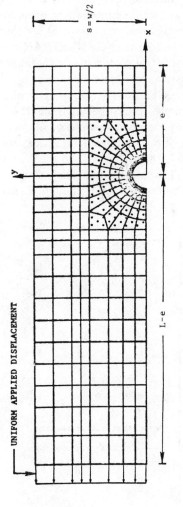

FIG. 5—*NASTRAN model for bolt bearing specimens [4].*

compares well with the commonly used cosine distribution as shown in Fig. 6 (see Ref *4*). The membrane model essentially considers a rigid frictionless bolt bearing on half the hole surface. Plate bending effects, interlaminar stress effects, bolt bending, and bolt shearing effects are assumed to be minimal. It should be noted that, in contrast to the assumptions in the analysis, the experimental setup transfers load through a single lap shear arrangement that could result in considerable bolt-bending effects (predominantly in the upper range of N) that are not accounted for in the analysis.

A preprocessor generates the mesh (Fig. 5) for the bolted joint and the necessary input data. The results of the NASTRAN stress analysis are stored on disks and are accessible to a postprocessor that determines the location of maximum stress, integrates the stress value over a "constant" length, and invokes the average stress failure criterion to determine the strength and the corresponding failure mode. It must be noted that the magnitude of the point stresses in the neighborhood of the geometric discontinuity is a function of the level of refinement of the finite-element mesh. The finer the mesh near the source of singularity, the higher will be the stresses predicted using an elastic stress analysis. Theoretically, the elastic stresses tend to infinite magnitudes at the singular points. In reality, though, a small region near the dis-

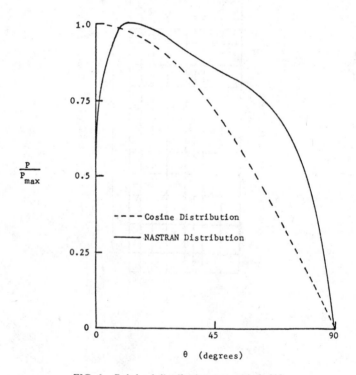

FIG. 6—*Bolt load distribution around hole [4].*

continuity suffers a permanent deformation on exceeding the elastic limit values, causing a blunting effect on the elastic stress concentration. An elastic stress analysis cannot account for this physical phenomenon, and the stresses get larger near the singular points for finer meshes. To minimize the variability effect inherent in any finite-element modeling and to indirectly account for the stress concentration blunting effect due to local inelastic deformations, the average stress failure criterion averages the computed point stresses over a finite length from the source of singularity. The tensile, compressive, and shear strengths of the basic unflawed laminate are required as input data. A tensile failure across the width of the laminate (and through the hole), a shearout of the bolt in the loading direction, and bearing failure at the bolt/laminate interface are the failure modes considered in the analysis.

A fatigue failure analysis can be carried out by dividing the spectrum fatigue loading into a large but finite number of time intervals, and carrying out a static failure analysis for each interval with degraded property inputs [14]. On fatigue loading a notched laminate, high axial and shear stresses in the vicinity of the notch cause the material therein to degrade more rapidly than in the rest of the laminate. This spatial variation of the property degradation in the laminate is a result of the different stress states in the individual laminas, and the large stress concentrations around the notch in the neighboring laminas. An analytical prediction of the instantaneous stress state in a fatigued notched laminate is a formidable task, and only an experimental evaluation of the resultant effects is presented in this paper.

Results

Experimental Results

Tables 2 and 3 summarize the results of the experimental program. The average tensile strength of a specimen with no hole, from Table 2, is 530 MPa (76.9 ksi). The average tensile strength of the specimen with an unloaded 4.76-mm-diameter (3/16 in.) hole is about 310 MPa (45.0 ksi) (from Ref 15 and 16). Strain levels at failure were measured to be approximately 11 800 μmm/mm and 6900 μmm/mm, corresponding to specimens without and with a central hole, respectively. The unloaded hole data [16] were obtained from tests on specimens with a fastener present at the hole during loading. A tensile strength loss of about 42 percent is induced by the hole. The average static compressive strength of the unflawed specimen is 524 MPa (76 ksi) and that for the same specimen with an unloaded 4.76-mm-diameter (3/16 in.) hole is 480 MPa (69.6 ksi) [15]. A compressive strength loss of about 8 percent is induced by the hole. It is noted that the strength loss is more significant in tension than in compression.

Figure 7 shows plots of gross tensile and compressive strengths of a set of tests conducted on 38.0-mm-wide (1.5 in.) specimens with 4.76-mm (3/16 in.)

TABLE 2—Summary of static tension and tension-dominated fatigue tests.

| Test Series | Specimen Description[a] | | | | Environmental Exposure | Load Description | | No. of Replicates | Static and Residual Strengths, ksi |
	Configuration No.	Width,[b] W, in.	Hole Type	No. of Holes		Bearing to Total Load Ratio	Fatigue Exposure		
1	1	1.50	···	0	RTD	···	none	2	73.57, 82.28
2	1	1.50	···	0	RTD	···	none	2	74.50, 77.07
3	1	1.50	···	0	RTW	···	none	2	72.998, 64.15
4	1	1.50	···	0	RTW	···	none	2	61.78, 88.81
5	2	1.50	3/16 TH	1	RTD	0.10	none	2	36.02, 37.66
6	2	1.50	3/16 TH	1	RTD	0.20	none	2	35.50, 36.08
7	4	1.50	3/16 TH	1	RTD	1.00	none	2	17.06, 13.24
8	2	1.50	3/16 TH	1	RTW	0.10	none	2	37.21, 37.87
9	2	1.50	3/16 TH	1	RTW	0.20	none	2	37.42, 37.87
10	4	1.50	3/16 TH	1	RTW	1.00	none	2	14.50, 13.62
11	2	1.50	3/16 PH	1	RTD	0.10	none	2	39.74, 38.03
12	2	1.50	3/16 PH	1	RTD	0.20	none	2	36.13, 36.45
13	4	1.50	3/16 PH	1	RTD	1.00	none	2	15.84, 16.50
14	2	1.125	3/16 TH	1	RTD	0.133	none	2	34.39, 33.36
15	2	1.125	3/16 TH	1	RTD	0.267	none	2	29.65, 29.59
16	4	1.125	3/16 TH	1	RTD	1.00	none	2	20.02, 19.19
17	3	1.50	3/16 TH	2	RTD	0.20	none	2	31.80, 31.57
18	3	1.50	3/16 TH	2	RTD	0.40	none	2	31.41, 32.57
19	2	1.50	3/16 TH	1	RTD	0.10	2 LT	2	33.72, 36.07
20	2	1.50	3/16 TH	1	RTD	0.20	2 LT	2	32.997, 33.97
21	4	1.50	3/16 TH	1	RTD	1.00	2 LT	2	14.29, 14.64
22	2	1.50	3/16 TH	1	RTW	0.10	2 LT	2	36.24, 38.60
23	2	1.50	3/16 TH	1	RTW	0.20	2 LT	2	37.10, 34.99
24	4	1.50	3/16 TH	1	RTW	1.00	2 LT	2	11.80, 10.14

[a] [±45/0/90]$_{3s}$ AS./3501-6 specimens.
[b] 1 in. = 25.4 mm.

TABLE 3—*Summary of static compression and compression-dominated fatigue tests.*

Test Series	Specimen Description[a]				Environmental Exposure	Load Description			Static and Residual Strength, ksi
	Configuration No.	Width,[b] W, in.	Hole Type	No. of Holes		Bearing to Total Load Ratio	Fatigue Exposure	No. of Replicates	
1	1	1.50	...	0	RTD	...	none	2	−38.93, −53.59
2	1	1.50	...	0	RTD	...	none	2	−52.16, −49.95
3	1	1.50	...	0	RTW	...	none	2	−52.56, −58.48
4	1	1.50	...	0	RTW	...	none	2	−48.43, −44.07
5	2	1.50	3/16 TH	1	RTD	0.10	none	2	−70.85, −71.56
6	2	1.50	3/16 TH	1	RTD	0.20	none	2	−55.82, −61.29
7	4	1.50	3/16 TH	1	RTD	1.00	none	2	−10.58, −11.73
8	2	1.50	3/16 TH	1	RTW	0.10	none	2	−63.33, −65.40
9	2	1.50	3/16 TH	1	RTW	0.20	none	2	−51.54, −53.04
10	4	1.50	3/16 TH	1	RTW	1.00	none	2	−10.91, −12.57
11	2	1.50	3/16 PH	1	RTD	0.10	none	2	−73.96, −74.25
12	2	1.50	3/16 PH	1	RTD	0.20	none	2	−56.80, −55.23
13	4	1.50	3/16 PH	1	RTD	1.00	none	2	−15.62, −15.48
14	2	1.125	3/16 TH	1	RTD	0.133	none	2	−71.69, −67.73
15	2	1.125	3/16 TH	1	RTD	0.267	none	2	−70.07, −55.59
16	4	1.125	3/16 TH	1	RTD	1.00	none	2	−15.06, −16.13
17	3	1.50	3/16 TH	2	RTD	0.20	none	2	−65.65, −67.45
18	3	1.50	3/16 TH	2	RTD	0.40	none	2	−59.30, −51.50
19	2	1.50	3/16 TH	1	RTD	0.10	2 LT	2	−71.20, −71.17
20	2	1.50	3/16 TH	1	RTD	0.20	2 LT	2	−61.898, −51.48
21	4	1.50	3/16 TH	1	RTD	1.00	2 LT	2	−13.32, ...
22	2	1.50	3/16 TH	1	RTW	0.10	2 LT	2	−69.79, −68.16
23	2	1.50	3/16 TH	1	RTW	0.20	2 LT	2	−55.07, −57.66
24	4	1.50	3/16 TH	1	RTW	1.00	2 LT	2	−12.70, −12.13

[a] $[\pm 45/0/90]_{3s}$ AS/3501-6 specimens.
[b] 1 in. = 25.4 mm.

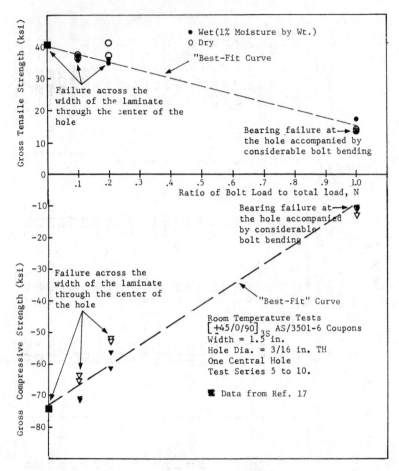

FIG. 7—*Comparison of tensile and compressive strengths under wet and dry conditions (1 in. = 25.4 mm).*

central holes countersunk for titanium fasteners. Unloaded hole data taken from Ref *17* and test results corresponding to $N = 0.1, 0.2,$ and 1.0 are plotted in Fig. 7. As the ratio of the bolt load to the total applied load (N) increases, the tensile and compressive strengths of the bolted joint decrease in magnitude. The decrease in the strengths with N seems to follow a linear path as shown by the dashed "best fit" line in Fig. 7. Only two data points were obtained for each value of N. The linear variation of the tensile and compressive strengths with N is very interesting. This is especially so because of the change in the mode of failure as N increases. For small values of N, the bolted joint fails like an unloaded hole specimen in tension and compression. The failure path in tension extends across the width of the specimen, transverse to the loading direction, and through the center of the hole. Due to the presence of the bolt, the failure path in compression goes around the boundary of the hole between two diametrical points in the transverse direction. For large values of N, close to 1.0, the bolted joints suffer bearing failures due to the large loads transferred by the bolts directly onto the laminate near the contact region. While the transverse mode of failure in the lower range of N is dictated predominately by fiber properties, the bearing failures in the higher range of N are precipitated by matrix failures that are primarily influenced by matrix properties. The seeming indifference of the rate of degradation of the tensile and compressive strengths, with respect to N, to the change in the failure mode is interesting. Strain levels at failure corresponding to $N = 0.0, 0.1, 0.2,$ and 1.0 in tension were measured to be approximately 6900, 5680, 5490, and 2300 $\mu mm/mm$, respectively.

An alternative presentation of the preceding results involves the replacement of N by the bearing stress at the hole boundary. To facilitate use of these curves by designers, the bearing stress at the hole is computed in the conventional manner as the bolt load divided by the product of the hole diameter and the specimen thickness. The results of Fig. 7 are presented in this form in Fig. 8. It should be noted that a linear best-fit curve in Fig. 7 transforms into a quadratic best-fit curve in Fig. 8.

Test results from wet specimens are superimposed on the dry specimen results in Fig. 7 for comparison. The effect of moisture is more pronounced

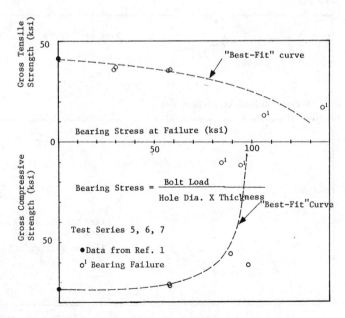

FIG. 8—*Room temperature tests on 1.5-in.-wide [±45/0/90]₃ₛ specimens (dry) with 3/16-in.-diameter countersunk fasteners. (1 in. = 25.4 mm).*

in the compression tests than in the tension tests. This is believed to be due to the matrix-governed compression failure modes and fiber-dominated tension failure modes. While the fiber properties are indifferent to the presence of moisture, the matrix (resin) properties are highly influenced by moisture content. Figure 7 shows a compression strength loss of about 12 percent at $N = 0.2$ due to the presence of moisture.

The replacement of the countersunk fasteners by protruding head fasteners resulted in small changes in the strengths under room temperature dry conditions. A design curve corresponding to the results obtained with protruding fasteners is presented in Fig. 9. A reduction in the width of the specimens from 38.1 mm (1.5 in.) to 28.6-mm (1.125 in.), retaining countersunk fasteners, results in a reduction in the tensile and compressive strengths for small values of N and an increase in the strengths for large values of N. A design curve corresponding to these results is presented in Fig. 10. Results from tests on dry specimens with two holes in tandem in the loading direction are presented in Fig. 11. Results from Test Series 19 to 24 are plotted in Fig. 12. The results correspond to 38.1-mm-wide (1.5 in.) wet and dry specimens with 4.76-mm (3/16 in.) central holes countersunk for titanium fasteners. All the specimens are subject to two lifetimes of tension (compression)-dominated F-5 fatigue spectrum loads before residual tensile (compressive) strengths are

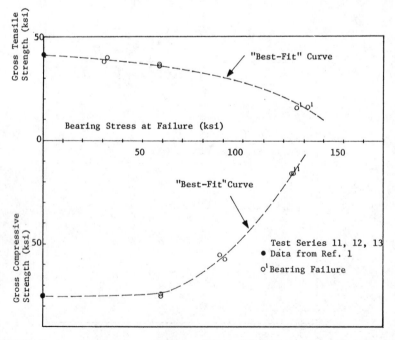

FIG. 9—*Room temperature tests on 1.5-in.-wide [±45/0/90]$_{3s}$ specimens (dry) with 3/16-in.-diameter protruding head fasteners (1 in. = 25.4 mm).*

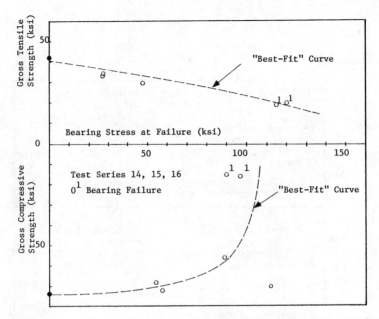

FIG. 10—*Room temperature tests on 1.125-in-wide [±45/0/90]₃ₛ specimens (dry) with 3/16-in.-diameter countersunk fasteners (1 in. = 25.4 mm).*

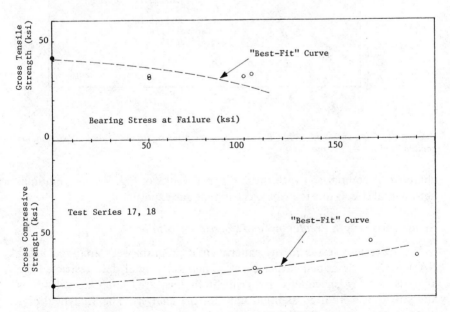

FIG. 11—*Room temperature tests on 1.5-in-wide [±45/0/90]₃ₛ specimens (dry) with two 3/16-in.-diameter holes (countersunk) in tandem (1 in. = 25.4 mm).*

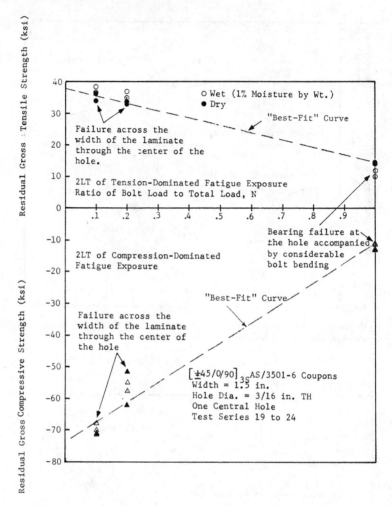

FIG. 12—*Comparison of residual tensile and compressive strengths under wet and dry conditions (1 in. = 25.4 mm).*

obtained. A comparison with the static test results of Fig. 7 shows minimal detrimental effects due to prior spectrum fatigue exposure.

Analytical Strength Predictions and Correlation Studies

The specimens tested in the experimental program were analyzed using NASTRAN as described earlier. Lack of necessary input data restricted the analysis to a few specimens. Experimentally measured laminate stiffnesses and strengths were used as input data for the analysis. The analytical predictions are presented in Table 4 along with the corresponding experimental data. The analytically predicted failure modes coincide with experimentally

TABLE 4—Analytical predictions.

Test Series	Load Type	N (Ratio of Bolt Load to Total Load)	Analysis[c]				Experiment	
			σT,[a] ksi	σB,[a] ksi	σ^{SO},[a] ksi	Predicted Failure Mode	Strength,[b] ksi	Observed Failure Mode
5	tension	0.1	36.95	61.85	177.26	T	36.84	T
6	tension	0.2	36.90	62.19	176.19	T	35.79	T
7	tension	1.0	14.60	15.93	22.63	T/B	15.15	B
	compression	1.0	...	11.63	...	B	11.16	B
10[d]	compression	1.0	...	12.14	...	B	11.74	B
14	tension	0.133	36.66	61.51	176.45	T	33.88	T
15	tension	0.267	36.62	61.85	175.30	T	29.62	T
16	tension	1.0	18.67	21.42	31.30	T/B	19.61	B
17	tension	0.2	36.12	59.82	169.84	T	31.69	T
18	tension	0.4	36.01	60.37	167.64	T	31.99	T

[a] Laminate strengths (based on gross cross-sectional area) for the tension (T), bearing (B), and shearout (SO) modes of failure.
[b] Average of two tests.
[c] Analysis assumes $a^{ot} = 0.035$ in., $a^{ob} = 0.245$ in., and $a^{os} = 0.45$ in.
[d] Room temperature wet.

observed failure modes for most of the test cases. The analytically predicted strength values are also in good agreement with the experimentally measured strength values. Results for Tension Test Series 7 and 16, though, do not exhibit similar correlations. While the analytical strength values are close to the experimental measurements, the analysis does not predict the proper failure mode. A possible reason for the lack of correlation could be the choice of the "constants" a_{ot}, a_{ob}, and a_{so} defining the stress concentration regions for the three failure modes. The analytical predictions point to a tension mode of failure for both tests, though the bearing strength predictions exceed the tensile strengths only by a small magnitude. So, a bearing mode of failure could precede the tensile mode as observed in the laboratory.

The use of a two-dimensional stress analysis precluded the quantitative evaluation of interlaminar stresses that precipitated delaminations in a few compression test specimens. Enhancing the analytical capabilities to include predictions of interlaminar failure modes and the corresponding strengths would mean the use of a costlier, high-order (three-dimensional) stress analysis and a reliable failure criterion to predict delaminations.

Conclusions

This paper summarizes the results from an experimental-analytical program that studied the effect of loaded holes on quasi-isotropic AS 3501-6 graphite/epoxy laminates. Specifically, various proportions of the total applied load were transferred directly through the bolt to the test specimens, and the corresponding strengths and failure modes were compared. Experimental results were cast in the form of useful design curves relating specimen strength to the local bearing stress at the hole, caused by the partial transfer of the applied load directly through the bolt. An alternative presentation of the results revealed a linear variation (decrease) in the specimen strength with N, the ratio of the bolt load to the total applied load. This is an interesting observation since the failure mode changes from a transverse mode to a local bearing mode as N approaches a value of unity.

In the analytical part of the program a two-dimensional finite-element stress analysis of the specimen under the various static loading situations was incorporated into average stress failure criteria for transverse (tensile), shearout, and bearing modes of failure. A fair correlation was demonstrated between analysis and experiment. Establishment of similar agreement between analysis and experimental observations on additional test specimens of different configurations and loading conditions could lead to use of the analysis as a tool to generate useful design curves.

Acknowledgment

The work reported herein was performed as an Independent Research and Development Program in the Structural Mechanics Research Organization at Northrop.

References

[1] Ramkumar, R. L., "Bolt-Bearing/By-Pass Study on Composite Laminates," Northrop Corp. Report NOR 78-154, Hawthorne, Calif., Dec. 1978.

[2] Nusimer, R. J. and Witney, J. M. in *Fracture Mechanics of Composites, ASTM STP 593,* American Society for Testing and Materials, 1975.

[3] Agarwal, B. L., "A Computer Program for the Strength Prediction of Double Shear Composite Bolted Joints and Laminates with Holes," Northrop Corp. Report NOR 78-49, Hawthorne, Calif., May 1978.

[4] Agarwal, B. L. in *Proceedings,* 20th Structures, Structural Dynamics, and Materials Conference, St. Louis, Mo., April 1979.

[5] Van Siclen, R. C., "Evaluation of Bolted Joints in Graphite/Epoxy," paper presented at the U.S. Army Symposium on Solid Mechanics: Role of Mechanics in the Design of Structural Joints, Sept. 1974.

[6] Hart-Smith, L. J., "Bolted Joints in Graphite/Epoxy Composites," NASA CR-144899, National Aeronautics and Space Administration, June 1976.

[7] Waszczak, J. P. and Cruse, T. A., *Journal of Composite Materials,* Vol. 5, July 1971, p. 421.

[8] Oplinger, D. W. and Gandhi, K. R., "Stresses in Mechanically Fastened Orthotropic Laminates," AFFDL-TR-74-103, paper presented at the 2nd Conference on Fibrous Composites in Flight Vehicle Design, Dayton, Ohio, 21-24 May 1974.

[9] Eisenmann, J. R., "Bolted Joint Static Strength Model for Composite Materials," NASA-TM-X-3377, National Aeronautics and Space Administration, 1976.

[10] Waddoups, M. E., Eisenmann, J. R., and Kaminski, B. E., *Journal of Composite Materials,* Vol. 5, Oct. 1971, p. 446.

[11] Harris, H. G., Ojalvo, I. U., and Hooson, R., "Stress and Deflection Analysis of Mechanically Fastened Joints," Technical Report AFFDL-TR-70-49, Air Force Flight Dynamics Laboratory, Dayton, Ohio, May 1970.

[12] Harris, H. G. and Ojalvo, I. U., "Simplified Three Dimensional Analysis of Mechanically Fastened Joints," paper presented at U.S. Army Symposium on Solid Mechanics: Role of Mechanics in the Design of Structural Joints, Sept. 1974.

[13] Oplinger, D. W. and Gandhi, K. R., "Analytical Studies of Structural Performance in Mechanically Fastened Fiber Reinforced Plates," paper presented at U.S. Army Symposium on Solid Mechanics: Role of Mechanics in the Design of Structural Joints, Sept. 1974.

[14] Ramkumar, R. L., Kulkarni, S. V., and Pipes, R. B., "Evaluation and Expansion of an Analytical Model for the Fatigue of Notched Composite Laminates," NASA-CR-145308, National Aeronautics and Space Administration, March 1978.

[15] Verette, R. M. and Labor, J. D., "Structural Criteria for Advanced Composites," AFFDL-TR-76-142, Air Force Flight Dynamics Laboratory, Dayton, Ohio, 1976.

[16] *Advanced Composites Structural Manual,* Structural Mechanics Research Department, Northrop Corp., Hawthorne, Calif., Vol. 2, Section 5, 1978.

T. R. Porter[1]

Environmental Effects on Composite Fracture Behavior

REFERENCE: Porter, T. R., **"Environmental Effects on Composite Fracture Behavior,"** *Test Methods and Design Allowables for Fibrous Composites, ASTM STP 734,* C. C. Chamis, Ed., American Society for Testing and Materials, 1981, pp. 396–410.

ABSTRACT: This paper describes the results of an experimental program investigating the effects of temperature and moisture on the damage development and fracture response of composite materials. Fracture data were obtained for graphite/epoxy composite materials that had been moisture-conditioned. The fracture response data were obtained for both room temperature and immediately after the application of a temperature transient causing the specimen to attain 150°C (300°F). Three 20-ply laminates were investigated in the tests: one typical angle-ply laminate, one laminate representative of polar/hoop-wound pressure vessels, and one laminate containing four plies of S-glass fibers representative of engine fan blades for propulsion systems. From the results it is shown that elevated temperature degrades the notched strength of some laminates while not degrading others. The unnotched strength of all laminates tested at elevated temperature was degraded. In addition to the fracture data, the influence of preloading on damage growth and moisture absorption, the effect of autoclave pressure during laminate cure on fracture, and the variation of interlaminar shear properties with layup, moisture exposure, laminate material, and cure pressure were investigated.

KEY WORDS: composite materials, graphite/epoxy, fracture, cracking, fatigue, flaw, moisture absorption

The objective of this program was to develop data for evaluating the integrity of fiber composite components. In particular, the static performance of three potential composite laminate designs was investigated. The results of these tests addressed the following topics:

1. the effects of moisture and temperature on laminate fracture strength,

2. the effect of defect type and size on fracture strength and failure mode,

3. the influence of moisture and temperature on failure mode and defect growth,

4. the effects of preloads on moisture absorbtion and damage growth, and

---

[1] Research engineer, Boeing Aerospace Co., Seattle, Wash. 98178

5. the feasibility of proof-loading composite structures that are subjected to moisture and elevated temperature.

Data were obtained on the effects of six different types of stress concentrations or flaws simulating potential defects anticipated in composite laminates. The test specimens were conditioned to near maximum moisture levels prior to static testing. Selected specimens were preloaded prior to the moisture conditioning to investigate the potential effects of proof loading on subsequent environmental sensitivity. The damage development due to the loading and exposure was evaluated to assess the mechanism of degradation of composite laminate materials. From these results the potential effects of preloading, moisture content, and temperature were assessed.

Specimen Design and Manufacture

The materials used for the program were Thornel 300 (T-300) graphite fiber, S-glass fiber, and Fiberite 934 epoxy resin. The three different layups used are listed in Table 1. All laminates are balanced, symmetrical, and 20 plies thick. Laminate L1 is a layup representative of many aerospace applications. The second layup (L2) is representative of a filament-wound pressure vessel, fabricated using both polar and hoop wraps. The third layup (L3) is representative of turbo-engine fan blades or, possibly, tubular support struts. The S-glass fiber is used in fan blade layups to increase the impact damage tolerance.

The laminate materials configuration is shown in Fig. 1. The specimen size was selected to minimize the width effects on the test results. Woven fiber glass grip tabs are bonded to the specimen.

The test specimens were layed up and cured in large panels to minimize specimen variability due to differences in layup practice and cure. After cure, the tabs were bonded to the panels and then individual specimen blanks were cut from the panels.

The potential effects of defects were assesed by testing laminates containing induced stress concentrations. Figure 2 illustrates the stress concentration types. These defect types are representative of the potential

TABLE 1—*Test laminates.*

Designation	Material	Layup	Application
L1	Thornel 300/fiberite 934 (T300/934)	$[(0/\pm45/0/90)_s]_2$	general structure
L2	T300/934	$[(0_3/\pm80)_2]_s$	pressure vessels
L3	T300/934 with 901-S	$[(0/\pm30/0^a/-30/0)_2]_s$	turbine engine fan blades or support struts

[a] Plies that are replaced with S-glass.

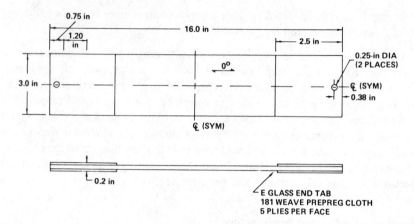

FIG. 1—*Specimen configuration.*

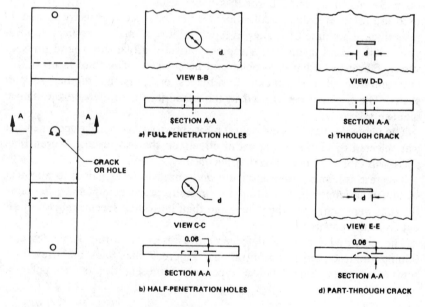

FIG. 2—*Stress concentrations.*

tension-critical defect configurations. The defect categories represented are full penetration cut or broken filaments and partial penetration defects such as scratches.

The hole and slit sizes selected for testing were 3.18 mm (0.125 in.), 9.52 mm (0.375 in.), and 15.87 mm (0.625 in.). These sizes represent a range of sizes near the threshold of detectable damage sizes. The sizes are identified as 1/8, 3/8 and 5/8 defects, respectively.

The normal cure cycle includes autoclave pressurization of 689 kPa (100 psi). Three panels from Laminate L2 were cured at reduced pressures of 345, 172, and 86 kPa (50, 25, and 12.5 psi). The lowest and perhaps the second lowest curing pressures had an effect on the quality of the laminate as can be seen from the C-scan result shown in Fig. 3.

Test Procedures

As shown in Fig. 4, four test profiles were followed: (1) static room temperature test, (2) preload-static room temperature test, (3) static elevated temperature test, and (4) preload-static elevated temperature test. In all cases the specimens were moisture-conditioned prior to static testing. Preloading was performed at room temperature and prior to moisture-conditioning. The preload was representative of a proof load that would be applied before service exposure. The preload stress was about 80 percent of the static strength for each specimen.

Moisture conditioning was by soaking the specimens in a water bath at 82°C (180°F) for eight weeks. The test specimens were removed from the water bath periodically to monitor weight gain. Additional untabbed laminate material was also included in the water soak for weight checks and for interlaminar shear specimens. After moisture conditioning the specimens were stored at room temperature until test.

The elevated temperature tests were performed after the application of a thermal spike. The thermal spike was produced by heating the specimen on one face with radiant heaters. The heatup rate was selected to produce a 17°C (30°F) degree front-to-back surface temperature gradient and required about 60 s to attain the test temperature. Test specimen loading was initiated 180 s after the start of heatup, allowing temperature stabilization through the thickness at the time of maximum load. The temperature survey results indicated that through-the-thickness temperature differences after stabilization were less than 3°C (5°F).

Test Results

All test specimens were C-scan inspected initially, after preload and after moisture-conditioning to monitor damage growth. In general, the preload did not influence the base laminate material C-scan result. The exception was the unnotched Laminate L2 tests. Figure 5 shows the change in C-scan results due to preloading. Subsequent fractographic examination has shown there is microcracking in the 80-deg plies at strain values applied to the unnotched specimens. Examination of the C-scan of the wet test specimens did not reveal any additional damage due to the moisture-conditioning. However, because of the water "fill in" of the voids or delaminations, some C-scans show less damage after moisture exposure.

● All specimens have 5/8-in full-penetration hole

SPEC L2-24-1
12.5 lb/in² CURE

SPEC L2-23-4
25 lb/in² CURE

SPEC L2-22-1
50 lb/in² CURE

BEFORE PRELOAD

FIG. 3—*Effect of autoclave pressure.*

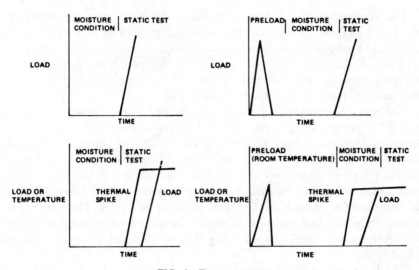

FIG. 4—*Test sequences.*

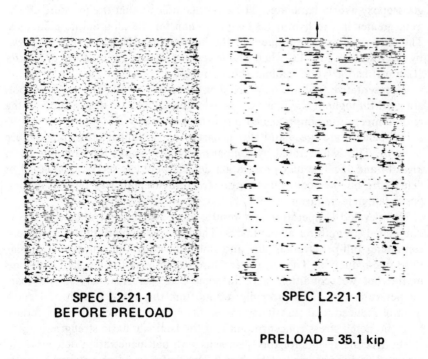

SPEC L2-21-1
BEFORE PRELOAD

SPEC L2-21-1

PRELOAD = 35.1 kip

FIG. 5—*Effect of preload on C-scan damage in unnotched L2 laminate.*

The application of the preload produces significant damage around the defect. In general, through defects propagate by splitting at the edge of the defect, and for part-through defects there is delamination between the plies. The extent of damage growth varies with the laminate type. Typical C-scan-detected damage for half-penetration defects from each of the three laminates is shown in Fig. 6 for comparison.

The moisture-conditioning procedures resulted in the laminates attaining about 1.5 percent weight gain. The weight gain experienced in each of the laminates during exposure is shown in Fig. 7. The weight gain in the pre-loaded test specimens was compared with the weight gain in the nonpre-loaded for potential loading effects. The result is masked somewhat by the moisture absorbtion in the tabs, but the comparison is valid. The weight gain at one week for each of the laminates is shown in Fig. 8. These results indicate that the initial moisture absorption is accelerated by preloading. For the results at eight weeks shown in Fig. 9, however, there is a possible reverse trend.

The interlaminar shear strength for the laminates was affected by both moisture and temperature, with the temperature effect being the most significant. The interlaminar shear data are shown in Figs. 10 and 11 for the graphite/epoxy laminates. In Fig. 12 similar data are shown for the graphite-glass/epoxy hybrid laminates. These results indicate that the moisture effects were greater in the hybrid L3 laminate than for the all-graphite laminates. The cure pressure also influenced the interlaminar shear strength as shown by the results in Fig. 13, where the lowest autoclave cure pressure of 86 kPa (12.5 psi) had low interlaminar shear strength.

Representative fracture data are shown in Figs. 14–19. The same symbols are used throughout the figures. The connecting lines are drawn through the room temperature nonpreloaded test results.

Full-penetration holes and slits generally had a similar effect on fracture strength. This is illustrated by comparing the test results for Laminate L1 in Figs. 14 and 15. The half-penetration defects were less severe. The effect of defect size on the room temperature fracture strength was typical of composite laminate behavior.

The elevated-temperature test results show a significant reduction in the unnotched strength for all laminates. The notched strength of test specimens containing full-penetration slits and holes, however, do not always display this same reduction. That is, the fracture strengths of the notched laminates in the hot wet condition were in some cases as great as for the room temperature tests. This probably results from the mixed benefit and detriment of reduced load transfer from one filament to another. This can reduce both the notch stress concentration and the laminate basic strength.

The fracture results for the specimens with half-penetration defects, as illustrated in Fig. 16, generally show a less direct trend of reduced fracture strength with increased defect size. The defect size refers to surface length.

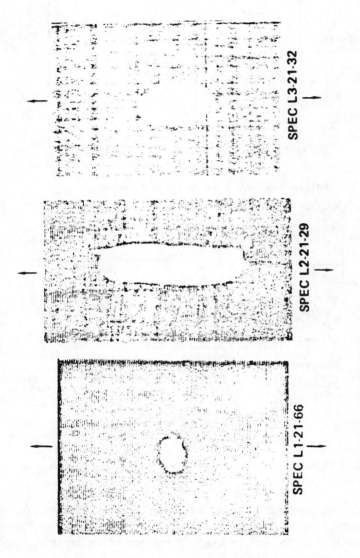

SPEC L3-21-32

SPEC L2-21-29

SPEC L1-21-66

5/8-in HALF-PENETRATION SLIT

FIG. 6—*Effect of preload on C-scan damage.*

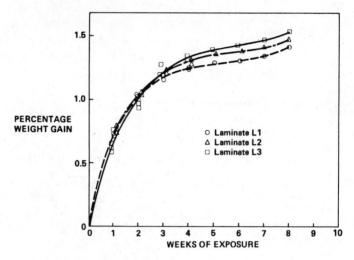

FIG. 7—*Weight gain in untabbed laminate specimens.*

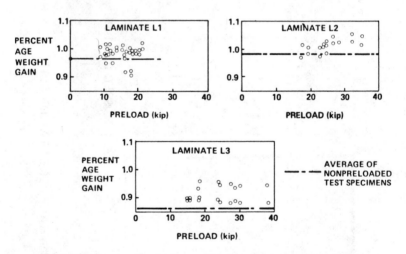

FIG. 8—*Influence of preload on weight gain after one week of exposure.*

The depth of the partial penetration defects was 1.5 mm (0.060 in.) for all surface lengths. It might be expected that depth would be a more significant flaw dimension, with the surface length of only secondary importance and therefore not producing definitive trends.

The test data indicates there is not a direct effect of the preloading on the residual fracture strength. This can be seen by comparing solid and open symbols in the figures. This result indicates that a prior preload is not detrimental to the residual tensile strength of environmentally exposed com-

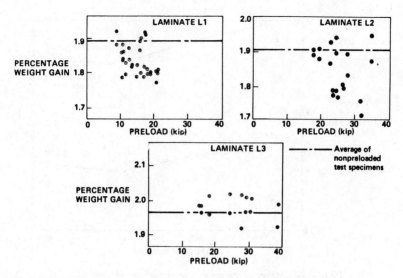

FIG. 9—*Influence of preload on weight gain after eight weeks of exposure.*

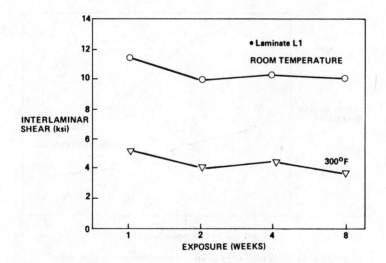

FIG. 10—*Influence of moisture exposure on interlaminar shear.*

posites. Damage growth does occur on preloading, however, and the potential for degrading effects for other residual structural properties (such as compression strength) is likely.

The low cure pressures were found to have only a minimal effect on fracture strength as shown in Fig. 19. This was true even for the lowest cure pressure, where C-scan and interlaminar shear tests revealed significant effects.

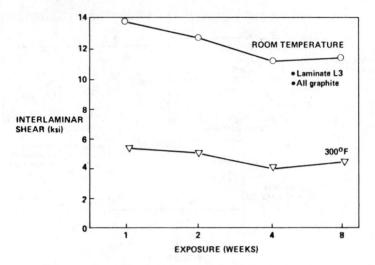

FIG. 11—*Influence of moisture exposure on interlaminar shear.*

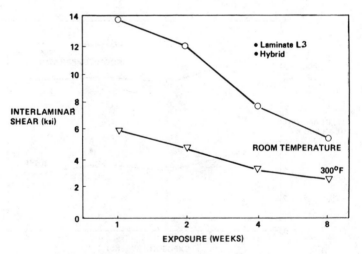

FIG. 12—*Influence of moisture exposure on interlaminar shear.*

Conclusions

These results indicate there are potentially degrading effects of moisture, temperature, and proof loading on composite laminates. This was determined by the damage development and by the reductions in interlaminar shear strengths. However, when considering the specific requirements of tolerance to defects under tensile loads as tested, the effects of preloading, moisture, and temperature were minimal. The environmental effects were

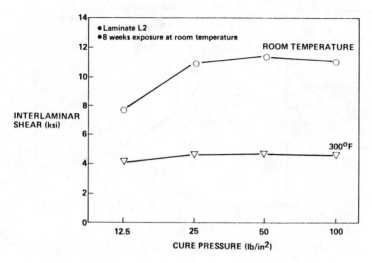

FIG. 13—*Influence of cure pressure on interlaminar shear.*

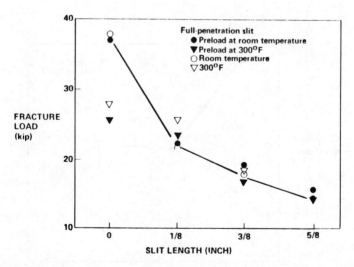

FIG. 14—*Slit specimen static fracture data for Laminate L1.*

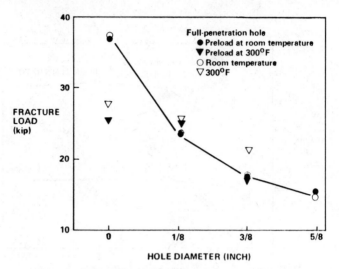

FIG. 15—*Hole specimen static fracture data for Laminate L1.*

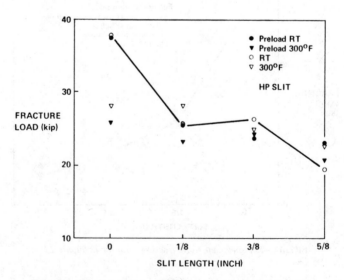

FIG. 16—*HP slit specimen static fracture data for Laminate L1.*

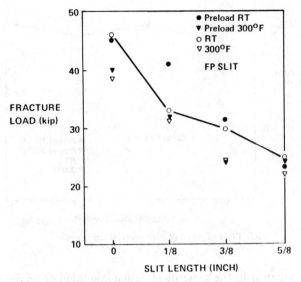

FIG. 17—*FP slit specimen static fracture data for Laminate L2.*

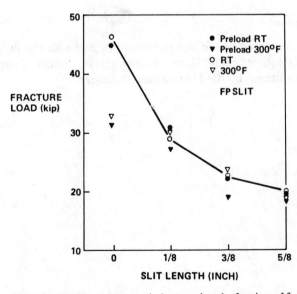

FIG. 18—*FP slit specimen static fracture data for Laminate L3.*

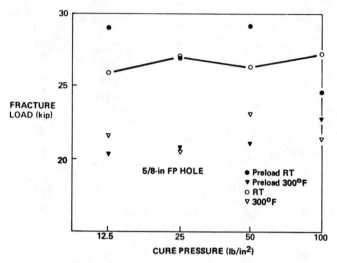

FIG. 19—*Influence of cure pressure on fracture.*

generally the greatest in the noncritical design condition of no defect. These results suggest that the application of proof-loading methods in the room temperature dry condition could by viable in tension-loaded composite structures exposed to environmental conditions.

Acknowledgment

The work presented herein was performed in part with the support of the National Aeronautics and Space Administration under Contract NAS 3-20405 administered by the Lewis Research Center.

S. V. Hoa[1]

Effects of Liquids on the Stress-Rupture Lives of Fiber Glass-Reinforced Plastics

REFERENCE: Hoa, S. V., "**Effects of Liquids on the Stress-Rupture Lives of Fiber Glass-Reinforced Plastics,**" *Test Methods and Design Allowables for Fibrous Composites, ASTM STP 734,* C. C. Chamis, Ed., American Society for Testing and Materials, 1981, pp. 411–419.

ABSTRACT: The results of a preliminary investigation on the effect of liquid environments on the stress-rupture lives of fiber glass-reinforced plastics are investigated. Specimens made from sheet molding compounds were subjected to stresses while being immersed in water and in isooctane. At stresses greater than 50 percent of the ultimate tensile strength of the material, the stress rupture lives show a large amount of scatter. Immersion of the polymer in the liquids indicates considerable liquid absorption. Immersion of the specimens in water also decreases the amount of scatter. Microscopic observation reveals a multitude of irregularities of glass fiber protrusions at the surface of the specimens. The investigation is limited and serves as an indication for future research.

KEY WORDS: fiber glass-reinforced plastic, sheet molding compounds, water absorption, stress-rupture lives, notch sensitivity, composite materials

Fiber glass-reinforced plastics have found wide applications in the automotive industry. In their applications, these materials are subjected simultaneously to loads and various environments such as water and hydrocarbons. It is known that the mechanical behavior of thermoplastics such as PMMA, polystyrene, and nylon is severely affected by the combination of loads and liquid environments and that thermosetting plastics such as polyester and epoxy are affected by the absorption of water. Work on the effect of environments on fiber-reinforced plastics has focused mainly on aircraft materials such as graphite/epoxy composites [1,2] [2], and much of this work is on the effect of fluids on the unstressed composite [2–4]. The effect of natural weathering on the creep behavior of glass fiber-reinforced polyester

---

[1]Assistant professor, Department of Mechanical Engineering, Concordia University, Montreal, Que., Canada.
[2]The italic numbers in brackets refer to the list of references appended to this paper.

laminates for building applications has also been studied [5,6]. In automotive applications, sheet molding compounds have been utilized for making various components such as auto body, door panels, and hoods. In these applications, the materials are simultaneously subjected to loads and various environments such as water, salted water, gasoline, and ice, and the mechanical response of the materials under these conditions is not well understood. This report presents the preliminary findings of the current research program on the effect of liquids on the mechanical behavior of fiber-reinforced plastics. These findings show only a qualitative trend that suggests possible research directions in the future and they should be used with caution since the data are observed to show a large amount of scatter.

Materials and Equipment

The materials tested were sheet molding compounds of Material No. G-1005-3 supplied by Somerville Belkin Ltd. They consisted of a matrix of polyester resin reinforced by short glass fibers distributed randomly. The material consists of 28 to 30 percent glass fiber, 27 percent polyester resin, 41 percent calcium carbonate filler, 1 percent thickener, and the balance is made of internal release and catalyst. The density of the material is 1.87. Stress-rupture specimens (Fig. 1) 20 cm long with a gage length of 6.25 cm were cut from 0.312-cm-thick sheets using a tool steel milling cutter. These specimens were tested in a constant-load lever loading machine (Fig. 2). An environmental chamber is incorporated into the machine from below the loading jaws and the specimen. This chamber can be easily applied or withdrawn so that different intervals of immersion and nonimmersion are possible. The deformation of the specimen is recorded using a dial gage or strain gage or both mounted on a cantilever beam. Tests in air were performed at 27 ± 3°C and 50 percent relative humidity. The liquids used were distilled water and laboratory-grade isooctane. The ASTM Test for Tensile Properties of Plastics (D 638-76) was used to obtain tensile strengths.

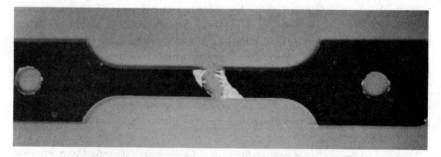

FIG. 1—*Stress rupture specimen.*

FIG. 2—*Stress rupture machine.*

Experimental Results

Absorption

Unstressed specimens of the polymer were immersed in both water and isooctane. The percent weight uptake after different intervals of time was recorded and is shown in Fig. 3. Both water and isooctane show considerable absorption.

Tensile Strength

Tension tests were performed on an Instron machine. The tensile strengths of specimens subjected to different environmental treatments are given in Table 1. There is considerable scatter in the results. The tensile strengths obtained from six specimens tested vary from 32.1 to 82.2 MPa. For presoaked specimens, the scatter is reduced and the standard deviation is about 10 percent of the average tensile strength. A presoak time of 20 min seems to give more consistent results than no presoak. There is a general decrease in tensile strengths as the time of presoak increases from 20 to 62 days. The effect of the application of 41.5 MPa in water reduces the tensile strength by 18 percent.

Stress-Rupture Lives

Stress-rupture tests were performed at the nominal stresses of 41.0, 48.0, and 62.0 MPa. The stress-rupture curves are shown in Figs. 4-6. In Fig. 4, at

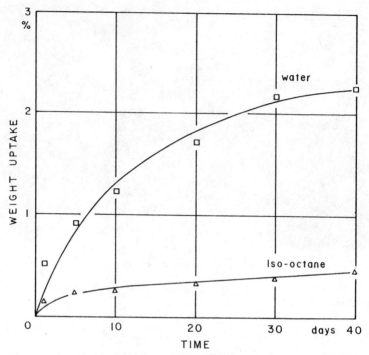

FIG. 3—*Absorption of water and isooctane into fiber glass-reinforced polyester.*

TABLE 1—*Variation in tensile strengths.*

Conditions	% Weight Increase	Tensile Strength, MPa
Soaked in water for 20 days	1.07 ± 0.03	74.9 ± 7.5
Soaked in water for 40 days	1.45 ± 0.03	73.9 ± 7.2
Soaked in water for 62 days	1.82 ± 0.03	62.85 ± 6.9
Presoaked in water for 20 min, tested in air (different batch)	...	64.13 ± 5.4
After creep test: 41.5 MPa to 10^6 s	...	61.0
Tested in air, no soaking treatment	...	32.1 to 82.2

41.0 MPa, the specimens tested in air failed shortly after load application. There is a considerable amount of scatter in the stress-rupture lives. The scatter varies from 200 to 1800 s for six specimens tested at 41.0 MPa. The stress rupture lives of specimens tested in water are longer than those tested in air. This is in contrast to what is usually expected in terms of environmental stress cracking of plastics. Usually, fluid environments tend to shorten the useful lives of the materials [7]. Figure 4 also shows the effect of presoak on the rupture behavior. Different specimens were soaked in distilled water for

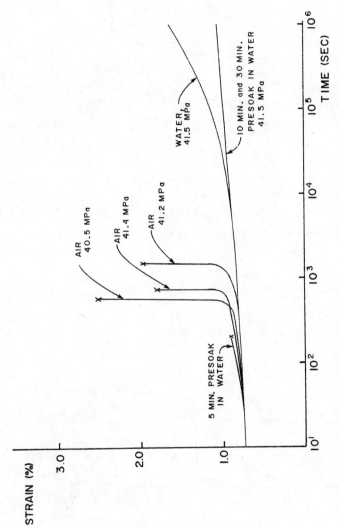

FIG. 4—*Stress rupture of fiber glass-reinforced plastics at 41.5 MPa in air, water, and with presoaks.*

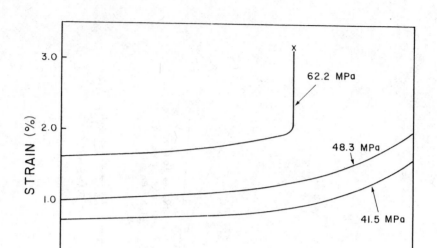

FIG. 5—*Stress rupture of fiber glass-reinforced plastics in distilled water.*

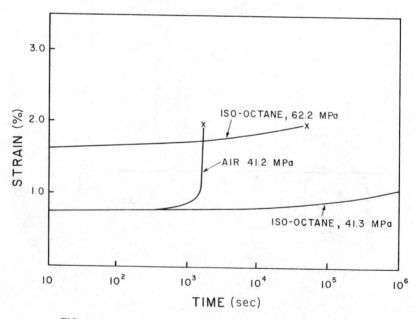

FIG. 6—*Stress rupture of fiber glass-reinforced plastics in isooctane.*

5, 10, and 30 min, they were then blotted dry and tested in air. Specimens that are presoaked for 5 min fail abruptly like those that have no presoaking. Specimens presoaked for 10 or 30 min show significantly longer rupture lives. The creep strains of the specimens with 10 and 30 min of presoaking do not increase as much as the creep strains of specimens tested in water.

Figure 5 shows the rupture response of fiber glass-reinforced plastics in water at different stresses. Under a stress of 62.2 MPa, the specimens fail after about 10^4 s, whereas at 48.3 and 41.5 MPa, the specimens are still holding at 10^6 s.

Figure 6 shows the rupture response of fiber glass-reinforced plastic in isooctane. Similarly to the stress rupture response in water, the specimens tested in isooctane last longer than those tested in air at 41 MPa.

Microscopic Observation

Observation of the thickness area (area formed by the thickness and the length) of the specimen reveals many regions of irregularities. In some regions of this area, partial lengths of glass fibers are observed (upper part of Fig. 7). In other regions, the ends of glass fibers are observed (lower part of Fig. 7). These irregularities are inherent in the formation of the material since the glass fibers are distributed randomly. Usually, the roughness and irregularities at this surface in materials such as steel can be improved by improving the method of machining. For sheet molding compounds, however, changing the method of machining may not help. Studies are underway to investigate the correlation between the different distributions of these irregularities and the tensile strength of the material.

Discussion

At this very preliminary state of knowledge on sheet molding compounds, a lot more data need to be obtained before any firm conclusion can be reached. However, a few trends can be observed from this limited investigation and can be used as an indication of the direction of future research. Even though the results are few (two specimens were used for each test unless otherwise specified), and the scatter is large, it is seen that soaking the sheet molding compounds in water reduces the scattering in the results. One of the hypotheses on the effect of liquid absorption on the mechanical behavior of polymers is that the liquid tends to behave like a plasticizer, therefore increasing the molecular mobility in the polymer [8]. Sheet molding compound is made up of hard glass fibers imbedded in a polyester matrix. These fibers have different orientations and some of them may be lying perpendicular to the loading axis. For simplicity in representation, a model for the fiber matrix system is shown in Fig. 8. The fracture toughness of this material

(a) x 205

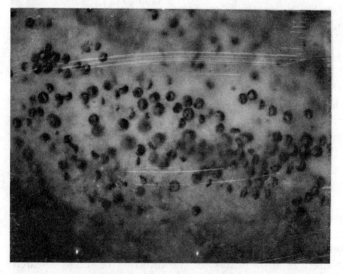

(b) x 205

FIG. 7—*Microscopic observations of the cut surface.*

depends on the plasticity of the matrix. Based upon the hypothesis that the absorption of water makes polyester more ductile, this increase in ductility can result in reducing the notch sensitivity of the material, and therefore less scatter is observed.

Conclusions

Based upon the results and the discussion of this limited investigation, the following conclusions are reached.

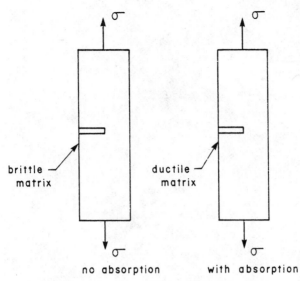

no absorption   with absorption

FIG. 8—*Effect of liquid absorption on the notch sensitivity of the composite.*

1. Sheet molding compounds SMC-R30 show a large amount of scatter in tension and stress rupture tests.

2. Presoaking in water reduces the scatter considerably.

3. Creep strains of specimens tested in water and in isooctane increase much more than creep strains of specimens tested in air (with presoak).

Acknowledgments

The financial assistance of the National Research Council of Canada through Grant No.A0413 is appreciated. The author also expresses his appreciation for the materials supplied by Somerville Belkin Ltd., and the technical assistance of Mr. M. Quinlan and Mr. P. Ouellette.

References

[1] Kerr, J. R., Haskins, J. F., and Stein, B. A. in *Environmental Effects on Advanced Composite Materials, ASTM STP 602,* American Society for Testing and Materials, 1976, pp. 3-22.

[2] Cainahort, J. L., Rennhack, E. H., and Coons, W. C., in *Environmental Effects on Advanced Composite Materials, ASTM STP 602,* American Society for Testing and Materials, 1976, pp. 37-49.

[3] Trabocco, R. E. and Standler, M. in *Environmental Effects on Advanced Composite Materials, ASTM STP 602,* American Society for Testing and Materials, 1976, pp. 67-84.

[4] McKague, E. L., Jr., Reynolds, J. D., and Halkias, J. E., *Transactions,* ASME, *Journal of Engineering Materials and Technology,* Jan. 1976, pp. 92-95.

[5] Norris, J. F., Crowder, J. R., and Probert, C., *Composites,* July 1976, pp. 165-172.

[6] Jain, R. K., Goswamy, S. K., and Asthana, K. K., *Composites,* Jan., 1979, pp. 39-43.

[7] McCammond, D. and Hoa, S. V., *Polymer Engineering and Science,* Vol. 17, No. 12, Dec. 1977, pp. 869-872.

[8] Bernier, G. A. and Kambour, R. P., *Macromolecules,* Vol. 1, No. 5, 1968.

Summary

Summary

The papers in this volume have been grouped into four major areas. The first area deals with new and special test methods. The second deals with special test methods and analysis. The third area covers procedures for establishing design allowables in general. The fourth area deals with design allowables and procedures for special applications. Specific subjects covered in these major areas are summarized in the following.

New and Special Test Methods—Research continues for improved test methods to measure transverse and sheer properties. A single-ply test method (short and very wide specimen one ply thick) is described by Foye to measure the transverse strength of fiber composites. The main advantage of this test method is that moisture conditioning is accomplished expeditiously since the single ply is thin and a short time is needed to reach saturation equilibrium. Use of the method, however, requires care and usually finite-element analysis to minimize possible tensile buckling.

A four-point load ring test method has been investigated by Greszczuk to measure the shear modulus of unidirectional composites. The test method is useful for determining shear moduli of cylindrical-type shell structures subjected to elevated and cryogenic temperatures. The equations required to extract the shear modulus from deflection measurements are based on the bending and torsion of the ring but do not include the through-the-thickness shear deformation. Use of the test for metals and for several composite systems yielded results which were in good agreement with available data from different test methods.

Three different test methods were evaluated by Clark and Lisagor for measuring compression properties of fiber composites. These methods included the Illinois Institute of Technology Research Institute (IITRI) method, the face-supported method, and the end-loaded coupon. All of these methods were used to investigate the effects of specimen size, specimen end support, load introduction, and ply configuration. The results showed that the IITRI test method provided the most consistent stress-strain data to fracture for unidirectional and quasi-isotropic angle-plied laminates, while the face-supported test method was more suitable for $[\pm 45]_s$ angle-plied laminates. The results also showed that the IITRI compression test method is sensitive to specimen alignment and buckling when thin specimens are used and to specimen width when thicker specimens are used. The face-supported compression test is sensitive to delaminations and local instabilities.

Low-velocity impact damage, due to tool dropping, for example, is receiving considerable attention. A test method to determine the low-velocity impact resistance of fiber composite structures is described by Sharma. The method consists of sandwich specimens with faces from three different angle-plied laminates. Preload and impact energy combinations to cause catastrophic failure as well as residual stength after impact damage can be determined in both tension and compression. Results presented are for $([\pm 45/0_4]_s, [\pm 45/90/0]_s, [90/\pm 45/0]_s)$ from graphite-fiber epoxy matrix angle-plied laminates. The experimental data were used to determine a lower-bound failure threshold for each of the laminate configurations studied in both tension and compression. The results show that both failure threshold and residual strength depend on laminate configuration and fiber/matrix.

Special Test Methods and Analysis—Fracture toughness testing for short fiber reinforced thermoplastics was investigated by Mandell and coauthors. Linear elastic fracture mechanics testing concepts were used. The results showed that the cracks propagated in a "fiber avoidence" path. The fracture toughness parameter depended on location and direction of the crack relative to the fiber direction and fiber length but not on the fracture toughness of the resin. This would imply that the fracture toughness may be related to fiber strength and therefore is not an independent parameter.

Comparison of off-axis and angle-plied predicted and measured fracture stresses in both tension and compression continues to receive attention. A major reason for the continuing interest is the difficulty associated with compression testing of thin specimens. Kim found that dog-bone-type specimens with side supports can be used to measure compressive strength. These same specimens without the side supports can also be used to measure tensile strength. Data presented in the paper show that these specimens fail in the test gage region, are in good agreement with predictions using a tensor polynomial failure criterion, and the predicted failure modes are consistent with fracture surface characteristics.

The fracture characteristics of quasi-isotropic graphite/epoxy panels with cracks and subjected to biaxial tensile loads were investigated by Daniel. The effect of crack length on fracture was also investigated. The strain intensity field in the crack-tip vicinity was measured using strain gages, birefringent coatings, and moiré grids. The results showed that the strain field near the crack tip becomes nonlinear at relatively low loads compared with failure loads. However, the crack-opening displacement and the crack shearing (forward sliding) remain linear until the crack starts propagating. The biaxial stress field reduced the fracture load by about 21 percent compared with uniaxial load for all crack sizes considered. The experimental data for failure stress near the crack tip correlated with predicted values using a tensor polynomial failure criterion for the ply and a progressive degradation model for the laminate.

A comparative study was performed by Herakovich and his colleague using finite-element analysis to assess the stress distribution in commonly used in-plane shear specimens. The specimens studied were slotted coupon, crossbeam, double V-notch coupon, and the rail shear. The results showed that the crossbeam, rigid rail shear, and rigid double V-notch specimens provide regions of uniform pure shear in the test section. Elastic supports or load introduction fixtures or both affect the stress distribution in the rail shear specimen. Stress concentrations are present in all four shear specimens. Slotted specimens are not recommended. Thermal stresses affect the stress distribution in the rigid rail shear specimen.

The four-point bending sandwich beam test method continues to receive attention. Shuart evaluated this method to assess its use for determining the compressive properties of graphite/polyimide composites at a wide range of temperatures, 117 to 589 k (-250 to $600°F$). The evaluation included experimental and 3-D finite-element analysis studies. The results showed that the four-point bending sandwich beam test method can be used to determine the compressive properties of graphite/polyimide composites in a wide range of temperatures. A uniform compression stress field exists in the top cover between the two load points. Fracture may initiate by cover debonding or at stress concentrations under the load points. The honeycomb core material, the laminate configuration, and the temperature affect the failure stress. In a parallel analytical study using finite difference, Salamon found that the interlaminar stresses for a $(\pm45)_s$ angle-plied laminate are confined to boundary regions near the free edges and extend about 1.3 times the face thickness. The in-plane stresses converge to the laminate theory solutions in the interior of the face. The analytical results also showed that the face/core interface stresses have negligible influence on the in-plane stresses in the faces. Therefore, both studies (Shuart's and Salamon's) indicate that the sandwich beam is a viable test method for measuring unidirectional composite and angle-plied laminate properties.

Design Allowables—Developing generic test methods and procedures to establish design allowables in general is a difficult task since most of the effort is special case directed. However, several investigators are tackling this difficult task and have had various degrees of success. Procedures for developing composite material static and fatigue design allowables are described by Rich and Maass. The design allowables are established at the ply level and account for a wide range of hygrothermal conditions. For different conditions, the corresponding allowables are established by interpolation and checked by selective testing. It is useful to represent the static design allowables for environmental effects as percentages of the average room temperature dry conditions data. Fatigue design allowables for environmental effects are represented by percentages of a constant life diagram of room temperature wet static tensile strength. The tension-compression and compression-compression range effects are shown in the constant-life fatigue diagram. Statistical evaluation of experimental data indicates normal distribution and, therefore, relatively small sample sizes can be used to establish reliable design allowables. Investigators testing for composite design allowables in fatigue should be aware that proven test methods for metals are not directly applicable to composites. They should also be aware that the most economical level for establishing design allowables for composites is the ply level with selective testing at the laminate level. Additionally, compression fatigue appears to degrade composite integrity rapidly and is aggravated by hygrothermal environments.

Test procedures for obtaining design allowables for composite bolted joints are described by Wilson and his co-authors. Specifically they investigated the static strength bolted joints in graphite-fiber/PMR-15 celion polyimide composites. Temperature effect (21 to 315°C), strain rate (0.002 and 1.00 sec^{-1}), and creep were investigated. The results showed that laminate configuration, temperature, strain rate, and time at load affect the joint strength as well as the failure modes. All these factors lead to the unavoidable conclusion that composite joints need special design considerations in order to assure the cost-effective usage of composites in structures.

Cost-effective procedures for establishing design allowables for stiffness controlled designs are described by Suarez. The procedure consists of conducting a small testing program of key property characterization and simulated component testing. The key properties are obtained using simple test methods. The simulated component is designed using these key properties. The simulated component is subjected to anticipated service life loading conditions to establish the margins between the simple test and the component structural behavior. The key properties from the simple tests and the margins established for the simulated component testing constitute the design allowables base for the types of structures represented by the simulated components. This approach for establishing design allowables has the advantage in reflecting fabrication and geometric scaling-up effects.

Research in statistical approaches to establish design allowables for composites static and fatigue strengths continues. Tenn examined four distributions and the goodness-of-fit for six samples of composite static strength data. The distributions considered included the normal, log normal, 2-parameter Weibull, and 3-parameter Weibull. The goodness-of-fit was determined using the chi-square and Kolmogorov-Smirnov tests and the correlation coefficient from linear regression analysis. Results showed that all four distributions and the two goodness-of-fit tests can be used to establish A and B design allowables with approximately the same level of confidence. Sendeckyj describes a new procedure for fitting fatigue strength models. The procedure consists of a deterministic equation to define the shape of the S-N curve and a probabilistic description of the fatigue data scatter. The procedure is general; it will fit any fatigue model; it will work with a minimum set of fatigue data. The procedure has added advantages: (1) it may be used with only four noncensored fatigue data points for the wearout model, and (2) it can handle runouts and tab failures through progressive censoring.

The use of intraply hybrids to obtain balanced design properties to meet diverse and competing design requirements is receiving increased attention. Test methods for generating data to establish design allowables are described by Chamis and coauthors. The test methods are based on the use of thin laminates (about 8 plies thick) and only one laminate to determine all the properties. Results presented are for tensile properties, intralaminar properties, and flexural, short-beam-shear, and Izod impact. Though not included in the paper, face-supported compression tests can be used to determine compression properties from the same thin laminates. Simple equations can be used in conjunction with normalization to assess the translation efficiency from constituent composites to the intraply hybrid. The translation efficiency serves also as an indirect means to assess the fabrication quality and the composite action of the intraply hybrid as well as any synergistic effects.

A major concern over the past few years has been the severity of compression fatigue on fiber composites and suitable test methods for establishing design allowables. The results of an extensive experimental investigation of compression-compression fatigue are reported by Grimes. Several laminate configurations and hygrothermal conditions were used in the investigations. A new face-supported test method was developed for compression-compression fatigue. The results showed that unidirectional longitudinal [0] specimens have compression fatigue runouts about 60 percent of their static strength. $(1.25 \times 10^6$ cycles) for room-temperature-dry (RTD) and room-temperature-wet (RTW) conditions. Unidirectional transverse [90] specimens have (RTD) compression fatigue runouts about 50 percent and RTW about 45 percent of their respective static strengths. The $[\pm45]_s$ angle-plied laminates have runouts of about 50 percent and the $[(\pm45)_5/0_{16}/90_4]_s$ angle-plied laminates about 80 percent. One encouraging implication of this investigation is that a viable compression fatigue test method may have been developed. Also, the so-called compression fatigue severe degradation may be more a shortcoming of the test method used rather than composite material weakness.

Design Allowables for Special Applications—Buckling of cylindrical shell components under combined loads generally presents difficulties in generating experimental data to verify theoretical predictions. Herakovich and Johnson subjected composite cylindrical shells to combined axial compression and torsional loads. They obtained good comparisons between experimental results and theoretical predictions using an available computer program based on Flugge's cylindrical shell buckling theory. The test data generated were obtained from cylindrical shells made from boron/epoxy and graphite/epoxy composites with symmetric and unsymmetric laminate configurations. The experimental data are presented in normalized interaction diagrams. These diagrams can be used to establish design allowables in conjunction with a reliable buckling theory. A conservative approach to this combined loading is a linear interaction curve connecting the "pure" axial compression buckling load with the "pure" torsional buckling load. The effect of possible initial imperfections on the axial buckling load may require proper attention if it is known that they are present.

The integrity of internal pressure vessels is usually established by proof pressure testing. Hahn describes a convenient procedure which can be used to determine the proof pressure to assure the design life of Kevlar-49/epoxy composite pressure vessels. The method is developed by assuming that the lifetime distribution is exponential. Also, the characteristic lifetime of the pressure vessel is related exponentially to the applied pressure. Use of the method is illustrated by two typical examples.

The fatigue strength of composite bolted joints was investigated by Ramkumar using an experimental/analytical program. Various portions of the applied load were transferred through the bolt to the composite, which was made from quasi-isotropic AS/3501 graphite/epoxy. The specimens were tested at room temperature dry and 1.2 percent moisture conditions. Joint fatigue strength plotted versus local bearing stress provides a basis for establishing design allowables. The local bearing stress may be determined using two-dimensional finite-element analysis.

Environmental effects on composite fracture behavior are reported by Porter. The environmental conditions included room temperature dry, room temperature wet, and wet at 150°C (300°F). The laminates investigated included an "almost" quasi-isotropic laminate, a pressure-vessel-type laminate, and an engine fan blade interply hybrid. The defects included through-holes and slits, half-penetration holes and slits, and internal delaminations. The results showed that the environmental conditions degrade the strength of specimens without defects, while these effects are minimal in composites with defects. Also, preloading showed similar trends. A useful observation from these data is that composite structural components free of defects designed for environmental degradation effects would lead to conservative designs when defects are present or induced.

Creep rupture tests conducted by Hoa on sheet molding compound showed that the data scatter is reduced when these tests are conducted in liquid environments compared with those in air.

It appears that test methods and procedures for establishing design allowables will be developed and adopted through an evolutionary process. From the data available to date and apparent trends, the following comments appear reasonable and instructive.

1. Design data and allowables should be generated and set at the ply level using thin unidirectional laminates with suitable validation at the laminate level.

2. Available sets of design data and design allowables should be used only as a guide and in the preliminary designs. Selective testing will still be required for the case under consideration.

3. Compression testing is difficult especially for compression-compression fatigue. The major difficulty centers on what is "true" material compression property and what is a shortcoming of the test specimen/method used. This difficulty will be resolved as more researchers obtain hands-on experience with compression testing and designers obtain field data feedback. The face-supported test appears to be the most attractive to date.

4. Fracture toughness testing at the laminate level should be done only to validate predicted response based on ply level properties.

5. Statistical methods for quantifying the goodness-of-fit of strength data provide a definitive approach for establishing A and B design allowables. Those which are based on deterministic models with probabilistic descriptions of key parameters or data scatter or both have the added advantage of requiring smaller sample size to obtain the same level of confidence.

6. Design allowables for bolted or adhesive joints should be examined and set, again, at the ply level accounting for the associated or participating failure modes. At this level, the joint can be designed to be cost-effective and also preserve the cost-effective use of the composite.

7. Analysis methods, and especially finite-element analysis, provide a direct means of assessing whether the test results are those that were anticipated or, if not, what do these results represent. All tests are physical events representing load-induced composite response. And considerable insight into composite behavior can be gained by proper analysis. One way to ascertain that the test method used provides the desired results is to establish benchmark-type cases which have been validated by thorough analysis for these test methods.

The collective discussion of the papers in this volume and the summary provide considerable and timely information for test methods and for composite analysis of these methods, as well as tests and procedures for establishing design allowables. The collective test data in the papers provide a valuable source which can be used for preliminary design of composite structures, including impact, fatigue, environmental effects, and joints. The collective data also provides a reasonable, reliable base for comparing the merits of new test methods or new composite systems, or both relative to those already in use.

C. C. Chamis

Aerospace and composite structures engineer,
NASA-Lewis Research Center, Cleveland,
Ohio 44135; editor.

Index